EXPLAINING THE COSMOS

EXPLAINING THE COSMOS

THE IONIAN TRADITION
OF SCIENTIFIC PHILOSOPHY

Daniel W. Graham

PRINCETON UNIVERSITY PRESS

PRINCETON AND OXFORD

ISBN-13: 978-0-691-12540-4

LIBRARY OF CONGRESS CONTROL NUMBER 2006922883

BRITISH LIBRARY CATALOGING-IN-PUBLICATION DATA IS AVAILABLE.

THIS BOOK HAS BEEN COMPOSED IN SABON

PUP.PRINCETON.EDU

1   3   5   7   9   10   8   6   4   2

For Sarah and Joseph

CONTENTS

PREFACE

I CANNOT SAY EXACTLY when I began to work on this project. It grew out of a study of the foundations of Greek science, which I pursued while I was a visiting fellow at Clare Hall, Cambridge, in 1988–89. There I was privileged to sharpen my understanding of the issues in the company of Geoffrey Lloyd, Malcolm Schofield, David Sedley, and Myles Burnyeat, among others. Previous to this time I had been working mainly on Aristotle. It seemed to me that Aristotle was carrying out a scientific program that had its roots in Plato, who was following Socratic principles in reaction to an earlier conception of science embodied by the Ionians. I had intended to deal with scientific developments from the Milesians to Aristotle, and I even wrote a draft of this study, but it seemed weak and unpersuasive. I came to believe that more work needed to be done to clarify the positions of the Presocratics who pioneered the scientific outlook that in many ways anticipates our present attitude toward the world.

As I worked through the Presocratics and scholarship on them, I was struck by the fact that at one time, scholarly opinion seemed to be converging on something like the view that I advance in this work. But in recent years that view has been not so much refuted as ignored, as scholars tended to go back to the received opinion of the middle of the twentieth century—which happens to coincide in many respects with the interpretation of Aristotle and his school. What I argue for here is a revisionary view that owes much to Harold Cherniss and his criticisms of Aristotle's historiography, especially as developed by Michael C. Stokes.

The kind of interpretation I criticize in the present study is embodied in two of the leading scholarly works of the English-speaking world: the second edition of G. S. Kirk and J. E. Raven's classic work *The Presocratic Philosophers*, with Schofield as an added author, and Jonathan Barnes's book by the same title. These two books, so different in many ways—the former an advanced textbook with an emphasis on philology, the latter a philosophical study of arguments with an emphasis on logical analysis—both support a similar line of interpretation that has been orthodox for most of the twentieth century. The early Ionians were Material Monists whose assumptions were undermined by Parmenides and his Eleatic school. Thereafter, philosophers with an interest in cosmology desperately tried to answer the Eleatics by proposing pluralistic theories which, alas, failed because they begged the question—all except the atomists, who begged the question so grandly that they solved the problems (at least the physical problems). All of this is present in some form already in Aristotle, except the historical role of Parmenides in changing the focus of cosmological theory. My studies have convinced me that every feature of this interpretation is deeply mistaken, except for the general point that Parmenides in criticizing his predecessors had a profound impact on the pattern of cosmological explanation.

One might have expected the two volumes on *Presocratic Philosophers*, both of which are full of insights and novel interpretations in many details of their studies, to usher in a renaissance of Presocratic studies in the English-speaking world. Unfortunately, the two decades since their publication have been marked by a dearth of work on the Presocratics. To be sure, valuable translations, monographs, and editions have appeared in recent years, but the number of scholars working in the field remains small and their influence on other areas of philosophy limited. The reason for the falloff has been the booming interest in Hellenistic philosophy, which requires many of the same scholarly skills as study of the Presocratics, and has attracted the attention of many erstwhile scholars of the Presocratics, rather than any intrinsic defects of the Presocratics themselves.

I hope that this study will at least contribute to a revival of interest in the Presocratics. New works are emerging challenging many features of the standard interpretation by scholars such as Alexander Mourelatos, Patricia Curd, Carl Huffman, and John Palmer. I hope that the future will see a lively debate about the Presocratics, on foundational questions as well as on details of their theories and influence; and I hope that this study may present in a more attractive light some options that have been neglected.

I would like to thank my mentors at the University of Texas, Alexander Mourelatos and Paul Woodruff, for their continuing influence on me, as well as my friends from Cambridge. I have been helped by ongoing conversations with Patricia Curd and Victor Caston, as well as (former fellow students) Carl Huffman and Herbert Granger. I have profited from consultations with other leading scholars in the United States, including Alexander Nehamas, Charles Kahn, James Lesher, and Richard McKirahan, and in Canada, T. M. Robinson and Brad Inwood. I have also been privileged to receive feedback from a lively community of scholars in Mexico, including Enrique Hülsz and Maria Teresa Padilla. I benefited from a meeting and correspondence with Jonathan Barnes at an early stage of this project. I have also been helped by discussions at international conferences in Lille, France; Mykonos, Greece; and Istanbul, Turkey.

My work in Cambridge was supported by a grant from Brigham Young University, and I have received paid leaves of absence from my university to pursue my research. As ever, I am grateful for the support of my wife, Diana, and my children (now grown), Sarah and Joseph, as we have traveled around the country and the world in search of knowledge.

In the text of this book I have tried to keep the argument accessible to nonspecialists. On the rare occasions in which I cite Greek, I provide a transliteration and translation. The footnotes, on the other hand, contain references in Greek, Latin, and some modern languages, for the benefit of scholars with specialized interests. In a few places I have presumed a basic knowledge of logic, but I have tried to avoid technicalities. I have published a number of more specialized studies on topics related to this book, which are cited in the references. All translations are my own except as otherwise noted.

My thanks to Joshua Gillon for preparing the index.

ABBREVIATIONS AND BRIEF REFERENCES

DG	Diels, Hermann. *Doxographi Graeci.* Berlin 1879. Reprint, Berlin: Walter de Gruyter, 1976. For reconsiderations of and corrections to Diels's treatment of the doxographical tradition in general and his reconstruction of Aëtius in particular, see Jaap Mansfeld and David Runia, *Aëtiana: The Method and Intellectual Context of a Doxographer*, vol. 1, Leiden: E. J. Brill, 1997.
DK	Diels, Hermann. *Die Fragmente der Vorsokratiker.* Edited by Walther Kranz. 6th ed. 3 vols. Berlin: Weidmann, 1951–52. References to source material often include the DK testimony numbers (section A, e.g., Heraclitus A25 = testimony 25 for Heraclitus) or fragment numbers (section B, e.g., Heraclitus B30 = Heraclitus fragment 30). The author number is included only in cases of possible ambiguity; e.g., DK 22B30 = Heraclitus fr. 30.
Hippolytus *Refutation*	Hippolytus. *Refutatio omnium haeresium.* Edited by Miroslav Marcovich. Berlin: Walter de Gruyter, 1986.
KR	Kirk, G. S., and J. E. Raven. *The Presocratic Philosophers.* Cambridge: Cambridge University Press, 1957.
KRS	G. S. Kirk, J. E. Raven, and M. Schofield. *The Presocratic Philosophers.* 2nd ed. Cambridge: Cambridge University Press, 1983.
LSJ	Liddell, H. G., and R. Scott. *A Greek-English Lexicon.* Revised by H. S. Jones. 9th ed. Oxford: Clarendon Press, 1940; with a supplement, 1968.
Simplicius *Physics*	Simplicii *In Aristotelis Physicorum Commentaria.* Edited by Hermann Diels. 2 vols. Berlin: G. Reimer, 1882, 1895.

EXPLAINING THE COSMOS

1

THE IONIAN PROGRAM

This man [Thales] is supposed to be the originator of philosophy, and from him the Ionian school gets its name. It became the longest tradition in philosophy. (Ps.-Plutarch *Placita* 1.3.1)

TODAY MILETUS is a mound rising above a flat plain dotted with olive trees. On the crest of the mound stands a Roman-era theater, and off to the east some stately marble facades line a swampy depression that is the remainder of a once proud seaport on the Aegean, from which little merchant ships sailed to far-off colonies in the Black Sea, the central Mediterranean, and the Nile laden with *amphorai* of olive oil from the ancestors of today's orchards. With her three harbors and a numerous progeny of daughter colonies, Miletus was the "jewel of Ionia,"[1] and she counted among her citizens not only wealthy traders but also wise men whose names have long outlived their native city. For it was here that Western philosophy and science were born, in the first days of the sixth century BC. The little ships carried with their perishable cargoes words that would echo across the Mediterranean Sea and eventually around the world.

Miletus was the most illustrious of a chain of city-states dotting the eastern shore of the Aegean Sea, colonies of Greeks from the Ionian tribe, who gave the name Ionia to their coastline.[2] The first prose books were written in their alphabet and dialect, and their culture combined the best of a resurgent Greek civilization, recently emerged from a dark age,[3] with borrowings from Egypt and the Middle East. Themselves great traders and colonizers, they had daughter cities in the south from the Nile and Libya, west to the coasts of Sicily and southern Italy, France, and Spain, and north to the Black Sea. Thus they were in touch with almost the whole Mediterranean world including three continents. They

---

[1] Herodotus 5.28, the inspiration for a recent book: Gorman 2001.
[2] For surveys of Ionian culture, see Huxley 1966; Emlyn-Jones 1980.
[3] See Snodgrass 1971, esp. 328.

had trading posts in the Levant and served as mercenary soldiers in Egypt. Like the European voyagers of the Age of Exploration, they looked at lands of less advanced cultures as ripe for their own taking, but they traveled to the more advanced civilizations of Egypt and Mesopotamia to learn their secrets.

These great civilizations had managed to organize kingdoms and empires under the direction of a single autocratic ruler. Vast bureaucracies ran complex operations from fielding armies to taxing produce. In Babylonia temple priests kept detailed observations of the skies in order to report—and, wherever possible, anticipate—ominous phenomena. Handbooks of omens were kept from about 1700 BC and records of eclipses from around 747 BC. The Babylonians developed a powerful if complex system of mathematics based on the number sixty, which they eventually (in Hellenistic times) used to track the motions of the sun and moon. The most important element of their calendar was the lunar month, which being of variable length, caused them to make minute observations. The Egyptians in their bureaucracy used skilled scribes who had a good knowledge of basic arithmetic on which to base practical questions of ordering supplies and the like. They used a simple but highly practical year of 365 days and made simple astronomical observations.

Both of these great civilizations developed some powerful tools for scientific research, but neither had the concept of a scientific research program. For the Babylonians, astronomical observations served astrology, while for the Egyptians they served both to determine religious festivals and to anticipate the Nile floods and the agricultural seasons. Both civilizations furnished textbooks to teach mathematical procedures and the solution of practical problems, but neither had a system of proofs.[4] The Greeks learned highly developed crafts and skills from their neighbors,[5] but could have found no real sciences to borrow. Babylonian archives contained vast stores of mathematical and astronomical data on cuneiform tablets, and Egyptian archives contained vast collections of practical documents on papyrus rolls, which could be used in the service of science. But there was nothing recognizable as an institute or association or organized practice of scientific research.

What the Ionians themselves accomplished, and what their contribution to Western knowledge was, have been the subject of ongoing scholarly debate. Some partisans argue that they fairly invented science, others that they merely speculated about the world in a manner incapable of

[4] See Neugebauer 1957.
[5] See Burkert 1992.

producing scientific knowledge.[6] Recently most commentators have been willing to grant to them a modest status as forerunners of scientific thought, part of a complex combination of activities that were destined to contribute to scientific thought and method, including mathematics, medicine, technology, and public speaking.[7] While it is surely true that the Ionians provided only one of several ingredients necessary for the creation of natural science, there remains a sense in which they deserve a special place in the history of Western thought.[8] For a good deal remains to be said about what they accomplished, and how, that will show their contribution as definitive of a new approach, both theoretical and practical, to the world, and in that sense genuinely revolutionary.

In this work I propose to address the Ionians' contribution in a fairly straightforward manner: to retell the story of their intellectual development in a roughly chronological order. But this story will not be just like

[6] A classic statement of the scientific character of Presocratic philosophy is found in Burnet 1930, 24–30. F. M. Cornford changed from seeing Ionian philosophy as scientific (Cornford 1912) to seeing it as non- or even antiscientific (Cornford 1942, 1952). See the famous reply to him in Vlastos 1955b and later in Vlastos 1975, 86–97. Also for the scientific value of the Presocratics is Popper 1958, disputed by Kirk 1960, with a rejoinder in Popper 1963, reviewed in Lloyd 1967. For a recent assessment of scientific claims, see Barnes 1982, 47–56. For a recent reaffirmation of the scientific character of Ionian philosophy, see Longrigg 1993, ch. 2. For analyses of Presocratic methods, see Stannard 1965; Hussey 1995. An early modern criticism of Greek science is found in Francis Bacon's *The Great Instauration*, preface: "[T]hat wisdom which we have derived principally from the Greeks is but like the boyhood of knowledge, and has the characteristic property of boys: it can talk, but it cannot generate; for it is fruitful of controversies but barren of works."

[7] E.g., Lloyd 1979. These activities are not sharply distinguished, nor do they coincide with modern categories: Lloyd 2002.

[8] For instance, though there were mathematicians who were not philosophers, there were also philosophers who were leading mathematicians; Thales in particular is credited with being the first Greek mathematician on the authority of Eudemus (Proclus *Commentary on Euclid* 157.10–11, 250.20–251.2, 299.1–4, 352.14–18 = A20). Philosophers deeply influenced the medical writers, even when the latter criticized them, e.g., "Hippocrates" *Ancient Medicine* 1, 2, 13, 15, 20; *The Nature of Man* 1, 2; Longrigg 1993, 26 et passim; Heidel 1941, 17–19; Jones 1979, 3–6, stressing the influence of Alcmaeon. Although technology no doubt had a life of its own, Thales is famous for combining science and technology, e.g., Herodotus 1.75.3–5 = A6; recently it has been stressed that Anaximander was in touch with the latest architectural technology of his time: Hahn 2001; Couprie, Hahn, and Naddaf 2003. Likewise, public speaking had a life of its own, often taught in the fifth century BC by sophists. Yet the sophist Hippias wrote a history of philosophy: Diogenes Laertius 1.24, cf. authorities in n. 50 below; and Gorgias took up basic ontology in his treatise *On What Is Not* (B3, Pseudo-Aristotle *On Melissus, Xenophanes, Gorgias* 979a12–33, b20–980b21). In general, Ionian philosophy seems to have helped shape all the intellectual currents of Greece (see ch. 11). The thesis I shall defend here is similar to that of Burnet 1930, introduction; however, his historical account is now obsolete. For a brief but powerful recent statement of the originality and independence of Greek science, see Kahn 1991.

other histories of the Presocratics. In the first place, I will tell the story exclusively from the point of view of the scientific or proto-scientific researchers, those who participated in what I shall call the Ionian tradition, as is rarely if ever done. In the second place, I shall maintain, contrary to assumptions common since ancient times, that Presocratic philosophy is not a mere patchwork of different schools or styles of thought; that the Ionian tradition is the dominant current in Presocratic history; and that even those schools of thought that seem most tangential or most opposed to Ionian philosophy—the Pythagoreans and the Eleatics—are deeply indebted to and even parasitic on the Ionian tradition.[9] In the third place, I shall argue that several of the key doctrines and positions commonly attributed to the Presocratics—continuously since the time of Aristotle—result from fundamental misunderstandings of the Ionians and their principles. When we get the doctrines of the Ionians right, we may be in a position to see relationships that were not evident before. The result will, I hope, be a more coherent picture of Presocratic development than is usually attained. It will, in any case, be different in important respects from standard accounts which have prevailed for most of the twentieth century.

1.1 Anaximander's Project

The story properly begins, as has been recognized for the last forty years, with Anaximander.[10] Although the first of the Milesians was Anaximander's mentor, Thales, we do not have enough reliable information about Thales to know how the various elements of this thought and practice fit together.[11] He viewed water as the source of all things, saw the magnet as somehow animated, studied the stars and allegedly founded geometry; he provided political advice to the Ionians and was famed for engineering feats; he seems either to have traveled widely or to have learned from those who did—or both; he may have brought to his countrymen Egyptian surveying techniques and a curiosity about why the Nile floods; he may have been inspired in his choice of a source by Near Eastern myths;

---

[9] I shall not, however, argue the primacy of the Ionian tradition in detail, but rather illustrate it by showing how the tradition developed.

[10] As demonstrated by Kahn 1960, still the standard work on Anaximander.

[11] See now O'Grady 2002, the first book-length study of Thales, and Marcacci 2000. Yet I tend to think that these studies are too optimistic in a number of ways. For criticisms of alleged accomplishments of Thales in astronomy and engineering, respectively, see Graham 2002b, 2004b; for one area in which Thales seems to have made an important contribution, see Graham 2003c. For a reasonable view of what he accomplished in astronomy, see White 2002.

he seems to have borrowed from Phoenician sailors a knowledge of the Little Dipper, which, by its proximity to the pole star, provided a point of reference for navigation and orientation. But apparently he left no writings, and consequently he became an enigma even to Aristotle and the researchers of the fourth century BC. Today we are forced to project back interpretations derived from what we can learn from the later tradition. The first Ionian to leave a written record from which his theory could be reconstructed with some confidence is Anaximander. It is indeed likely that Anaximander inherited both his general assumptions about the world and his approach to it from Thales. But we can get a foothold in the Ionian intellectual world only with Anaximander.

1.1.1 The Pre-philosophical Background

The idea of explaining the world in a scientific way is so common to us now that we should pause to consider its novelty in an earlier time. Many cultures have creation myths which in some way tell the community that shares the myth how it came about that there is a world, and that explain the present order of things, often including the present religious and political arrangements. The Greeks seem to have had creation myths which shared features with other cultures of the Near East and Middle East.[12] At the very beginning of the historical period, when alphabetic writing was new, these stories were organized and unified by Hesiod, around 700 BC, in his *Theogony*.[13]

In this epic poem, Hesiod tells the story of the origin of the world:

> Hail, children of Zeus! Give to me desirable song,
> and proclaim the holy race of immortals who ever are,
> who were born from Earth and starry Heaven
> and Dark night, and whom salty Sea nourished.
> Tell how first the gods and Earth came to be,
> and rivers and boundless sea, with raging swell,
> shining stars and wide heaven above
> [and the gods that came from them, givers of good things];
> and how they distributed their wealth and divided their honors,
> and how first they laid hold of many-folded Olympus.
> Tell me these things, Muses, who dwell on Olympus,
> from the beginning; and say who was first born of them.
> Indeed, first was Chaos born, but then
> broad-bosomed Earth, a steadfast seat always of all

---

[12] Most notably the Hittites' Kumarbi Epic, the Babylonian *Epic of Gilgamesh*, and others; see West 1966, 20–31, KRS 45–6.

[13] West 1966, 40–48.

[the immortals, who hold the peaks of snowy Olympus],
and misty Tartarus in a recess of the wide-wayed earth,
and Eros, who fairest among the immortal gods,
looser of limbs, of all gods and all men
overcomes the thought in their breast and their wise counsel.
From Chaos Erebus and black Night were born,
And from Night Aether and Day were born,
whom she bore being with child after mingling in love with Erebus.
And Earth first bore equal to herself
starry Heaven, that he might cover her all around,
that he might be a steadfast seat always for the blessed gods.
And she bore long Hills, lovely haunts of the divine
Nymphs, who dwell on the woody hills.
And she bore the fruitless deep, with raging swell,
Sea, without desirable love. But then
lying with Heaven she bore deep-swirling Ocean,
Coeus, Crius, Hyperion, Iapetus,
Theia, Rhea, Themis, Mnemosyne,
golden-crowned Phoebe and lovely Tethys.
After them was born the youngest, wily Cronus,
most terrible of her children. And he hated his flourishing sire. (104–38)

Beginning with an invocation to the Muses, the daughters of Zeus who inspire the poet with truths he cannot know for himself, Hesiod recites the beginnings of the world. In an account that has the form of a genealogy, he names the first beings as Chaos, signifying a gap or open space, Gaia or Earth, and Tartarus, or the Underworld. Then dark and light conditions are born from Chaos, while Earth bears Heaven and hills, and in a sexual union with Heaven, Ocean, the body of water that occupies the edge of the earth disk where earth meets heaven. Further, Heaven and Earth beget the race of Titans, who beget the Olympian gods.

Hesiod's story is not a scientific account. In it cosmic beings beget other cosmic and divine beings, who eventually produce human beings. Yet there is an order to the events he describes which provides a kind of systematic explanation for the world, its structures, its inhabitants, and its processes. A begets B, who begets C. Each divine being has its powers and dominion. Earth provides a place for humans to live, Heaven a place for (most of) the gods. Night and Day take turns traveling abroad on earth; Zeus hurls thunderbolts, Poseidon shakes the earth from below the sea. The divinities form alliances, foment plots, and go to war with one another. A succession of divine potentates leads up to the present state of affairs in which the Olympian gods, led by Zeus, control the world. In short, the world has a history which accounts for the way things are now.

How does Hesiod know the history of the world? He calls on the Muses, daughters of Zeus, to inform him of things he does not know. No doubt he has heard similar stories many times told by other bards, who in turn invoked the Muses. Hesiod's stories are not mere inventions of his own: they agree with stories told by Homer and are similar to stories of the Hittite Kumarbi Epic.[14] Yet Hesiod's authority for his story—his version of the truth—is the inspiration he receives from the goddesses of poetry. He tells a story of divine things inspired by divine beings.

1.1.2 Anaximander's Account

Writing in the sixth century BC, Anaximander seems to have also told a story about how the world came to be. But he did not recount it as a tale of divine beings interacting. Rather, he talked about things coming to pass in a natural way, starting with an initial undifferentiated state:

> [Anaximander] says that that part[15] of the everlasting which is generative of hot and cold separated off at the coming to be of the world-order and from this a sort of sphere of flame grew around the air about the earth like bark around a tree. This subsequently broke off and was closed into individual circles to form the sun, the moon, and the stars. He also says that in the beginning man was generated from animals of a different species, inferring this from the fact that other animals quickly come to eat on their own, while man alone needs to be nursed for a long time. For this reason man would never have survived if he had originally had his present form. (Pseudo-Plutarch *Miscellanies* 2 = A10)

According to Anaximander, the original state of affairs consisted of some everlasting stuff, which he elsewhere calls "the boundless." From this primordial stuff some seedlike substance was, as it were, secreted, which gave rise to differentiated things such as hot and cold. From this arose a mass having an earthy nucleus surrounded by a layer of air, surrounded by a shell of fire. The mass burst, producing concentric rings of fire enclosed in air, surrounding a cylindrical earth. The rings are invisible because of the air surrounding them, but a hole allows the fire inside to be seen.[16] The outer ring is that of the sun, the middle that of the moon, and the inner ring, or, presumably, set of rings[17] are those of the stars. We

---

[14] See n. 12 above.

[15] Reading τι with Kahn.

[16] Hippolytus *Refutation* 1.6.3–5 = A11, Aëtius 2.20.1, 2.21.1 = A21, and ibid. 2.25.1 = A22.

[17] Hippolytus, ibid. (see previous note), 5, describes plural rings for the stars. In a recent reconstruction, Couprie 1995; Couprie, et al. 2003, 224–26, hypothesizes a "virtual cylinder." (However, he seems to withdraw or qualify the notion at 227–228.)

see also that he pursues his account all the way to the formation of human beings out of other creatures.

The earth gradually dried out:

> Some say the sea is what is left of the original moisture. For the region about the earth was first moist, and then part of the moisture was evaporated by the sun, and winds arose from it and the turnings of the sun and moon, because their turnings are produced as a result of these vapors and exhalations; and where there is an abundance of moisture for the winds, the turnings take place. And what is left of the original moisture in the hollow places of the earth is sea. Accordingly it continually diminishes as it is dried out by the sun and finally some day it will be dry. Anaximander and Diogenes held this view, as Theophrastus reports. (Alexander of Aphrodisias *On the Meteorology* 67.3–12 = A27)

The muddy earth gradually became drier and perhaps eventually will be completely dry. Life arose in the primeval seas and moved to land:

> Anaximander said the first animals were generated in moisture surrounded by a prickly bark or shell, and as they matured they moved onto land and breaking out of their shell they survived in a different form a short while. (Aëtius 5.19.4 = A30)

Because human offspring must be nourished for a long time before they are self-sustaining, they must have been born as adults.[18] Anaximander explains that they were nourished inside fish until they reached maturity. When the fish burst open, adult humans emerged to populate the land.[19]

The fiery bodies of heaven are, as we have said, holes in the heavenly rings through which the inner fire shines out. The phases of the moon result from the periodic opening and closing of the aperture on the middle ring. Moreover, eclipses result from a sudden blocking of the apertures of the sun and the moon. Winds bursting out of clouds cause lightning and thunder, and other winds cause the "turnings" of the sun, or solstices, which govern the seasons.[20]

1.1.3 The Structure of the Account

We find in this account a kind of pattern which will be important for later accounts.

(1) There is a source from which everything arises:

---

[18] As Ps.-Plutarch explains in the passage quoted above.

[19] Censorinus 4.7, cf. Plutarch *Symposium* 730e = A30; Hippolytus *Refutation* 1.6.6 = A11.

[20] A23, A24, A27.

Anaximander was the son of Praxiades, of Miletus. He said the source and element was the boundless [*to apeiron*], not defining it as air or water or anything else. And the parts change, but the totality is changeless. (Diogenes Laertius 2.1)

In Anaximander the source is incurably vague: the boundless or the ever-lasting. Obviously there is something there which seems always to have been there, and to be everywhere. But what kind of thing it is and what kind of qualities it exhibits are not expressed. Presumably it is some indeterminate stuff that cannot by itself be described or identified. But there is certainly no thought of ex nihilo creation.

(2) There is a process by which the constituents of the world arise out of the originative stuff. First a seedlike stuff appears, then contraries. We next meet with earth, air, and fire. Their precise relation to the emergent contraries is not made clear. It is possible that some further stage of organization takes place by which the contraries give rise to the stuffs. But it is also possible that the contraries are just an abstract way of referring to the stuffs needed to make the world.

(3) The constituent stuffs of the world are organized into the material layers of the world, the *maxima membra mundi*, as Lucretius will call them. Thus the fire, air, and earth form the cylindrical earth surrounded by the rings of heaven.

(4) The structures and materials of the world are stabilized into the state of affairs we are familiar with in the world. The earth dries out so that it consists mostly of dry land in the hollows of which seas, the residue of primeval moisture, are found. Water is evidently one of the cosmic materials, even though it is not mentioned explicitly in reports of Anaximander's cosmogony. The present world is stable, though over a long period of time it may yield to a sort of greenhouse effect and dry up completely.

(5) Living things emerge. They are formed in the seas when moisture prevails, and some things migrate to land by emerging from shells or from aquatic life-forms that produce them.

(6) A wide variety of phenomena are explained by the model. The phases of the moon, lunar and solar eclipses, thunder, lightning, seasonal variations are all accounted for.

This list identifies the content of Anaximander's account. At first sight his account appears impressive in its breadth of conception and the extent of its application. But content alone is not sufficient to establish the scientific character. We may see this by comparing Hesiod's mythological account in the *Theogony* (supplemented by a few points from his *Works and Days*).

In Hesiod we find (1) a source of all things, Chaos, the great yawning gap, the womb in which the world takes shape. (2) There is a process by which the parts of the world appear, namely birth. Indeed, we may distinguish asexual reproduction, which takes place a number of times in the early stages of the cosmogony, and sexual reproduction, which takes place under the influence of Eros, one of the first figures to be born. (3) The structures of the world are born so as to constitute the existing world, as Hesiod's generation understood it, consisting of a flat, circular earth topped by a vaulted heaven and supported by a vaulted underworld, Tartarus.[21] (4) Through dynastic struggles among the leading deities a political/religious order emerges with Zeus as the ruling figure.[22] (5) The human race is created in several successive generations, leading up to the ill-starred present generation.[23] (6) Many phenomena of the world are accounted for by the presence of titulary deities who govern them. Zeus throws thunderbolts, Poseidon shakes the earth, Iris produces the rainbow, and so on.[24] A mythological account can, then, cover the same territory as the proto-scientific account of Anaximander. Indeed, Anaximander and his followers do in a sense continue the tradition of Hesiod in important ways.[25] But there are important differences in the two kinds of narrative.

1.1.4 Scientific Features of Anaximander's Account

What makes Anaximander's account fundamentally different from mythological accounts is not the content and scope of the explanation, but the method. In Hesiod the items that do the explaining, the *explanantia* (singular: *explanans*), are divine persons; explanation consists of showing how a divine person or pair of divine persons begot another divine person. Or, in later stages of the story, it consists of showing how these persons interacted with each other in quasi-human interactions—for instance alliances, ambushes, and battles—to bring about the present state of the world. In Hesiod, to explain the world is to show how it arose from the interactions of supernatural persons.

By contrast, (A) in Anaximander the explanantia are natural events or things: fire, air, earth, wind in clouds, hot and cold. Even the more inscrutable items in the narrative—the boundless, and what is productive of hot and cold—are portrayed as natural entities.

---

[21] The symmetry of Heaven and Tartarus is seen at *Theogony* 721–6.

[22] *Theogony* 666–720.

[23] *Works and Days* 174–201.

[24] E.g., *Theogony* 501–6 tells how Zeus was given thunderbolts as a reward for services rendered.

[25] See Stokes 1962–1963; Diller 1946.

(B) Similarly, the things to be explained, the *explananda* (singular: *explanandum*), are all natural events or things. In Hesiod, on the contrary, it appears that the poet is out to explain not only natural events but also mythological events such as the consignment of the Titans to the underworld, and religious practices and beliefs such as the worship of the Olympian gods. In Anaximander there are no corresponding mythological explananda. Only observable events seem to come in for consideration.

(C) In consequence of (A) and (B), we may ascribe to Anaximander a *closed system of natural explanation*, in which one privileged set of natural phenomena explains other natural phenomena. There is no room for a second set of items apart from those found in nature. There is no "Two-World" theory as in Plato, for whom sensible things are to be explained in terms of nonsensible Forms which exist outside of the natural world. Much less does Anaximander appeal to supernatural beings or their actions to account for phenomena. The world of nature is autonomous and, in a certain sense, self-sufficient for explanation. Consequently, the system leaves no room—or rather, only a limited role—for the supernatural.

(D) In Anaximander's system of explanation, explananda are accounted for by showing how they arose out of or are constructed out of simple materials. Roughly, the kind of explanation Anaximander uses is material cause explanation. The concept of material cause comes from Aristotle; I shall have occasion to modify his particular account of it as it applies to the Ionians. But it will serve as an initial approximation: Anaximander explains the world by deriving it from a certain kind of matter or stuff. In Anaximander there is also a strong component of genetic explanation: showing how the world and its contents arose in a kind of unique historical development.

(E) Anaximander develops his account in prose (even if the prose is enriched with poetic terms).[26] His work is, in fact, a pioneering work in prose.[27] The medium invites an everyday understanding of events as it avoids the magical and marvelous dimension of poetry. The author's use of prose seems to underline the natural, commonplace character of the account, avoiding appeals to the supernatural. And it seems to allow for evidence and argument to support the claims made. Today, we take for granted the value of prose expression for communicating scientific information, but in a time when the alphabet itself was new and literacy limited, when written communications were few, and when the prestige of verse was unchallenged, the choice of prose marked an important departure from previous norms of communication.[28]

---

[26] Simplicius comments on the poetic diction of B1, to be discussed in the next chapter.

[27] Perhaps the very first prose treatise: Kahn 1960, 240.

[28] Greek epic poetry seems to have been composed orally and is adapted to memorization: Parry 1971; Lord 1960. There has been much work on the role of literacy in early thought;

(F) Anaximander appeals to the evidence of experience. If Hesiod's authority comes from the Muses,[29] or more generally traditional lore, Anaximander can claim only that he makes reasonable connections among phenomena. For instance, in his account of the human race (as seen in Aëtius and Pseudo-Plutarch above, A10 and A30), he *infers* that humans were born out of sea creatures because they cannot nourish themselves when they are young. He uses information about the nature and behavior of things (including human beings) to construct a rational account of how they must have arisen or how phenomena such as meteorological events take place. The sole grounds for his account are reason and experience (both taken in a fairly ordinary sense), not inspiration or tradition.

Roughly, characteristics (A)—(F) can be said to be shared by modern science.[30] There are, no doubt, significant differences in the standards of rigor between the Ionian conception and the modern practice of science—none so great as the notion of what constitutes experience in (F). I shall maintain that what the early Ionians have in mind is plain, everyday encounters with the world in which the objects around us are taken more or less at face value. By the end of the Presocratic period, under the pressure of philosophical criticisms, ordinary experience will become problematic.[31] Nevertheless, the first historical step in quasi-scientific explanation seems to arise from showing how remote and apparently portentous phenomena can be understood in terms of everyday events. Of course, modern science has taken the notion of experience to ever greater levels of precision as it tracks nanoseconds and nanometers and collects photons from rare collisions with neutrinos. What seems to be missing from Ionian scientific inquiries, from a modern point of view, is an effective way to test theories and adjudicate between rival hypotheses. The modern scientific community spends vast sums of money constructing particle accelerators, telescopes, satellites, and other powerful instruments designed to collect data that can determine the validity of theories. By contrast, the Ionians seem rich in hypotheses but poor in efforts to test or evaluate them. As an initial approximation, we may conjecture that the program of the Ionians embodies a necessary condition for science as we know it: naturalistic

---

see Robb 1983a; on preliteracy and the Presocratics, see Havelock 1966; on the general question of Greek literacy, see Harris 1989. An even more basic question than in what form to publish one's ideas is the question whether one should publish them at all. The early Pythagoreans may have forbidden publication of their ideas, and hence did not enter the intellectual mainstream until some of their members ignored the ban; see Burkert 1972, 218–38. On the titles and transmission of early cosmological works, see Schmalzriedt 1970.

[29] *Theogony* 104–15, quoted above; cf. Homer *Iliad* 2.484–93, where the Muses are Homer's informants for the catalog of ships because they were eyewitnesses.

[30] Cf. a list of characteristics in Long 1999, 13 (which is more general).

[31] Cf. Democritus B6–8, B9, B10, B11.

explanation of phenomena, but that it falls short of providing sufficient conditions for science insofar as it lacks a plan for testing hypotheses in relation to the world.

At this point we seem to see a strong contrast between a religious-mythological approach in Hesiod and a secular-rational approach in Anaximander. But is this too hasty?[32] Certainly Anaximander did not shun the religious dimension cosmology:

> For that reason, as we say, there is no source of the infinite, but this seems to be a source of everything else and to contain all things and to steer all things, as everyone claims who does not posit some cause beyond the infinite, as for instance mind or love. [B3] And this is the divine, for it is deathless and imperishable, as Anaximander says, and most of the natural philosophers. (Aristotle *Physics* 203b10–15 = A15)[33]

> Anaximander declared the countless heavens to be gods. (Aëtius 1.7.12 = A17)[34]

Aristotle rightly notes that Anaximander assigns divine attributes to his boundless, and he treats it as somehow having the ability to control the world; the world itself, or the several successive worlds, if there are such, seem also to have divine attributes.[35] Yet we must not overlook the vast difference between Homeric or Hesiodic divinities and Anaximander's divine beings. Homer's Zeus asserts his ascendancy over his fellows on the basis of his ability to win a cosmic tug-of-war—though he can be waylaid if he is not careful.[36] Hesiod's Zeus rules in part by wisdom, but he too must be on the lookout against palace coups.[37] Zeus's rule is a very personal one in which he punishes wicked deeds with penalties which he metes out on the basis of the particular crime, as in the case of Prometheus: because Prometheus steals fire for man, Zeus chains him to a rock to have his liver eaten by a bird, while he sends Pandora with a box of evils for men (not least of which is the female sex embodied in Pandora herself).[38]

---

[32] For a discussion of the question of myth and rational discourse, see Lloyd 1987, ch. 1.

[33] Cf. Kahn 1960, 43–44.

[34] See other texts in A17.

[35] Although the sources attribute plural worlds to Anaximander, they may be confusing states of the world with the world: Kahn 1960, 46–53, cf. Cornford 1934. On the religious language of the Presocratics, see Deichgräber 1933. The religious dimension of *to apeiron* is stressed by Sinnige 1971, 1–14.

[36] *Iliad* 8.18–27; 1.396–406.

[37] *Theogony* 886–900; though he is also powerful with his thunderbolts: 687ff.

[38] *Theogony* 521–5, 565–616; *Works and Days* 50–105. Debate continues about how amoral Zeus is in the administration of justice, but all parties concede that Zeus is seriously concerned with his personal honor: Dodds 1951, chs. 1–2; Adkins 1960, chs. 2–3; Lloyd-Jones 1983, chs. 1–2; Vlastos 1975, 10–18.

With Anaximander, by contrast, "it is essential to remember that the 'justice' and 'reparation' of fragment 1 operate simply through the self-regulative periodicities of a mechanical equilibrium."[39] Though there is much more we need to say about Anaximander's theory, we can note at the outset that the kind of regulation that the boundless seems to exercise is an impersonal one that is based on lawlike cycles rather than a personal one based on arbitrary interventions.[40] To assign divine names to the boundless is thus not merely to rearrange the pantheon, it is in part to reconceive what it is to be divine and how divinity operates.[41] In fact, the boundless seems to steer all things through enforced regularities rather than through the willful use of powers such as thunderbolts and earthquakes. But if that is the case, the influence of the boundless can recede into the background as we study the regularities themselves which become the foreground of research. The kind of divinity Anaximander recognizes seems compatible with natural order.

1.2 Anaximander's Project as a Scientific Program

Anaximander's explanation of the world can be seen as providing a pattern for future explanations. It exemplifies a kind of project for anyone who wishes to engage in naturalistic explanation of the world. If we look forward to the last of the Presocratics, we see the project embodied in Anaximander's account still in use. A rough contemporary of Socrates, Democritus applied the atomic theory of his predecessor Leucippus to account for the natural world.

(1) Everything in the world comes from microscopic atoms and empty space.[42] (2) Atoms have an everlasting and irreducible motion by which

---

[39] Vlastos 1952, 115.

[40] "The notion that the primary physical elements are alive and divine is as old as Greek philosophy itself. While we can connect the living earth and air of the early cosmologists with earlier ideas of a divine Gaia in Hesiod, for instance, the more important point is again that in cosmology earth and air, while still divine, are not personal gods. They have no will. They are left with one and only one of the properties of the living, namely the capacity for self-movement" (Lloyd 1975, 203).

[41] "Anaximander's *apeiron* is in no way conscious or personal and, if it guides all things, it does so in no voluntary sense. . . . If [the lawlike control of the boundless] is to be considered in relation to theology, it must be admitted to be a complete rejection of all that was traditional in Greek religion. It is the denial that natural order can be suspended by any supernatural being or force, the denial in fact that any supernatural being can exist, and the assertion that, if the divine means anything at all, it can mean only the system of nature ordered according to infrangible law" (Cherniss 1951, 327). It is possible that predicates associated with divinity were more fluid than at a later time; cf. Clarke 1995 on Thales.

[42] Diogenes Laertius 9.30 = Leucippus A1.

they dart about in the void. At certain times and places in infinite space, atoms by chance collide in such a way as to create a whirling motion that draws in more atoms. A kind of cosmic storm takes place in which a membrane is produced about the whirling mass. (3) Inside the membrane, the vortex motion sifts different kinds of atoms into different places, forming the great masses of the world. A flat earth disk forms in the center which gradually dries out. The heavenly bodies are dried out and ignited by friction.[43] (4) The earth takes shape with the seas in its hollows as residual water from the swampy proto-earth, with the heavenly bodies taking their customary place and order.[44] (5) Living things emerge from the moisture, come to inhabit land, and form the present biosphere. The human race emerges, develops language and culture, and dominates the earth.[45] (6) The multifarious phenomena of heaven and earth are explained by the actions of different kinds of atoms in different sorts of arrangements.[46]

In this account (A) the atoms and empty space, alleged basic features of the world, account for all events, processes, and things. (B) Democritus's account explains all natural phenomena, now including not only events of nature but even human culture and history, which are subsumed under natural processes. (C) Democritus constructs a closed system of natural explanation in which one set of items (the microscopic ones) explains all others. (D) Explanation takes place as a reduction of things and events to pieces of matter in interaction with one another. (E) Democritus writes numerous prose treatises to develop his theory, producing a new vocabulary for some of his ideas.[47] (F) He uses examples from everyday experience to illustrate and support his theory.[48]

The content of the account and the method by which it is arrived at are evidently the same as in Anaximander. The one striking variation we find is that in Democritus the explanantia are nonperceptible and indeed in principle nonperceptible. Democritus posits certain unseen elements from which the world is theoretically constructed, without being able to appeal to the explanantia as familiar items of our environment. (Anaximander did start with an original source that was somewhat mysterious: the boundless, and some generative material that arose from this. Hence there is a sense in which he too depends on imperceptible posits. But Anaximander appears to explain events in the present cosmos in terms of famil-

[43] Diogenes Laertius 9.31–2.
[44] Diodorus Siculus 1.7.1–2.
[45] Ibid. 7.3–5, 8.1–7.
[46] E.g., lightning and thunder in Leucippus A25, Democritus A93.
[47] von Fritz 1938, 12–38.
[48] E.g., Democritus B164; for a recent account of the scientific character of Ionian philosophy, see Hussey 1995.

iar materials.) But even though Democritus's first principles are not sensible objects, they are allegedly natural beings, to be understood by analogy with everyday particles of matter.

Perhaps there is some value in making a further comparison to present-day scientific explanation. (1) More than thirteen billion years ago, a sudden explosion of energy occurred. (2) In the first few instants, a great profusion of subatomic particles was produced, leading to the formation of protons, neutrons, and electrons. The laws of nature we are now familiar with began to apply to the new matter. (3) About three hundred thousand years after the initial burst of energy, protons, neutrons, and electrons joined to form hydrogen atoms. In places hydrogen gathered together, drawn by gravity, to form ever more dense clouds of gas, some of which collapsed into dense spheres which began to produce energy by nuclear fusion, generating helium nuclei as a by-product. In time moribund stars exploded as supernovas and novas, generating a wide range of elements by fusion, which were scattered by the force of the explosions.

(4) Galaxies, systems of stars, formed throughout space, held together by the forces of gravity, often with a massive black hole in the middle. One galaxy, the Milky Way, formed. Twenty-eight thousand light-years from the center, a cloud of dust and gas with the right ratio of metals to gases contracted to produce a middle-sized yellow star, the sun, surrounded by a disk of matter. The matter collected into planets, the third of which, with the right mass and distance from the sun, held an atmosphere and a sea of liquid water. About four and a half billion years ago, the impact of a planetary body the size of Mars sent debris from the planet into orbit; the ejecta coalesced by gravity into a satellite.

(5) Lightning strikes in the sea, rich in hydrocarbons, produced increasingly complex organic molecules, which gave rise to living cells, which evolved into multicellular organisms. Eventually mammals were differentiated from reptile ancestors. As a result of a devastating meteor strike sixty-five million years ago, most large reptile species were driven to extinction, and mammals emerged as the dominant life-form. In Africa hominids emerged among primates, and eventually homo sapiens appeared. Endowed with a large brain, members of the species developed language and culture and spread throughout the world, coming to dominate it and drive to extinction many species of fauna.

(6) All objects are constructed of molecules compounded of atoms, made of protons, neutrons, and electrons. Protons and neutrons, in turn, are composed of two kinds of quarks, while three fundamental kinds of force are carried by photons and bosons. By appealing to matter and the laws of physics, as well as laws of chemistry and biology, phenomena from light radiation to star formation to diseases to volcanoes to earthquakes can be accounted for.

(A) The explanantia now are various and complex, but they are all natural particles or elements. (B) A complex array of laws of nature accounts for interactions, from the quantum level to the redshift of stars. Laws make use of elaborate mathematical relationships. (C) Scientific explanation includes many subjects with different areas of research, but there is at least an assumption embodied in the Unity of Science principle, that a basic set of principles can ultimately account for all phenomena, however difficult it may be in practice to carry out the unification. (D) Explanations ultimately presuppose simplification from organisms to organs to cells to molecules to atoms to subatomic particles, which are themselves pieces of matter or of energy, operating according to basic laws of physics. (E) The many accounts of scientific research are composed in prose treatises giving empirical and mathematical justifications for inferences made. (F) Science draws its evidence from experience, now highly refined in the form of controlled experiments and observations aided by elaborate physical and conceptual instruments such as optical and radio telescopes, electron microscopes, and spectroscopes, mathematical formulas, and computer processing.

It would be naive to think of modern science as a straightforward successor to the theory of Anaximander. Before the twentieth century, scientists could not identify a determinate temporal beginning to our universe, or recognize that there were galaxies or quarks or black holes or many of the entities and events that now populate the standard account of science. There were major historical detours from the Ionian scientific program, advocated by Plato and Aristotle, which dominated the intellectual landscape until the early modern period.[49] Nor is modern science simply a revival of ancient models. Nevertheless, before there was modern science, there was Ionian philosophy, and the Ionian project formed, as we shall see, a sine qua non of all ancient science. And there has been an unbroken historical succession of physical theories from the Ionians to the present. Moreover, the vortices of Descartes and the atoms of Newton were echoes of ancient models from the Ionian tradition. Thus, while modern science is not a pure descendant of Ionian philosophy, Ionian philosophy remains a remarkable anticipation of modern science and unquestionably served as an inspiration and model to early modern scientists.

Anaximander's project, in any case, proved in the hands of his successors a program capable of endless development and, in light of its modern

---

[49] This is not to say that Plato and Aristotle were unscientific, or that they acknowledged no debts to the Presocratics, but that their conception of science in some key respects had less in common with modern science than did Presocratic conceptions. See Vlastos 1975 on Platonic science. Yet even in diverging from Presocratic methods, Plato roughly follows the scheme of a Presocratic cosmogony in his *Timaeus*, and Aristotle, himself ethnically Ionian, in many ways embodies the best principles of Ionian *historia*.

incarnation, productive of the greatest advances in knowledge the world has known. In a sense his private project has become the grand quest for knowledge of the world. In its conception it owes something to Hesiod's myths; but in its execution it draws on a wholly naturalistic method of describing and explaining the world. We may quibble about whether the Ionians were scientists. Certainly they had no scientific method per se, no adequate device for testing their own bold speculations or for adjudicating between competing theories. But that the program was in essence a scientific program is established by parallels to its modern counterpart. And a commitment to the program itself seems to have been the one essential ingredient of science as we know it. If this is right, the West had a scientific worldview before it had science, and the former was an indispensable precursor to the latter.

1.3 Toward an Understanding of the Ionian Tradition

1.3.1 Questions of Classification

Thus far I have spoken very generally about an Ionian tradition; I have not attempted to define the term or to delimit the Ionian tradition from any other movement, school, or tradition. In fact, it is difficult to identify this or any other particular movement among the Presocratics, largely because the interpretive tradition gives a muddled picture of the various schools and stages of thought. Aristotle laid the foundations for historiography of Presocratic philosophy by his at least partially historical account of them in *Metaphysics* I. His aim in that work was not to give a general characterization of Presocratic philosophy but to show that there are just four causes, namely the ones Aristotle himself recognizes. He set out to demonstrate this by showing that his predecessors gradually came to make use of at least some of the four causes. In the process of showing how the causes were used by the Presocratics, Aristotle provided a story of their development from one very specialized point of view. Aristotle's story became the basis of Theophrastus's multivolume history of doctrines, which in turn became the bible of Presocratic history throughout antiquity. Aristotle's own views were not completely original but apparently drew on previous attempts to pigeonhole early philosophers.[50] The interpretation of Presocratic philosophy presented by Aristotle and expanded by Theophrastus has severe limitations, as has been emphasized by twentieth-century commentators.[51] Yet paradoxically, it continues to

[50] As shown by Snell 1944; Classen 1965; Patzer 1986; Mansfeld 1990, 22–83.
[51] Cherniss 1935 provided a detailed survey and argument for Aristotle's failures; McDiarmid 1953 extended the criticisms to Theophrastus. Guthrie 1957 attempted to defend

provide not just the data but also the principles of contemporary interpretation, as we shall see.

What emerged from ancient accounts was a view that there were several schools and movements of philosophy (which overlap each other, as we shall see shortly). The Ionians (consisting in the first place of the philosophers of Miletus) were *physiologoi*, or natural philosophers who focused on cosmology and explained the origin of the world in terms of some fundamental matter. The Pythagoreans appealed to the formal cause and said that everything was number. The Eleatics (Xenophanes, Parmenides, Zeno, and Melissus) argued for the One as against the many appearances and rejected any kind of change. The pluralists (chiefly Anaxagoras, Empedocles, Leucippus, and Democritus) argued for a plurality of fundamental beings that would allow for change and save the appearances from Eleatic arguments. Heraclitus was an important but isolated Ionian thinker who argued for universal flux and the identity of opposites and consequently transgressed the Law of Non-contradiction.

The one thing that twentieth-century historiography added to this story was a clear recognition that Parmenides was the great watershed of Presocratic thought: his arguments forced some kind of a fundamental reaction to his criticisms. Unfortunately, who he was arguing with and why remained unclear, indeed became more uncertain with time. From the late nineteenth to the mid-twentieth century, the role of the Pythagoreans was emphasized: there was an important Pythagorean school in southern Italy from the late sixth century, according to the story, which dominated the philosophical scene of the western Greeks. Parmenides and even Zeno were reacting to members of this school, arguing for their One against the Pythagorean Many, embodied in a kind of mathematical atomism (according to one version).[52] In the fifties and sixties this picture collapsed: there was no evidence for such a Pythagorean doctrine—it was all a mirage generated by the wishful thinking of scholars.[53] This left Parmenides with no dialectical opponent to criticize. About this time analytic philosophy diverted attention away from the historical and dialectical situation to questions about the semantics of the verb "to be," and so the question

---

Aristotle, but his argument is weak: Stevenson 1974. Recently Collobert 2002 has reexamined the question, but she concludes that Aristotle is less than a historian of philosophy. While we are deeply indebted to Aristotle for the information he provides, and while recognizing Aristotle can often distinguish the views of a predecessor from his reactions to them, we need to be careful of his global interpretations. In this study I shall give reasons for thinking Aristotle's overall reading of Presocratic thought is untenable.

[52] Sometimes called "number-atomism." See ch. 6 below.

[53] E.g., Vlastos 1953b, a review of Raven 1948, which sees Eleatic philosophy as closely connected to Pythagoreanism; and Burkert 1972 [1962].

was left unanswered and often unaddressed.[54] There was near unanimity about what happened after Parmenides: the pluralists saw in Parmenides and his followers the threat to scientific philosophy and desperately developed pluralistic schemes designed to allow for phenomenal change without presupposing coming to be and perishing, which Parmenides had ruled out. Unfortunately for them, they ignored other equally damaging arguments against differentiation and any kind of change in general, which precluded the possibility of phenomenal change.

Although today we can claim to have fairly coherent pictures of individual Presocratic philosophers, the details of how philosophy developed in the sixth and fifth centuries BC remain obscure. Once scholars had rejected Pythagoreans as the dialectical targets of the Eleatic school, they lost any direct link between the early Ionians and the rest of the tradition. Who were the Eleatics reacting to—if anyone—and why? On the other side of the divide, why were the natural philosophers after Parmenides so helpless and inept at replying to him?

The obvious place to look for developments in the Presocratics is in the realm of what Aristotle called their causes and principles. According to Aristotle, the Ionians (or most of them) adhered to Material Monism, a doctrine which posited one kind of matter—for instance, water or air—as the source and essence of the world and everything in it. Not only did the world arise *from* the one substance in the distant past, but also all the different stuffs of the world were always composed *of* that substance. For it was the only substance in the world, all other alleged substances being merely modifications of the one stuff. After Parmenides, virtually all the natural philosophers are pluralists, who posit of plurality of stuffs or elements, which combine to form the features of the world. Thus we have a kind of ontological shift that takes place between the pre-Parmenideans and the post-Parmenideans. Aristotle, however, did not exploit this possibility but undermined the scheme by suggesting, sometimes at least, that Anaximander was to be classed with the pluralists because he relied on contraries to make up the stuff of the world and by identifying the post-Parmenidean figure Diogenes of Apollonia as a Material Monist.[55]

Although some leading modern interpreters have proposed alternative ontological schemes, the majority still ascribe Material Monism to the early Ionians, or at least to Anaximenes and Heraclitus.[56] But modern adherents of the traditional view have not resolved the problems of inter-

[54] Starting with Owen 1960.

[55] He is ambivalent about Anaximander, whom he sometimes treats as a monist, sometimes as a pluralist, e.g., *Physics* 187a12–23, where lines 12–15 seem to allude to him as a monist (cp. *On Generation and Corruption* 332a19–25), 20–23 refer to him as a pluralist. Diogenes is classified as a monist like Anaximenes, *Metaphysics* 984a5–7.

[56] Perhaps most forcefully Barnes 1982, chs. 2–3.

pretation any more than did Aristotle. Indeed, modern interpretation is still surprisingly dependent on the schemes of Aristotle and Theophrastus. For all their advances in the study of individual philosophers, interpreters of the late twentieth century have been conservative to a fault. Even some of the most novel recent interpretations of details have been advanced within an old-fashioned, even reactionary, framework of interpretation.[57] In light of gaps and deficiencies of the standard interpretation, I believe it is time to reconsider the development of Presocratic philosophy. The position I will argue for in the present work is not original, but can be seen as a revival (with some significant modifications) of a classical interpretation of Harold Cherniss, reasserted in a revised form by Michael Stokes[58]—an interpretation whose time has come to be taken as the best account of the data.

The set of interpretations proposed in the twentieth century is sufficiently unified in its overall assumptions and its continuity with traditional interpretations that we may call it the Standard Interpretation of Presocratic philosophy. I take it that this view comprises the following claims:

The Standard Interpretation

1. The early Ionian philosophers, or some subset of them, were Material Monists.

2. Parmenides attacked the cosmological program of the Ionians by attacking the foundations of their program.

[57] In their preface to the second edition of KR (= KRS), Kirk and Schofield say, "There has been a spate of publications on the Milesians, Xenophanes and Heraclitus over the last quarter-century, but the effects have been minor compared with those of work on Pythagoreans and Eleatics, and on Empedocles" (x); consequently they made only minor changes in the first part of the book. While this remark correctly describes the received opinion, it also indicates how impervious scholarship has been to several radical criticisms that have been made of explanations of early Ionian philosophy. Barnes 1982, in many ways the most philosophically sophisticated book on the Presocratics in recent times, is most innovative in treatments of arguments and details. In areas of global interpretation, however, such as of Milesian theory, Heraclitus's philosophy, and the pluralists' response to the Eleatics, Barnes follows traditional, even Aristotelian, models almost slavishly, as will be pointed out in subsequent chapters. For more recent reaffirmations of the standard view or parts of it, see Wright 1995, 78–79; Schofield 1997, 68; Taylor 1997b, 209; Hankinson 1998, 19–20; Algra 1999, 57–58.

[58] Cherniss 1935; Cherniss 1951; Stokes 1971. My account of Ionian physical theory is closer to that of Stokes than Cherniss. But Stokes's account loses some force because he rejects Heraclitus as a link between the Milesians and the Eleatics, which Cherniss relies on; I shall defend a view like Cherniss's at the beginning of ch. 6. Strokes also fails to give a positive characterization of Milesian theory. My account of the pluralists' relation to the Eleatics, however, will be fairly different from Cherniss's. I stand by Cherniss's conclusion concerning the early Ionians: "So we find no 'material monism' in the Ionic theories all of which Aristotle reduced to this formula" (1935, 382).

3. The post-Parmenidean pluralists tried to rescue the cosmological program from Parmenides' attack by ascribing Eleatic properties to a plurality of beings.
   a. The early pluralists were unsuccessful because they did not provide a theoretical foundation for plurality or change.
   b. The atomists were successful because they denied a key principle of the Eleatics.

It is my contention that (1) is false, that (2) has not been and probably cannot be connected to (1) by proponents of the Standard Interpretation, and that (3) seriously misrepresents the historical situation: contra (a) there is no evidence that the early pluralists were even trying to answer Parmenides' attacks; contra (b) the case typically made by the Standard Interpretation should entail that the atomists were even more unsuccessful than earlier pluralists.

One general problem for the Standard Interpretation is the scope of the theory: how much does it explain? It seems to leave out at least three major cosmologists as irrelevant: Xenophanes, Heraclitus, and Diogenes.[59] Xenophanes and Heraclitus are philosophical dead ends, while Diogenes is a throwback. On my reading, all three will be integrated as important voices in the ongoing debate. In general, according to my interpretation, Parmenides' criticisms are all the more telling because they address not Material Monism but another type of theory; on the other hand, his criticisms are taken as constructive criticisms by the early pluralists, who see themselves as disciples of Parmenides, so that there is significant continuity between the early cosmologists and the later cosmologists. Parmenides plays a pivotal role but not the role of spoiler he is traditionally accorded.

Against the Standard Interpretation, I shall offer something like the following (in a point-by-point contrast with the Standard Interpretation):

Revisionary Interpretation

1. The early Ionian philosophers were generating-substance theorists (to be explained later).

2. Parmenides attacked the cosmology of the Ionians by attacking their ontology and theory of change.

---

[59] In the present *status quaestionis* Pythagoras gets left out of both the Standard Interpretation and my own, for want of evidence that he had a cosmology. The Pythagorean Philolaus, however, fits into my scheme, though I shall not focus on him. Philolaus has increasingly been accepted as the source of early cosmological reports about the Pythagoreans: Burkert 1972, chs. 3–4; Huffman 1993 (see further sec. 6.1 and n. 16). There have been attempts to rehabilitate early Pythagorean cosmology independent of Philolaus: Zhmud 1997; Kahn 2001, with Kahn 1974, but I remain skeptical. Even if Pythagoras had a cosmology, it must have been indebted to Ionian models in a significant way (see Zhmud 289, 292).

3. The post-Parmenidean pluralists saw themselves as followers of Parmenides, accepting his cosmology as paradigmatic.

    a. The early pluralists carried out the program implicit in the second half of Parmenides' poem.

    b. The atomists modified Parmenides' assumptions to accommodate later Eleatic challenges, but continued to work within an Eleatic framework.

The aim of this book will be to explicate the several theories of the Ionian philosophers, and to elucidate the relationships among the philosophers of the Ionian tradition and between them and their Eleatic critics. Yet I shall maintain that the Eleatics are, in a curious way, also heirs of the Ionian tradition, and connected with that tradition in a more complex way than is usually recognized.

One major consequence that emerges from this line of interpretation, if it is correct, is that the Ionian tradition is *the* great tradition of Presocratic philosophy. Even so harsh a critic of cosmology as Parmenides feels bound to produce a cosmology (or anti-cosmology) of his own—which willy-nilly becomes a paradigmatic cosmology in its own right. The Pythagoreans become participants in the conversation only by introducing Pythagorean religious elements into an Ionian-style cosmology—as does Empedocles[60]—or by founding a cosmology on Pythagorean-style principles—as does Philolaus. Later philosophers are free to introduce new religious, psychological, or ethical elements, and even to find new kinds of fundamental entities, but they must work within the framework of a cosmology—they must make Anaximander's project their program of research. There are, in this vibrant and creative conversation, anti-cosmologists. But they remain the minority view, the Loyal Opposition. Later cosmologists employ ever more sophisticated ontologies and achieve modest but important scientific advances in an ever-expanding domain of research. They always work within the Ionian program.

I shall argue, then, that when we replace inadequate interpretations, we may see the Ionian tradition as pursuing a common program by applying first one, then another system of explanation. Roughly, the two schemes of explanation embody two successive paradigms, in the terminology of Thomas Kuhn, for research. Perceived failures in the former—failures of theory rather than of fit between theory and fact—force the abandonment of one system and the adoption of another. We can, I shall argue, descry a kind of progress from the earliest to the latest cosmologi-

---

[60] Kingsley 1995, 2002 rightly argues for the importance of the religious dimension in Empedocles—a feature often downplayed by scholarship. But religion is integrated with cosmology in the Ionian style, as has become increasingly apparent with the publication of the Strasbourg Papyrus: Martin and Primavesi 1999.

cal theories. And to recognize progress is to allow the possibility that even in its infancy, the scientific program of the West was capable of making advances and in some measure of justifying its own existence. But to test these claims, we must look at the details of Ionian explanations and the controversies they generated.

1.3.2 The Ionian Tradition Characterized

One piece of unfinished business remains. I have yet to say what I mean by the Ionian tradition. Several schemes have been imposed on the Presocratics since ancient times. One is a geographical one, along a north-south axis or meridian: the philosophers from the East, especially from Ionia proper (roughly the Aegean coast of Asia Minor), but including the Ionian language and ethnic group (e.g., Democritus of Abdera, on the coast of Thrace in northern Greece), are cosmologists; the philosophers from the West—southern Italy and Sicily—are religious or metaphysical thinkers, for instance, Pythagoras, Xenophanes, Parmenides, and Empedocles.[61] Second, there is a numerical ontological scheme: monists vs. pluralists, so that, for instance the Ionian monist Thales would be classified differently than the Ionian pluralist Anaxagoras.[62] Third, Aristotle sometimes divides his predecessors into natural philosophers on the one hand, and anti-natural philosophers on the other.[63] The latter class is exhausted by the Eleatics, while the former consists of roughly everyone else. Finally, there is a classification strictly by local schools: the Milesians, the Pythagoreans, the Eleatics.[64]

Each scheme brings with it certain insights. The geographical scheme allows us to see certain differences of content and emphasis, and suggests cultural and historical reasons as their basis. The ontological scheme allows us to group philosophers by their principles, and hence by their philosophical commitments. The division into cosmologists and anti-cos-

[61] Diogenes Laertius 1.13–15. A distinction between Ionic and Italian schools is a major theme of Diogenes: Mejer 1978, 51. KRS follows this scheme in structure, even more closely than did KR, as can be seen in their respective tables of contents. Actually the notion of schools, *haireseis*, was originally applied only to those institutions organized by the Socratics or later (Diogenes Laertius 1.18), and was only gradually extended to the Presocratics (see Mansfeld 1992, 23–26).

[62] Plato *Sophist* 242c–e; Aristotle *Physics* 184b15–25; Isocrates *Helen* 3, *Antidosis* 268. Aristotle's scheme is further complicated by Theophrastus; see Wiesner 1989.

[63] Aristotle *Physics* 184b25ff.

[64] This scheme was followed especially by the composers of successions of philosophers, starting with Sotion or perhaps Ariston of Keos, both Peripatetics: *DG* 147; Wehrli 1967–1969, 65; Burnet 1930, 37. There is a further division of philosophers into dogmatists and skeptics, presumably made by members of the latter group (Diogenes Laertius 1.18); it does not have a strong application to Presocratic philosophy.

mologists allows us to classify philosophers in terms of their most fundamental project. And the school designations allow us to group philosophers in terms of their participation in an intellectual community.

Each scheme, however, also brings with it its own problems and obscu-
rities. The geographical scheme glosses over complex geographical relations. Pythagoras of Samos and Xenophanes of Colophon migrated from
Ionia to the West. Are they then westerners or easterners? Traditionally
the former is classified as a Pythagorean, the latter as an Eleatic. Melissus
of Samos in Ionia is classified as an Eleatic, although we do not know
that he ever left his native Ionia. Empedocles of Acragas in Sicily has much
in common with the Ionians, although he is also indebted to the Eleatics
and Pythagoreans. The ontological scheme breaks down when we see that
Aristotle seems to treat Anaximander sometimes as a monist, sometimes
as a pluralist. Xenophanes can be read as a monist, a dualist, or a pluralist.
And we find no correlation between geography and ontology: the easterners Thales, Anaximenes, and Diogenes are allegedly monists, but other
easterners—Anaximander (perhaps), Anaxagoras, and Democritus—are
pluralists, and so on. The distinction between cosmologists and anti-cosmologists breaks down when we realize that Xenophanes, allegedly an
anti-cosmologist, is arguably a cosmologist, while Parmenides the anti-
cosmologist devotes most of his poem to developing a cosmology. The
school classification projects onto the Presocratics institutional relationships that can be documented only from the fourth century on.[65] It further
creates problems by sometimes combining figures who arguably never
associated with each other, such as Xenophanes, Parmenides, Zeno, and
Melissus. Of these four Eleatics, only the second and third are likely to
have known each other. School successions regularly combine philosophers who, for chronological reasons, demonstrably could not have studied together. For instance, Anaxagoras and Diogenes of Apollonia are
alleged to have been students of Anaximenes; but both students, according to our chroniclers, were born after their teacher's death.[66] Meanwhile, important thinkers such as Heraclitus are left out of the schools.

[65] The one solid piece of information that seems to point in the direction of a teacher-
student relation is Alcmaeon B1, which mentions three disciples. Alcmaeon was a physician
as well as a philosopher, and disciples may have been apprentices. Problems with viewing
the Ionians as a school are laid out already by Grote 1881, 2.217: "Writers, ancient as well
as modern, have professed to trace a succession of philosophers, each one the pupil of the
proceeding, between these two extreme epochs [from Thales to the time of Socrates]. But
the appellation is in truth undefined and even incorrect, since nothing entitled to the name
of a school, or sect, or succession . . . can be made out."

[66] Diogenes Laertius 2.6; cf. Hippolytus *Refutation* 1.8.1. According to Apollodorus, Anaximenes died in the sixty-third Olympiad, 528–25 BC (DL 2.3), while Anaxagoras was
born in the seventieth Olympiad, 500–497 BC (DL 2.7). Diogenes of Apollonia "lived
around the time of Anaxagoras" (DL 9.57), but must be younger still.

Thus the traditional classes do not seem to be self-consistent or to reinforce each other.

Is there, then, any hope of meaningful classification of the Presocratics, and is any meaning to be assigned to terms like "Ionian"? The one scheme that seems to work fairly well is the most modern: the distinction between pre- and post-Parmenideans. This is a historical-dialectical distinction that does seem to account for differences in approach according to the thinkers' location in historical space. As to the term "Ionian," it admits of expansion. If we include Xenophanes and Heraclitus with the Milesians, we see similar patterns of explanation—or so I shall attempt to show. And after all, both Xenophanes and Heraclitus are from Ionia, and both are pre-Parmenidean.[67] The early (sixth-century) Ionians do embody an approach to cosmology that has a chance of being in some sense unified and hence instructive to us. And if we can establish that Parmenides is reacting to *their* style of cosmology, instead of to some shadowy Pythagoreans or to no one at all, we shall have established an important dialectical link for the history of philosophy.

But of course, after Parmenides there were cosmologists too. Barnes has dubbed their theories "neo-Ionian systems" in a happy phrase.[68] The authors of these theories include Ionians as well as non-Ionians (most notably Empedocles). These neo-Ionian thinkers continue, in broad outlines, the project of the early Ionians. To that extent, the fifth-century cosmologists are the intellectual heirs of the sixth-century cosmologists, who happen to be Ionians. Accordingly, I shall refer to those who pursue the goal of rational explanation of the cosmos, in either century, as adherents of the Ionian tradition. By calling it a tradition I do not wish to commit myself to the school scheme; by calling it Ionian I do not wish to commit myself to a geographical scheme. I wish only to pay respect to the great intellectual flowering of the sixth century that established in Ionia for the first time a program of the sort that could be emulated and pursued ever after by thinkers making similar assumptions.

One further caveat: it is a problem of philology how soon the term *kosmos*, "order" came to mean "world." According to different scholars, the latter sense could occur as early as Anaximander or as late as Plato.[69] But it is a fact that from Anaximander on, philosophers were attempting to explain the cosmos, whether they had a word for the world or not. Since

---

[67] The case for Heraclitus is controversial; I will argue for this point in ch. 6.

[68] Barnes 1982, 305 et passim.

[69] For the early date, Kranz 1938, 433; Kahn 1960, 188; for the late, Finkelberg 1998a. See Diller 1956; Kerchensteiner 1962; Vlastos 1975, ch. 1. In any case *kosmos* is thematically linked with the natural world, at least from the late sixth century: Heraclitus B30, Diogenes B2, in both cases marked by a demonstrative apparently referring to the actual state of affairs.

my thesis focuses on the type of explanation the early cosmologists used rather than on the language they used to describe that type of explanation, my thesis does not depend on the resolution of the philological question.

It is a contingent fact of history that a new style of explanation was invented in Ionia in the sixth century BC. But it is precisely the ability of this style of explanation to be exported, adapted, continued, and expanded that made it so valuable. A refugee from the Persian invasion of Ionia, Xenophanes could carry it in his head, and embellish with it the poems he recited to kings and symposiasts. It guided a seaman from Massalia (Marseilles) on the first recorded Atlantic voyage by a Greek, and it guided a historian from Ionian Halicarnassus on his voyage up the Nile in the mid-fifth century BC.[70] By such modest means it seems to have quickly traveled the length and breadth of the Mediterranean, wherever learned Greeks gathered and talked about the nature of things. It transcended schools and geographical regions, and ultimately historical epochs as well: decades and centuries, and, arguably, millennia. Indeed, it has become a possession for all time. Perhaps what gave the new style of explanation its impetus was the very process in which it was embedded: more than a set of results or a final conclusion, Ionian inquiry embodied an ongoing program of research. The program generated successes and apparent successes that justified its continued pursuit.

Ultimately, driven by a promising program of research, the Ionian tradition defined the world as a natural realm governed by lawlike regularities. The program promised to make the events of the natural world comprehensible in all their details, and to make the world itself an object of knowledge. The first steps in the tradition can be traced in Anaximander, to whom we now turn.

---

[70] A striking (but little-known) example of the early diffusion of physical theories is that of the explorer Euthymenes of Massalia, who seems to have used Thales' theory of the Nile floods in his observation of the waters of a major river on the Atlantic coast of Africa (perhaps the Senegal River), which he mistook as the source of the Nile (Seneca *Natural Questions* 4A.2.22). Euthymenes' date is uncertain, but could be as early as 530 BC: Jacoby 1909; Cary and Warmington 1929. Ionian theories guided Herodotus's inquiries in his own visit to the Nile *(Histories* 2.19–29); see Graham 2003c.

2

ANAXIMANDER'S PRINCIPLES

ANAXIMANDER STANDS AT the beginning of the Ionian tradition. There is much we do not know about Anaximander's theory. But we are fortunate to have a few valuable reports and, more important, one informative fragment from which we can glean at firsthand a few precious insights into the structure of Anaximander's world and his way of explaining it. What is most important for our story, and most difficult, is to understand the principles of explanation in Anaximander's theory.

2.1 Out of the Boundless

Most of what we need for a reconstruction of Anaximander's theory comes from two texts as follows:[1]

> After [Thales] Anaximander, who was his associate, said the boundless contained the whole cause of coming to be and perishing of the world, from which he says the heavens are separated and generally all the world orders, which are countless. And he declared perishing to take place and much earlier coming to be, all these recurring from an infinite time past. He says the earth is cylindrical in shape, and its depth is one third of its width. He says that a part of "the everlasting" which is "generative of hot and cold separated off" at the coming to be of the world-order and from this a sort of sphere of flame grew around the air about the earth like "bark" around a tree. This subsequently broke off and was closed into individual circles to form the sun, the moon, and the stars. He also says that in the beginning man was generated from animals of a different species, inferring this from the fact that other animals quickly come to eat on their own, while man alone needs to be nursed for a long time. For this reason man would never have survived if he had originally had his present form. (Ps.-Plutarch *Stromateis* 2 = A10

---

[1] Cf. Hippolytus *Refutation* 1.6.1–2, which provides some information about the boundless that will be discussed later.

Of those who say the source is one and in motion and boundless, Anaximander, the son of Praxiades, of Miletus, the successor and student of Thales, said the source and element of existing things was "the boundless" [*to apeiron*], being the first one to apply this term to the source. And he says it is neither water nor any other of the so-called elements, but some other boundless nature, from which come to be all the heavens and the world-orders in them: [B1] From what things existing objects come to be, into them too does their destruction take place, "according to what must be: for they give recompense and pay restitution to each other for their injustice according to the ordering of time," using these rather poetic terms. (Simplicius *Physics* 24.13-20 = A9)

In these passages I have put in quotation marks those words that are likely to have been Anaximander's.[2] In these passages *to apeiron*, "the boundless," is a prominent concept which, according to Theophrastus and the doxographical tradition, identifies the source and element of Anaximander's philosophy. The assumption that he has a source and element goes back to Aristotle's interpretation of the early Ionians, about which I shall have more to say later.[3] For now, suffice it to say that Aristotle and his followers regarded the Ionians as explaining the world by reference to a single source or principle from which it emerges.

2.1.1 Meaning of Apeiron

The term *apeiron* seems to go back to Anaximander himself. What does it mean? Several issues are to be considered: (a) etymology, (b) ordinary, nonphilosophical, usage, and (c) philosophical usage. In Homer, land and sea are sometimes given the epithet of *apeirôn*.[4] There is good reason to think this meaning may be illuminated by etymology, which suggests that the term originally meant (1) "uncrossable," related to *peraô* "pass over."[5] But this in turn comes to be understood as implying vast extension already in Homer, which suggests a meaning of boundless,[6] while the original meaning disappears from sight. By the fifth century BC it clearly is

---

[2] Following Kirk at KRS 117–18, Barnes 1982, 132–34, and earlier Burnet 1930, 52, n. 6; yet I am persuaded that Kahn 1960, 166–78, is right to say the first phrase of B1 is a close paraphrase of what Anaximander said. Dirlmeier 1938, on the other extreme, argues that all the wording is infected by later terminology and concepts.

[3] See below, sec. 3.2.

[4] The epic form of *apeiros*.

[5] See Kahn (1960, 231–39) for a detailed analysis. Cf. Aristotle *Physics* 204a2–6 with Solmsen 1962, 122–24.

[6] Cunliffe 1924, s.v. ἀπείρων.

contrasted with what is bounded or has a limit (*peras*).[7] Thus it seems appropriate to translate the term (2) "boundless" even if it did not originally mean that.[8] But scholars have also asserted that the term means (3) "indeterminate," implying some kind of nondescript stuff as the basis of the world. And it comes to mean (4) "infinite" in a mathematical context, the ramifications of which Aristotle explores in *Physics* III.

The argument for the meaning (3) "indeterminate" is that since Anaximander's colleagues supply a source that is definite, such as Thales' water and Anaximenes' air, Anaximander's failure to mention a specific source is tantamount to an assertion that the boundless is indeterminate. But, as recent critics have pointed out, "boundless" never *means* "indeterminate."[9] We may be justified in inferring from Anaximander's failure to define his boundless further that it is an indeterminate stuff, but that will be an inference from his use of the term, not a meaning of it. What he calls it is "boundless"; that it is indeterminate is a possible inference, but it is our inference, not his statement.

As to the meaning (4) "infinite," that seems too technical.[10] In the fifth century BC, Anaxagoras will assert infinite divisibility and Zeno will explore its problems; Philolaus will divide all reality into limiters, unlimiteds, and their products. Geometry will make major advances in dealing with conceptual space, and the atomists Leucippus and Democritus will propose a universe that is infinitely extended. In the fourth century Aristotle will explore the concept of infinity and deny that there can be an actually infinite extension.[11] But all of this remains in Anaximander's future. The conceptual distinctions needed to identify his boundless with infinity are not available, and consequently it seems anachronistic to understand his vague term with such precision. We are left, then, with "boundless" or "limitless" as offering the only viable translation for Anaximander's term.

A further question arises as to whether the boundless is to be taken as spatially unlimited, temporally unlimited, or both. Clearly Anaximander regards his boundless stuff as temporally unlimited, for it is "everlasting and ageless" (B2), "deathless and imperishable" (B3). But as one scholar has pointed out, *apeiros* usually means spatially unbounded, but never means temporally unbounded in its early occurrences, and Aristotle and

[7] The contrast is especially obvious in Philolaus B1, B2, B6. But see earlier Xenophanes B28 with Dancy 1989, 163–64.

[8] Kahn 1960, 233.

[9] Gottschalk 1965, 51–52; Dancy 1989, 151, 163ff., esp. 170–72. For a list of those who have held *apeiron* means "indefinite," see Dancy, ibid., nn. 58, 59, 61.

[10] Cf. KRS 109–10; Bicknell 1966b. Sweeney 1972, 55–73, too readily identifies *to apeiron* with infinity.

[11] *Physics* III.4–8.

others have to use terms other than *apeiros* to describe the temporally unbounded quality of the *arche*.[12] Thus we can take *to apeiron* as a spatially unlimited stuff.[13]

2.1.2 Function of the Boundless

Now even those who criticize "indeterminate" as a translation of *apeiron* agree that it is in fact an indeterminate stuff, according to the reasoning we have already examined.[14] The next question that arises is: what is the nature of this indeterminate stuff? Is it something neutral between opposites? For opposites such as the opposing sides referred to in B1 seem to arise out of the boundless (in A10). Or are the opposites present as in a mixture, from which they can be extracted? These sorts of worries are precisely those that appear in Aristotle's exposition of Anaximander, and they recur in modern commentators.[15] But whichever of the two views interpreters take, they tend to recognize in the boundless a neutral state of matter—neutral either because it is essentially a neutral kind of matter between hot and cold, wet and dry, and so on, or neutral because it is a mixture containing equal amounts of the contrary powers.

This allows the interpreters to appeal to one possible reason for positing the boundless, which was proposed by Aristotle long ago:[16] the boundless is a kind of reservoir which assures that the particular kinds of matter will never run out because it contains all kinds of matter in it, at least potentially. If, by contrast, the primitive state of the universe

[12] Dancy 1989, 165–67. Stokes 1976, 12–18, argues for a temporal understanding of the boundless. It is controversial whether Theophrastus claimed that Anaximander was the first to use the term *arche*: Burnet 1930, 54, n. 2; Jaeger 1947, 24–28; Kahn 1960, 30–32; KRS 108–9, on Simplicius *Physics* 24.15–16, 150.23–4, and Hippolytus *Refutation* 1.6.2. Surely Anaximander did not use the term in the technical way Aristotle later would (see n. 42 below), but it seems to me not unlikely that he used it in a nontechnical way, as Jaeger and Kahn argue. In any case the boundless *is* the source of the world in a fundamental way, and I shall so refer to it. For an early use of the term *arche* meaning "starting point," with perhaps an implication of a physical source, see Diogenes of Apollonia B1 with Laks 1983, 18–19.

[13] Solmsen 1962, 114, points out that whereas for Anaximander *apeiron* functions as a subject, for his successors it tends to function as a predicate.

[14] Dancy 1989, 172ff.

[15] Aristotle *Physics* 187a12–23 seems to present both possibilities. Alexander (*On the Metaphysics* 60.8–10) presents the notion of a substance between air and fire as being Aristotle's understanding of the boundless. In favor of an indeterminate stuff are McDiarmid 1953, 100–103, who argues that Theophrastus follows Aristotle's confusion; Gottschalk 1965, 50, and Dancy 1989, 179ff. Finkelberg 1993 further identifies the boundless as airy stuff; for the mixture view, see Cherniss 1935, 375–79; Vlastos 1947, 170–72. For effective criticisms of the mixture view, see Seligman 1962, 40–49.

[16] *Physics* 204b22–9, 203b18–20, cf. Barnes 1982, 30–31.

consisted of fire, for example, then there would be a constant danger of the world's being consumed by fire. Indeed, it would be difficult to account for the fact that any stuff containing properties opposite to those of fire, say something cold and wet like water, could arise in the first place. Thus by assuming the neutrality of the boundless, we can derive a kind of philosophical justification for positing the boundless as the primitive state of matter.

There are, then, some reasons for thinking the boundless is a neutral stuff. But before we accept this interpretation, we ought to ask whether the inference from the boundless to the indeterminate is in fact justified.[17] The move from the boundless to the indeterminate presupposes that Anaximander is trying to avoid saying something like his predecessor Thales says when he names water as the principle and element of things. But since we know so little about Thales and his project, we should be cautious here. Is Anaximander trying to avoid saying what his stuff is, or is he telling us something positive? Notice that in A10 he refers to the original stuff as "the everlasting," designating it as something having a temporal as well as a physical extension. Thus it has at least two positive characteristics: physical and temporal extension. Furthermore we learn that it "surrounds all things" and "steers" them.[18] So far there is nothing negative or privative about his conception of the original stuff.

But how does the boundless actually function? What does it do? The one (relatively) clear statement we get is about the role of the boundless in cosmogony. A part of the boundless which is generative of hot and cold separates off from the boundless and produces stuffs that form the parts of the world, according to A10. The term "generative," *gonimon*, suggests a seedlike material, and "separating off," *apokrinesthai*, is a term used in later medical and biological treatises to denote the body's act of secretion. As has often been noted, the language of A10 suggests a biological metaphor of generation.[19] The substances that make up the world: fire, air, and earth, arise from this creative process and form into a circular configuration in which the fire surrounds the earth like bark around a tree. Again, the biological references are obvious.

Anaximander's use of biological metaphors is, in a certain sense, both natural and historically appropriate. It is natural because generation is the closest parallel in the world of human experience to natural production of any kind. Indeed, in Greek the linguistic connections are inescapable: the

---

[17] Barnes 1982, 34–36, gives cogent reasons for rejecting the view.
[18] Ibid. 203b11–12.
[19] E.g., Kahn 1960, 156. The classic study is Baldry 1928. See also Lloyd 1966, 232ff., on the biological analogies in general, and 234–35 on Anaximander. Cf. Baeumker 1890, 11–14.

verb *gignesthai* expresses not only the abstract concept of coming to be, but also the specific action of being born, down to the latest ages of classical antiquity.[20] Furthermore, the Milesians' closest predecessor is Hesiod, who in his *Theogony* told the story of cosmogony precisely as a theogony: a genealogy of gods in which the earliest generations represent cosmological features: Earth, Underworld, Heaven, Hills, Ocean, (inland) Sea, Day, Night, etc. In noting this similarity we should not minimize the differences: Anaximander's coming to be is a natural process starting from a kind of impersonal matter and producing impersonal stuffs. The supernatural figures, their liaisons and personal prerogatives are all gone, in favor of a natural process of impersonal entities. The ontology and the etiology are utterly different. But both authors understand cosmogony as somehow analogous to natural generation.

But if we now apply the biological analogy seriously to our understanding of Anaximander's concept of matter, we observe problems. However natural the metaphor of birth and generation is for cosmogony, it remains mysterious at a certain level. The gestation process is a highly complex series of events that even now is difficult to comprehend; in the sixth century, before the earliest biological and medical studies, it was totally beyond comprehension. The important feature for Anaximander is no doubt the fact that it is a thoroughly natural process. Yet it would have remained to him and his contemporaries inscrutable. To appeal to the generation of the world as evidence for the role of the boundless seems perverse. First some part of the boundless with generative powers separates off from it; it generates opposites which form into a circular configuration and develop into the cosmos.[21] If Anaximander's point is the neutrality of the boundless, why tell such a complicated story? Why not have the boundless differentiate itself immediately into the opposites that are potentially or actually present? What the analogy suggests is not the neutrality of the boundless but its fertility and fecundity. The boundless is not the matter for the stuffs of our world, as Aristotle suggests, but the parent or matrix. First it must secrete some specialized material which then produces the several parts of the world not by a mechanical process but by a quasi-biological process. The boundless is the original matter out of which the world and its component stuffs come to be, but it is not itself the matter of the world, as Aristotle wants to claim.[22]

[20] Cf. Kahn 1960, 158.

[21] We cannot be certain that Anaximander referred to the world as *kosmos*, a world order. Yet it seems likely that he did (Kahn 1960, 33–35, 188–93, 219–30) or perhaps used the term *kosmos* in a more limited sense (Kerschensteiner 1962, 29–66). In any case, his concept of the world is such that it is adequately described by the term.

[22] Cf. Stokes 1971, 30–31.

What then is the ontological relationship between the boundless and the stuffs of our world? We simply do not know. The boundless remains outside the cosmos, surrounding and controlling it in some fashion, but it is not, so far as Anaximander tells us, in our world. It is forever inaccessible and mysterious, beyond empirical scrutiny. We know only its results, not its nature. As to events in this world, we must understand them in light of the powers and materials in it, and their own behavior.

2.2 Powers in Conflict

The interpretation of Anaximander's physics that has been generally accepted is that of Charles Kahn (1960), which is anticipated in important ways by the study of Vlastos (1947). According to this interpretation, the "existent objects" of B1 are the elements or powers. They come to be out of their opposites, and perish back into them, thus paying restitution to them for their injustice. For in the familiar cycles of the seasons, and of day and night, one power comes to predominate, for instance heat in summer, cold in winter. The power comes to trespass on the territory of the other, or to usurp its prerogatives. But it in turn loses its dominion and is displaced by its opposite in the succeeding time period. Thus there is a cycle of transgression and retribution, which in turn calls forth a corresponding retribution, and so on. This interpretation makes for a very satisfying account of the periodic cycles as embodying an order. But there have been some recent challenges to this account.

Vlastos supported the claim that the powers maintain an equilibrium over time by showing connections between this cosmic theory and the political concept of *isonomia*, which allowed citizens equal participation in government.[23] The value of *isonomia* as an explanatory concept is reflected by Alcmaeon of Croton, who regards health as *isonomia* of different powers in the body, sickness as *monarchia*, whereby one power dominates. Against Vlastos, Freudenthal (1986) has argued that we can see from related theories of disease in early medical writings that it is possible to have different states of equilibrium, including states of ill health, which require intervention to correct. On this analogy, the cosmos requires the intervention of the boundless, which, according to Anaximander, steers all things, to correct the excesses of each season or stage of a cycle. "The Boundless must alternately counteract the opposites, just like the physician" (208).

[23] Cf. Vlastos's 1947, 156–58, with his historical-political discussion of *isonomia* (Vlastos 1953a).

This criticism does raise some genuine questions about the received interpretation of Anaximander. (a) What guarantee, if any, is there for the maintenance of the cosmic equilibrium, on Anaximander's theory? (b) What role, if any, does the boundless play in maintaining the equilibrium? To the degree that (a) remains unanswered, we may be driven to positing a large role for the boundless in (b). Vlastos sees the boundless as merely enclosing the cosmos and hence as keeping it as a closed system.[24] Hence he seems to owe us a strong account of the mechanism, or whatever we can put in its place, for an autonomous system of changes. As to an explicit statement by Anaximander, there is only the statement about time. The phrase at the end of the fragment can be taken in a more figurative way than the relatively neutral translation I used above: "according to the ordering of time,"[25] where time could be a personification, representing a cosmic magistrate by whose decrees cosmic justice is maintained. This reading, however, only delays the question: what power and authority does time have to maintain the balance of opposites? For of itself what happens in time seems to result from other forces and factors. At one extreme, Anaximander may be offering his image only as a placeholder for some kind of process immanent in nature which assures the continued stability of the world.

At the other extreme, it is possible that Anaximander sees the structure of the world itself as the guarantor of stability. Looking forward to Aristotle, for instance, we see in him that the regular motions of the sun determine the advent of summer and winter, with their characteristic qualities.[26] In a certain sense, then, to depict a world with regular motions is to provide, at least tacitly, the conditions for cosmic justice. But is this reading too much into the first cosmologist, perhaps putting the cart before the horse? As always at the beginning of a tradition and with scanty evidence, it is impossible to know with certainty how much Anaximander is asserting and on the basis of what assumptions and evidence. But if we remember that time is closely connected with astronomical cycles in Greek thought—after all, we still measure time primarily in astronomical units: days, months, years—it seems plausible to think that the measures of time play a causal role. The sun rises, burns off the mist, at midday it heats the air, in the evening it sets, bringing cooler temperatures and moist dew. As the sun rises higher in the sky in spring, it warms the earth and dispels the fog and mist of winter, and so on. Thus the regular motions of heavenly bodies are responsible for the sequential increase and decrease of heat, of moisture, and of light. The circles of the heavenly bodies pre-

---

[24] Vlastos 1947, 173. Kirk (KRS 116) favors control by instituting the law of retribution.
[25] Or, more pointedly, "the assessment of Time," in Kirk's rendering (KRS 118).
[26] *On Generation and Corruption* II.10.

sent a structure for temporal cycles, and in that way guarantee a limited advance and retreat of contrary powers.

Somewhere between the image of cosmic justice as a statement of faith in the power of naturalistic explanation and a completely worked out cosmology lies the system of Anaximander. As to the specific criticism that different states of equilibrium are possible, this seems true but in a certain sense irrelevant. The medical analogy is at best a later invention, informed no doubt by the cosmic conception of justice as well as the political. But it appears at the earliest with Alcmaeon of Croton in the early fifth century.[27] Neither Vlastos nor Kahn wants to make the medical analogy the standard for Anaximander's justice. Vlastos stresses *isonomia*, or political equality, as at least helpful for understanding the issues at stake. One can question whether such a concept was sufficiently established in Anaximander's time, and especially in Ionia where aristocratic values were still prevalent.[28] But we should not get too far from Anaximander's fragment. The kind of justice envisaged in the fragment is not concerned directly with civic political arrangements, but with possessions. Anaximander does not have in mind primarily a constitutional crisis, but a dispute about ownership. My neighbor rustles my cattle or plants his crops on part of my land; in retaliation I demand my cattle back along with some of his as a compensation, or my land back along with some of his land. I take back my cattle with some of his—but I take more of his than I should, and he demands compensation from me. This private feud can, it is true, occur in the context of the whole polis, where one party or faction takes too much and is displaced in a revolution.[29] It can also occur in the context of an intercity dispute in which the army of one city occupies the territory of another.[30] But in no case is Anaximander's image one of stable constitutional rule: it is one of serial injustice. In a democratic constitution, such as envisaged by Vlastos, the officers do not perish when they lose power, nor is the exercise of their office an injustice. The injustice Anaximander has in mind is violent occupation or usurpation, not sanctioned by a constitutional system. In the personal sphere it is *hubris*, in the city *stasis* or partisan strife, between cities in a state of *polemos* or war. Heraclitus, who was able to read Anaximander B1 in context and as a near contemporary and neighbor, understood it this way, as we see in his criticism:[31]

[27] Alcmaeon B4.

[28] Hahn 2001, ch. 5, gives evidence that major public projects in Ionia were supported by wealthy private donors.

[29] Cf. Thucydides 3.82–4 on factional strife in wartime.

[30] Cf. Homer *Iliad* 18.509–40.

[31] That Heraclitus is in part criticizing Anaximander is widely recognized: "Heraclitus' statement (fr. 80) that 'strife is justice (the normal course of events)' is almost certainly a

We must recognize that war is common, and strife is justice, and all things happen according to strife and necessity. (B80)

Heraclitus identifies the systematic alternation of power as strife, which he identifies with war. And he notes that the system itself is an embodiment of justice, rather than injustice, as Anaximander would have it.

What ensures the stability of the whole system? In the case of a private feud, the authority of the city can at least put limits to the dispute. In the case of class strife or intercity warfare, the limited resources of parties and states can cause them to become overextended and vulnerable to counterattack. The image of alternating injustice makes a certain kind of sense in its own right and does not need to be eked out by theories of medical balance or political equality. While we can learn much about the interaction of different concepts of justice from such theories, we should not lose sight of the simple notion of injustice as trespass and justice as reoccupation, or more generally, injustice as overreaching, *pleonexia*, justice as getting what is one's own.[32] "To the Greeks, for whom the just is the equal, this pleonexy, or taking-too-much, is the essence of injustice. We must not think of civil and constitutional rights but simply of property rights—the daily quarrel over mine and thine."[33] If Anaximander's conception of justice is in some sense political, in a broad sense of "political," it concerns not the politics of constitutional rule but of justice, and indeed not distributive but retributive justice.[34]

One further challenge has recently been raised to Kahn's interpretation of cosmic justice. Engmann (1991) stresses that those of the doxographical tradition, including Simplicius (whom Kahn takes as supporting his reading), are unified in interpreting cosmic justice as resulting from destruction of the opposites into the boundless itself. On her interpretation, "When one opposite had increased too greatly over a period of time, it forfeited some of its gains by the resolution of these back into the infinite" (21). While this does seem to agree with at least many of the statements of doxographers, there is a problem noted long ago by Vlastos, Kahn, and others: in the fragment, Anaximander declares that the opposites pay restitution "to each other."[35] On the ancient reading of the doxographers, they pay restitution to the boundless, a third party. This may accord with

---

criticism of Anaximander's metaphor"; Kirk 1954, 401, cf. 240; Vlastos 1955a, 356–58; Kahn 1979, 206–7. It is disputed by Marcovich 1967, 139–40.

[32] Cf. Plato *Republic* 331e.

[33] Jaeger 1947, 35.

[34] See Aristotle *Nicomachean Ethics* V.1–4, who distinguishes between distributive and rectificatory (*diorthôtikos*) justice 1130b1–2, 1131b25. In the latter "the judge tries to equalize this kind of injustice, which consists of inequality" (1132a6–7).

[35] Vlastos 1947, 169–72; Kahn 1960, 180–83.

modern conceptions in which violation of a city ordinance results in a
fine to a third party, the city itself; but it does not accord with ancient
practices which specify paying compensation to the injured party directly.[36] In this case, Anaximander's own image is inconsistent with the
ancient interpretation of it, and we are justified in departing from the
doxographical reading, no matter how venerable it may be.

Let me add one further consideration to the argument. The first phrase
of B1, "From what things existing objects come to be, into them too does
their destruction take place," is often taken as a paraphrase by
Theophrastus based on an Aristotelian formula.[37] For instance, Aristotle
says, with special reference to the Ionians,

> that of which all existing things consist, and that from which they come to
> be first and into which they perish last . . . *this*, they say, is the element and
> source of existing things. (*Metaphysics* 983b8-11)

It appears, then, that Theophrastus is merely reciting a Peripatetic formula for the element and source of things. But this formula is not a pure
invention of Aristotle's. We find in Xenophanes, in the late sixth century,
the following statement:

> For from earth are all things and into earth do all things die. (B27)

This line embodies the same formula, but two centuries before Aristotle.
Aristotle does indeed give us a formula, but one based on early usages.
And it is at least possible that the earliest instance of the formula was one
found in Anaximander B1. The crucial point is that the one early parallel
we have describes a situation in which particular things arise out of a (or
the) major element, not one in which worlds come to be out of a primeval
stuff. For Xenophanes has no cosmogony as far as we know.[38] Thus the
context suggests that Anaximander is discussing the origin of particular
portions of stuff, not of the world itself, much less plural worlds. The
world itself, as we have seen, is generated by a quasi-biological process
out of the boundless.

The only consistent account of Anaximander's cosmos seems to be that
developed by Kahn. Although it is not clear how well-founded his conception of cosmic justice was, we can say that Anaximander did see cyclical
changes of opposites as coming to be out of each other in accordance with
some sort of corrective system.

[36] E.g., Homer *Iliad* 18.497–508.

[37] E.g., Kirk at KRS 118.

[38] Hippolytus *Refutation* 1.14.2–6 = A33. Although Xenophanes recognizes cyclical
changes in the world (presumably the *kosmoi* in v. 6 are arrangements of the world), the
world itself does not come into being or perish. For a similar statement about the origin of
things, cf. Diogenes B7.

2.3 Elements and Powers

Thus far I have discussed the entities of Anaximander's theory using a number of expressions such as "matter," "power," "stuff," "material," "opposite." Our doxographic sources use later terminology, much of it developed by Aristotle, to describe the things that are left vague in earlier writers. (Even Aristotle often uses neuter plural adjectives where we need some sort of substantives to make sense of his concepts.) Anaximander talks of his basic starting point as simply *to apeiron*, "the boundless," giving no characterization of its nature, and he probably talks of "existing things," *onta* (in Theophrastus's report, *eonta* in the Ionic dialect), without giving a specific characterization of them.[39] We are told by Simplicius that Anaximander's "contrarieties are hot, cold, dry, moist, and the rest" (*Physics* 150.24-5). But immediately we run into the problem of whether we are dealing with, in terms of later distinctions, types of matter or qualities or what.

As with other concepts of Anaximander's theory, our best recourse is to compare his concepts with other early conceptions of the tradition. We also need to be aware of later schemes used to explicate his theory. Briefly, Aristotle analyzes physical objects as composed of a formal and a material component. There are hierarchies of form and matter ranging from the complete organism, whose form can be viewed as the soul, and whose matter can be viewed as the body, to the four elements. The most rudimentary kinds of matter are the four elements, which in their own right have a formal and a material component. The formal component is one member of the pair hot/cold and one of the pair dry/wet. Each element is uniquely characterized by its forms—for instance, the hot, dry element is fire. The material component is a characterless substrate called "prime matter."[40] Since the four elements change into each other, they are not simple or irreducible, and are not really elements (simple components). Hence Aristotle often refers to them as "the so-called elements."[41] The identity of the four elements Aristotle inherits from Empedocles (who called them *rhizomata*, "roots"), though the analysis of them is his own. Aristotle uses the term *arche*, "beginning, source, principle," to designate

---

[39] See Kahn 1960, 80–82, who shows how the *onta* are often best taken as elements or primitive entities; cf. Diogenes B2, τὰ ἐν τῶιδε τῶι κόσμωι ἐόντα νῦν, γῆ καὶ ὕδωρ καὶ ἀὴρ καὶ πῦρ καὶ τὰ ἄλλα, Anaxagoras B17, ἀπὸ ἐόντων χρήματων, Hippocrates *On the Nature of Man* 7.

[40] Aristotle *On Generation and Corruption* 329a24ff. For prime matter, *Metaphysics* 1049a24–6. It is controversial whether Aristotle believes in prime matter, and the concept is certainly problematic philosophically, but I think Aristotle does believe in it: Graham 1987b.

[41] See Bonitz 1870, 702b2ff., for references.

anything from the starting point of a line to the source material of a change to a geometrical principle.[42] The second sense occurs frequently in his discussion of the Presocratics. The most basic ontological concept for him is that of "substance" (*ousia*), for which the paradigm examples (at least sometimes) are biological individuals.[43] He believes that all his predecessors were trying to understand substance, but they only managed to posit the existence of elements or stuffs ("homoeomeries," *homoiomerê*, or things whose parts are like the whole—e.g., the parts of flesh are portions of flesh).[44] They talk of contraries, but fail to understand clearly that contraries can only occur in some matter, and only certain contraries, the primary objects of touch, are appropriate as the characteristics of the elements. For Aristotle there are neat correspondences between the four primary objects of touch (hot, cold, dry, wet) and the four elements.[45] But what of the Presocratics?

Each philosopher has his own story to tell. But for Anaximander, on the one hand, it appears that some sort of contraries were necessary to his account of cosmic justice. On the other hand, his cosmogony makes use of ordinary stuffs that arise out of the cosmic seed. To go back to A10:

> He says that that part of the everlasting which is generative of hot and cold separated off at the coming to be of the world-order and from this a sort of sphere of flame grew around the air about the earth like bark around a tree.

Here instead of explicating his cosmogony in terms of hot, cold, and the like, Anaximander seems to have quickly introduced fire (or "flame"), air, and earth as the components of his cosmos. Apparently the concept of contraries does not preclude the existence of determinate stuffs (all of which will later be classified as elements). It seems likely, indeed, that the contraries are embodied in stuffs.[46] Anaximander's story of how the earth was dried out also implies the existence of water (found in seas), giving us the fourth element:

[42] *Metaphysics* V.1 gives Aristotle's several uses of the term.
[43] E.g., *Categories* 1b3–6, *Metaphysics* VII.4, 16.
[44] *On Generation and Corruption* 314a18–20, *Metaphysics* 1028b8–12, 1040b5ff.
[45] *On Generation and Corruption* II.2–4.
[46] Thus Hölscher 1944a, 1953b; Kahn 1960, ch. 2; Lloyd 1964, 95–97, cf. KRS 120; against this conception, see Fritz 1971, 21–22. It is perhaps significant that Simplicius, commenting on Anaximander B1, identifies the opposites with elements: "It is clear that, observing the change of the *four elements* into each other, he did not think it appropriate to make one of them the substratum of the others, but something else beside them. And he did not derive generation from the alteration of some element, but from the separation of *contraries* due to everlasting motion" (*Physics* 24.20–25, A9, continued from passage quoted at beginning of the present chapter). His talk of the four elements is, no doubt, anachronistic; yet he finds it easy to see the contraries functioning as components of elements.

Anaximander says the sea is the remainder of the original moisture, of which fire dried out the majority, and what was left changed its character owing to the heating action. (Aëtius 3.16.1 = A27)

It is not clear what the change of character comes to. Minimally it could be the change from fresh water to salt water. On the other hand, there may be something about moisture in its original state that is different from any water. In any case, what is left of the original moisture manifests itself as seawater now. It is possible that earth in its primal swampy state was not quite earth, but something of different character, but now it is earth. And there is no clear indication that the moisture ever existed apart from, or in abstraction from, moist things.

Accordingly, the contraries seem to enter into and manifest themselves as familiar stuffs of the world. We do not see any clear appeal to disembodied or elementary powers (*dunameis*) such as are sometimes invoked as primitive entities. Furthermore, before Aristotle there seems to be no clear-cut mapping of qualities or powers to elements such that the elements are reducible to or straightforwardly definable in terms of qualities.[47] As far as one can see in Anaximander, phenomena can be described either in terms of stuffs or in terms of powers indifferently. There does not appear to be a clear ontological distinction. If this is right, then when the hot of summer perishes into the cold of winter, the change may plausibly occur in events such as a portion of fire turning into a portion of water. Certainly when we get to the level of meteorological explanation, we find changes accounted for in terms of concrete substances:

Concerning thunder, lightning, thunderbolts, cyclones, and hurricanes: Anaximander says these all happen as a result of wind. For when wind surrounded by thick cloud breaks out violently owing to its rarity and lightness, the tearing action produces sound, the separation against the dark cloud the flash. (Aëtius 3.3.1–2 = A23)

Here wind, treated as at least a force, embodies the qualities of rarity and lightness. Its action is quite physical, a violent rending that produces a flash and a shock wave. Anaximander also explains wind itself:

Anaximander [says] wind is a rush of air when the most fine and moist parts of it are moved or dissolved by the sun. (Ibid. 3.7.1 = A24)

So wind is air in a state of motion caused by the action of the sun, itself a body composed of fire. Anaximander's method of explaining meteoro-

---

[47] Longrigg 1993, 220–26, finds the nearest anticipations of Aristotle's theory of elements with its associated pairs of qualities in the medical theory of Philistion, a contemporary of Plato's. Before this we cannot document any fixed correlation of powers and elements.

logical phenomena depends on determining the states and natures of physical stuffs and seeing how they interact with each other. Qualities and powers are an essential part of the story, but they do not act apart from the stuffs in which they occur.

Indeed, apart from the enigmatic boundless and its generative seed, there is nothing in Anaximander's ontology that we do not experience in everyday life: there is earth, water, fire, air. Air in motion is wind, which produces some meteorological phenomena, and cloud is some kind of moist vapor that produces winds. And winds account for the sometimes terrifying phenomena of lightning and thunder, and indeed even the "turnings of the sun," the solstices:[48]

> The whole region about the earth was moist at first, but being dried out by the sun, they say the vapor produced the winds and the turnings of the sun and moon, while what was left became the sea. Accordingly they think the sea is diminishing as it dries out and finally some day it will all be dry. (Aristotle *Meteorology* 353b6–11)[49]

Thus everyday events can account for portentous occurrences in the sky, and even for astronomical phenomena.

In Anaximander we see a closed system of explanation, in which a set of items, apparently including elemental stuffs and their contrary properties, accounts for all the phenomena of experience. To judge from our reports, Anaximander does not give laws of action for his explanantia, other than one very general one: excesses of one contrary are punished by succeeding excesses of the other. But he recognizes orderly interchanges between different elemental bodies, and a physical structure within which the interchange takes place. Indeed, if the winds control the solstices, they also control the seasons and the interchanges of hot and cold, dry and wet. It is possible, then, that cosmic justice is attained by the structure of the cosmos itself. The drying out of the earth may spell doom for future life if the process is irreversible. But whatever the case, the world attains a long-lasting regularity. From what we can see in the working out of Anaximander's cosmology, the world needs no intervention by the boundless. It is autonomous.

But what is the boundless? It is a spatially unlimited and everlasting stuff that exists outside the cosmos, whence it surrounds and "steers" the world we humans live in. From what we can see, it produces the world through a quasi-biological process of generation. What exists in our world is not the boundless itself, but contraries, or perhaps elemental bodies characterized by contrary qualities. Apparently the elemental powers or bodies con-

---

[48] Alexander *On Aristotle's Meteorology* 67.3–12 = A27.
[49] Alexander, ibid., and Aetius 3.16.1 confirm this as referring to Anaximander.

flict with one another in such a way that some of them perish to produce their opposites in cyclical interchanges, so that at one time (e.g., summer) the hot and the dry dominate, at another time (winter), the cold and the wet. The boundless itself does not seem to enter directly into these meteorological processes, but at most to contain them.[50]

On the other hand, to stress the absence of the boundless in our world seems to slight its importance. The boundless has a creative function and brings, with its intimations of immortality, hints of divinity, power, and providential control. It serves as parent and god of the world. Are we not misrepresenting Anaximander when we view him as a proto-scientist concerned with physical explanations of the world rather than with its divine parentage and purposiveness?[51] Indeed, there is a divine dimension in Anaximander's cosmology. His *arche* is endowed with immortality, power, fecundity, and, apparently, foresight.[52] Yet it is also, as far as we can see, bereft of individual character and particular motivations, and hence of the normal conditions of personhood. The boundless is virtue and power at once idealized and depersonalized. Anaximander's theology is more like deism than theism, with a creator, or rather begetter, that remains at the periphery of the world both literally and figuratively. In many ways it is the forerunner of Anaxagoras's enigmatic Mind (B12), which initiates the cosmic sorting process and comprehends its outcomes, without apparently directly accounting for anything that goes on after the initial impulse.[53] At best both Anaximander's boundless and Anaxagoras's Mind remain part of the explanatory background of everyday processes. And insofar as they sit as benign spectators on the margins of the cosmos, they can be ignored in explaining quotidian events. If Anaximander envisages teleology in his grand view, he does not seem to invoke it in his account of how the world operates. And if he has theological preconceptions, he does not draw on them for the details of his cosmol-

---

[50] It has even been argued that popular meteorological views inspired Anaximander's conception of justice (Shelley 2000). This may be so in the order of discovery, but as far as the order of justification goes, meteorology is surely the explanandum rather than the explanans.

[51] For a maximal view of Anaximander's religious theory, see Jaeger 1947, 23–37, who concludes, "In this so-called philosophy of nature we have thus found theology, theogony, and theodicy functioning side-by-side" (36). For an extreme and outdated view of Anaximander as a theologian, see Burch 1949. In general, see criticisms in Vlastos 1952, 113–17, who emphasizes the fact Anaximander and his tradition stressed the lawlike features of the divine.

[52] It would be appropriate to study teleology in early Greek philosophy at least from the time of Anaximander; unfortunately Theiler's monograph ([1925] 1965) effectively starts with Diogenes of Apollonia. But there are serious problems with Theiler's account even regarding Diogenes: Laks 1983, 250–57. See ch. 10 below.

[53] Plato *Phaedo* 97c–98b, Aristotle *Metaphysics* 984b8–22, 985a10–21; Lesher 1995.

ogy. We need not deny the obvious religious dimension of Anaximander's work to see him as making groundbreaking contributions to a new kind of discourse in which natural processes explain natural events.

In Anaximander the boundless is a divine, if inscrutable, being, the source in some quasi-biological sense of the world. As far as we can see, however, the boundless does not exist in our world, but only its offspring, the elementary powers or bodies that are subject to a kind of law of reciprocity, a *lex talionis*. The law itself appears to be as impersonal as it is inescapable, and if it is enforced by the boundless, the enforcer remains invisible. In any case, the boundless is a source of the world, but not its ground or substratum. That simple fact will have important implications for a study of the elements.

3

ANAXIMENES' THEORY OF CHANGE

THE THIRD PHILOSOPHER from Miletus, Anaximenes is often thought of as the least interesting member of the school.[1] Thales is known for grand schemes and great predictions, Anaximander for his sweeping vision of cosmology and his daring concepts. Anaximenes retreats from the exotic boundless to prosaic air as a source, and eschews indifference as an account for the earth's position in favor of a cushion of air. What Anaximenes is renowned for is his theory of change, which assigns a regular sequence of changes for air as well as a mechanism for action. Scholars have recently questioned even this relatively modest achievement as an invention of later commentators, as we shall see. In this chapter I wish to vindicate Anaximenes as offering an important theory of change. But in the process of defending his achievements, I will argue that the theory of change commonly attributed to him since ancient times has been misunderstood in crucial ways. Furthermore, Anaximenes' theory, understood in its historical context, is not only helpful for the Ionian project, but absolutely essential. It is this theory, more than any other, that put Ionian inquiry on the path toward scientific thinking.

3.1 The Theory of Change

As we have seen, Anaximander believed in a regular order of changes. There are natural cycles—days, years, months—in which opposites give way to opposites. The hot, bright, dry conditions of day turn into the cold, dark, moist conditions of night; the hot, bright, dry conditions of summer turn into the cold, dark, moist conditions of winter. There is a kind of balance in the cycle that is analogous to a system of justice in which the trespasses of one power are recompensed to the contrary power. The image of justice is powerful and suggestive, but it is merely an image. What actually controls the cycle, ensures the regularity, and enforces the assessments made? Our sources do not say, and it seems likely that the

---

[1] "last and least of the Milesian triad" (Schofield 2003b, 50).

silence goes back to Anaximander himself. He expresses faith in regularity, but does not explain why or how it comes about.

It is precisely here that we need some sort of theory to account for regularity. We need to know what elements and what forces are in play, and how they interact to maintain the balance that is critical to the continuance of the world. And here precisely is the innovation of Anaximenes. Simplicius extracts from Theophrastus Anaximenes' general account of change:

> Anaximenes, son of Eurystratus, of Miletus, was an associate of Anaximander, who says, like him, that the underlying nature is single and boundless, but not indeterminate as he says, but determinate, calling it air. It differs in essence in accordance with its rarity or density. When it is thinned it becomes fire, while when it is condensed it becomes wind, then cloud, when still more condensed it becomes water, then earth, then stones. Everything else comes from these. And he too makes motion everlasting, as a result of which change occurs. (Simplicius *Physics* 24.26–25.1 = A5)

Hippolytus adds some details:

> Being condensed or thinned [air] changes its appearance: when it is dispersed to become thinner, it becomes fire; when, on the other hand, air is condensed it becomes winds, and from air cloud is produced by "felting"; when it is condensed still more water, when condensed still more earth, and when it is condensed as much as possible stones. (*Refutation* 1.7.3 = A7)

The theory as reported here has three important features: (1) air is the source of everything by virtue of (2) being the underlying nature; (3)(a) there is a set of elements or basic stuffs: (i) fire, (ii) air, (iii) wind, (iv) cloud, (v) water, (vi) earth, (vii) stones, (b) ordered by their relative density, (4) which arise from each other precisely by their being rarefied or condensed; (5) the condensation process is analogous to the action of felting. The reference to felting is interesting because Anaximenes himself seems to have introduced the model on which the metaphor in this passage depends. Felting is the process of making felt out of wool by subjecting the wool to great pressure (as well as moisture, heat, and movement). This piece of simple technology illustrates the way in which compression can change the properties of things, and several references to felting in the doxography of Anaximenes seem to go back to the philosopher himself.[2]

Here we have a rich account of how change takes place in the world. Whereas Anaximander's source is mysterious and seems not to exist at

---

[2] In particular, the analogy of the heavens as a felt cap (*pilion*) is a striking image: "This image is scarcely likely to have been invented by anyone except Anaximenes" (KRS 156). Bicknell 1966a suggests that the object referred to may be a circular strip rather than a cap; but for our purposes the material is more important that the shape of the headgear.

all in our world, Anaximenes' source is familiar and ever-present in our world. Although Anaximander does not specify what his "existent objects" are, Anaximenes gives us a determinate list of stuffs. He also orders them with reference to a single principle, density, along a presumed continuum between the most rare and the most dense. Finally, he provides a principle by which one stuff turns into another. That principle is, moreover, completely mechanical: if certain conditions are met—for instance, air is adequately compressed—the outcome inevitably occurs: wind results. The mechanism leaves no room for events happening arbitrarily or mysteriously. They happen, when they happen, necessarily and predictably. Anaximenes has given us something like a scientific law. The law is not, to be sure, mathematically expressed in a formula, as we are accustomed to have since the time of Newton. Yet it has, in its crude way, a quantitative basis: the continuum of density is determined by the quantity of matter contained in a given space. There is, then, something ineluctably scientific about Anaximenes' explanation, however crude and speculative it may be.

But perhaps this view is hasty: like Anaximander, Anaximenes ascribes divine properties to air:

Anaximenes, son of Eurystratus, of Miletus, declared air to be the source of beings. . . . As[3] our soul, he says, which is air, controls us, so do breath and air encompass the whole world-order. (Aetius 1.3.4 = B2)

This "fragment," which is probably at best a paraphrase,[4] indicates that air has the power to control things; in other places we are told that air is a god.[5] If Anaximenes stresses the spiritual qualities of air, should we not take him as more a religious than a scientific thinker? As with Anaximander, we need to take the religious dimension of Anaximenes' thought seriously. But as with Anaximander, there are reasons to doubt that this dimension dominated Anaximenes' theory. In his famous digression on cosmology in the *Phaedo*, Plato considers, among others, the theory that we think by means of air (96b). But he has Socrates point out the folly of trying to explain why he is in prison on the basis of the disposition of his limbs and the constitution of his body (98d–99b). He singles out a cosmology like that of Anaximenes as inadequate to deal with the im-

---

[3] Or: "For example": Longrigg 1964.

[4] See, e.g., KRS 159; Alt 1973 argues persuasively that, although the "fragment" goes back to an early Peripatetic original, it paraphrases Diogenes of Apollonia rather than Anaximenes. For the sake of argument, however, I will assume that it tells us something about Anaximenes.

[5] Cicero *On the Nature of the Gods* 1.10.26, Augustine *City of God* 8.2, Aetius 1.7.13, all collected at A10.

portant questions of why the world is as it is.[6] The only thinker who seems to offer anything more than a mechanical account is Anaxagoras (97b–98b); yet disappointingly he does not follow through with his hints of a teleological account of the world (98b–d). Although Plato works more by allusions than by laying out a history of early philosophy, he clearly does not find in any of the early thinkers a teleological account of the world, such as is suggested by Anaximenes B2. It is often suggested that Plato is not acquainted with Anaximenes (or Anaximander),[7] so that his disappointment with early philosophical theories might be thought to result from his ignorance of some of them. But I will show below that he is indeed familiar with Anaximenes' theory. If that is so, then either Plato is very unfair to Anaximenes, or he does not find a suitable application in Anaximenes of the teleological and providential character the Milesian ascribes to air. As in Anaximander and Anaxagoras, the vitalistic or religious dimension seems to retreat into the background.

The mechanical dimension of Anaximenes' theory entails that change is no longer spontaneous or capricious. But what precisely are the assumptions that he is making? What is the status of air, and what is its relation to the other elements of the series? Here our ancient authorities have a story to tell that we must attend to carefully.

3.2 Material Monism

Thus far we have sidestepped the interpretation of the Milesians' *archê* or source. Ancient interpreters, starting from Aristotle, are quite specific about its ontological status in a complex theory of matter:

> Of the first philosophers, the majority thought the sources[8] of all things were found only in the class of matter. [I] For that of which all existing things consist, and that from which they come to be first and into which they perish

---

[6] *Phaedo* 99b–c describes a theory according to which the earth rests on air like a lid. This view is shared by Anaxagoras (Hippolytus *Refutation* 1.8.3 = A42) and Diogenes of Apollonia (Diogenes Laertius 9.57 = A1). But as we shall see below, Plato knows Anaximenes' theory as well as those of Anaxagoras and presumably Diogenes.

[7] Thus Gigon 1968, 43: "Die entscheidende Tatsache ist, dass zur Zeit Platons von keinem der drei Milesier mehr Schriften bekannt waren. Das darf man daraus schliessen, dass Platon von ihnen überhaupt nur Thales kennt." No source before Aristotle mentions Anaximenes by name: Wöhrle 1993, 31. But Plato does mention Thales' school: *Hippias Major* 281c (I am inclined with many others to accept this dialogue as genuine).

[8] We might well translate *arche*, "principle," in this passage, given Aristotle's sophisticated view of the concept (expressed in the next sentence). But I wish to stress what may be the common core of the Ionians' *arche*. For the range of meanings of the term, see Aristotle *Metaphysics* V.1—which, however, does not clearly include the sense Aristotle explains here.

last—[II] the substance continuing but changing in its attributes—*this*, they say, is the element and this the source of existing things. Accordingly [III] they do not think anything either comes to be or perishes, inasmuch as this nature is always preserved. . . . For a certain nature always exists, either one or more than one, from which everything else comes to be while this is preserved. [IV] All, however, do not agree on the number and character of this source, but Thales, the originator of this kind of theory, says it is water. . . . Anaximenes and Diogenes [of Apollonia] posit air as the simple body prior to water that is most properly the source. (Aristotle *Metaphysics* 983b6–13, b17–21, 984a5–7)

Here Aristotle identifies a type of theory presumably going back to Thales, according to which there is only one substance. Everything arises from this and perishes back into it. But even when it appears that new things have arisen from the source, there really are no new things, but only the source itself is really present. There is no coming to be or perishing because that would imply the coming to be of some new substance or reality, whereas everything really is identical with the one substance. According to Aristotle, different Ionians defend different sources as the ultimate reality: Thales says it is water, Anaximander the boundless,[9] Anaximenes air, and Heraclitus fire. Some later doxographers (but not Aristotle) say Xenophanes makes earth the source.[10]

The present interpretation clearly provides a strongly metaphysical reading of the theory of change, identifying a single reality as the ontological basis of the theory. This interpretation is known as Material Monism and may be summarized as follows:[11]

Material Monism

I. Everything arises from and terminates back into one source.

II. Everything is in essence identical to that source, which is a single substance.

III. There is no (unqualified) coming to be or perishing, but only alteration.

IV. The source of all things is (a) water or (b) air or (c) fire or (d) the boundless (?) or (e) earth (?).

The term "substance" in (II) is an Aristotelian term for an independently existing reality. Two particular things are the same in essence just in case they have the same definable nature.[12] The terms "coming to be," "per-

---

[9] As we have seen (ch. 1, n. 55), Aristotle wavers between making Anaximander a monist and making him a pluralist.

[10] See below, sec. 3.4.3 et passim, for a discussion of Xenophanes.

[11] This is a preliminary characterization only; in the next chapter a formal characterization will be given.

[12] Aristotle *Metaphysics* 1018a6–7, cf. 10–11.

ishing," and "alteration" in (III) are all Aristotelian terms denoting different kinds of change that are defined in terms of his categories. Coming to be and perishing are changes in the category of substance, i.e., the arising or destruction of a basic reality, whereas alteration is change in the category of quality, such as a change of color.[13] In other words, since there is only one independent reality in the world, any changes that occur are changes of its features, not cases of destruction of the reality itself. Point (IV) identifies the one substance with some particular stuff. In general Aristotle contrasts Material Monism with Material Pluralism as represented by, among others, Anaxagoras and Empedocles. There is a rough correlation between the early (sixth-century) monists and the later (fifth-century) pluralists, but it is not complete: Aristotle toys with treating Anaximander as a pluralist, and he sees Diogenes of Apollonia as a monist.[14]

This interpretation of the early Ionians as Material Monists is taken over by Theophrastus and the doxographical tradition, and is so prevalent that there is no rival account.[15] Virtually all interpreters from Aristotle on accept this reading of the early Ionians. Similarly, the great majority of modern commentators accept the view that the early Ionians are Material Monists.[16] To reject the view would be to take arms against over twenty-three hundred years of consensus.

3.3 Problems with Material Monism

3.3.1 An Argument for Material Monism

Yet there are reasons to question the consensus.[17] First, there is no fragment that unambiguously commits the early Ionians to such a view. This may, of course, simply reflect the scarcity of material we have. But it raises the question of what textual evidence Aristotle used as the basis for his

[13] Aristotle *Categories* 12, *Physics* III. 200b32–201a9. In *Physics* V.1–2 Aristotle limits the term "motion" (*kinêsis*) to accidental changes of place, quality or quantity; coming to be and perishing are changes (*metabolai*) but not motions. This nicety need not concern us.

[14] On Anaximander see below; on Diogenes, *Metaphysics* 984a5–7, quoted above.

[15] Simplicius *Physics* 24.29–31 in A5a quoted above is virtually identical to ibid. 149.29–31, showing that it is a quotation from Theophrastus.

[16] Including Gilbert 1907, 42–43; Burnet 1930, 73–74; Bailey 1928, 16–18; Cornford 1937, 180; Guthrie 1962–1981, 1.115–16; Lloyd 1970, 19–22; Lloyd 1979, 140–41; von Fritz 1971, 33–34; Hussey 1972, 27; Kirk in KRS, 145–46, Barnes 1982, 38–44; Hankinson 1998, 19. Often Anaximenes is seen as the first Material Monist, as in Lloyd and von Fritz. Against Barnes in particular see Graham 1997, 12–17.

[17] Guthrie 1962–1981, 1.64–65, criticizes MM as being inadequate in light of the Milesians' failure to distinguish material and spiritual aspects. They do indeed, but the relevant sense of "material" in "material monism" refers to Aristotle's material cause, not to matter in contrast to mind; the post-Cartesian debate is out of place here.

interpretation. Plato gives a reprise of early Greek philosophy in which he fails to attribute Material Monism to the Ionians.[18] Evidently he did not find an overwhelming case for making the Ionians monists of any kind. If we look at passages like that in which Theophrastus characterizes the views of Anaximenes (A5a, quoted above), we find no compelling reason in the exposition itself. The only place that the story appeals to the monistic metaphysics is in the statement that Anaximenes' "underlying nature," *hupokeimenê phusis* is air. But the adjective here is clearly Peripatetic terminology, and Theophrastus is merely identifying Anaximenes' *arche* with what he takes a priori to be a kind of single matter that underlies everything. In other words, Theophrastus is interpreting Anaximenes' source according to an Aristotelian schema. But one could perfectly well accept the whole subsequent narration without conceding that Anaximenes is a Material Monist (as will be seen more clearly below).

Besides tendentious readings of the Ionians, one argument seems to be compelling to many. This is given in a perspicuous statement by Jonathan Barnes:

> Now if Y comes to be *from* X by "thickening" or "thinning," by condensation or rarefaction, then surely Y is made *of* X. If ice is condensed water, if it is made from water by a process of condensation, then it is made of water. (Barnes 1982, 42, emphasis added)

Barnes applies this to Thales and notes that the evidence concerning Anaximander is problematic. Of Anaximenes he says,

> as I have argued, the inference from "Y was produced from X by a Φ process" to "Y is made of X" is eminently plausible and natural when the Φ process is one of condensation or rarefaction; and there is, I think, no cause to doubt that Anaximenes was a material monist in the standard Aristotelian sense. (44)

Barnes's overall point is that we can infer that Y is *made of* X if Y *arises from* X with the right kind of process.

---

[18] Plato *Sophist* 242c–d: "Each [theorist] seems to me to explain things to us in a story form, as if we were children, one saying that there are three things, and sometimes some of them are at war with each other in a way, then he has them make friends and marry, beget children, and raise their offspring; another says they are two, wet and dry or hot and cold, and gives them in marriage to each other." It is difficult to know who he has in mind here. Cornford 1935, ad loc., does not try to say, nor does McCabe 2000, 64–65. Kirk (KRS 71, n. 1) suggests Pherecydes for the theorist of three principles, Fehling 1994, 130–31, Pherecydes or Hesiod. Plato may have Anaximander in mind for the theorist of two principles. But he attributes monism only to the Eleatics in the following lines. Isocrates, *Antidosis* 268, another source independent of Aristotle, also fails to classify the Milesians as monists. It is likely that these alternative lists go back to Hippias' study of early philosophy (Mansfeld 1990, 22–83, esp. 29–30) and include the Ionian philosophers.

This argument seems to me just wrong.[19] Barnes uses the water example to persuade us: if ice is condensed water, then it is water. But we are inclined to accept this inference only because it has the weight of the whole theory of modern science behind it. We know ice is condensed water because we know that water is H_2O, that every chemical substance can exist in a solid, liquid, or gaseous state, that ice is the solid state of water. . . . In other words, we import the whole of our modern scientific worldview when we assent to this inference. The Ionians do not share our worldview, and we frankly do not know what they have in mind when they say that Y comes to be from X by a process of condensation. It could very well be that they think X undergoes a sea change in the process, so that is no longer X. Again, it is not the process of condensation that justifies our acceptance that if one thing arises from another it is made up of that other, but the process *as embedded* in a whole conception of how the world works. But that is just what is in question at present: how *does* the world work for an Ionian, and in particular for Anaximenes? We cannot appeal to our intuitions here, precisely because they are not Ionian intuitions.

3.3.2 The Case against Material Monism

A priori three kinds of consideration should support or defeat an interpretation of a historical philosophical theory: (A) The theory should be historically appropriate. It should be the kind of theory that would make sense in its time and place, to people with a certain level of culture and theoretical sophistication. (B) It should be philosophically coherent. That is, it should be a viable theory in its own right. (C) It should be dialectically relevant. Here I use "dialectic" in a broadly Hegelian way, to indicate an assumption that there is an ongoing conversation in which certain positions react, positively or negatively, to certain others. On the assumption (which drives historiography of philosophy) that there is a fruitful interaction between philosophers—one that supports at least the appearance of progress—a given interpretation will be preferable to another to the degree it makes better connections with contiguous philosophical movements and positions. I take it that these conditions in fact operate as tacit criteria in our evaluation of interpretations in the history of philosophy, and that they are relatively neutral as between different philosophies.

Relative to this background, I wish to argue in this section that the interpretation of Material Monism (MM) as applied to the early Ionians

---

[19] Indeed, it was refuted before Barnes: Stokes 1971, 43–48, replied to a similar argument suggested by Vlastos in correspondence.

is historically inappropriate, philosophically incoherent, and dialectically irrelevant.

3.3.2.1 Material Monism Is Historically Inappropriate. What would support or confirm MM as an interpretation of Anaximenes and some at least of his fellow Ionians would be some sort of evidence that they understood the world to function in the way described by the interpretation. In Simplicius's exposition of Anaximenes cited above, he describes the several substances coming to be from air, using the word *gignesthai*. The corresponding concept is ruled out by MM (III), which prohibits coming to be. At his most precise, Aristotle distinguishes three kinds of motion (*kinêsis*), all involving a change of accidents (incidental features): change of quality (alternation), change of quantity (increase and decrease), and change of place (locomotion); and two additional kinds of change (*metabolê*) which are not motions. The latter kinds are both changes in substance or being, coming to be (*genesis*) and perishing (*pthora*).[20] Aristotle does accommodate broader talk about coming to be by distinguishing between unqualified and qualified coming to be. Syntactically, this is the difference between coming to be (period) and coming to be *F*; ontologically it is the difference between coming to exist and coming to instantiate some property. Thus Socrates comes to be (unqualified) when he is born (comes to exist), but he comes to be (qualified) when he gets a tan (comes to be dark).[21]

Now (III), the denial of (unqualified) coming to be for the material source, follows from (II), the assertion that there is only one substance, which in Anaximenes' case is air. In A5a Simplicius speaks of air becoming other kinds of stuff. Is he speaking of unqualified or qualified coming to be? Since he speaks of, for instance, air becoming fire, i.e., becoming *F*, he may simply have in mind qualified coming to be. On the other hand, he and his source Theophrastus may be imitating the language of Anaximenes without the Aristotelian background. Another testimony from Hippolytus hints at some original language:

Anaximenes . . . said the source was boundless air, from which the things that are and were and will be, and gods and divinities *come to be*, the rest from the offspring [*apogonoi*] of these. (*Ref.* 1.7.1)

---

[20] *Physics* V.1–2; cf. *Categories* 14, which does not distinguish between changes which are and which are not motions—that distinction seems to be a later refinement.

[21] Thus Aristotle *Metaphysics* 983b13–17, omitted in the passage quoted above: "Thus we do not say that Socrates comes to be without qualification when he becomes fine or musical, nor that he perishes when he loses these qualifications, because the substratum, Socrates himself, remains; likewise none of these other [alleged sources comes to be or perishes]."

The pleonastic reference to things past, present, and future is suggestive of early phraseology.[22] The things of the world, including, presumably, other stuffs, are said to come to be. In this case, however, the unqualified sense seems implied: these things come to *exist*. They come *from* air. Is air present in them as a component, or is it merely the starting point for the change? A further generation of unspecified things are the *offspring* of the second generation. The image of offspring presupposes the connection between cosmology and birth, the most literal form of coming to be. What little we have of Anaximenes clashes with MM and its rejection of unqualified coming to be, based on a strict ontological monism.

What of Anaximenes' immediate predecessors? In the last chapter, we saw that what we know of Anaximander suggests that his source, the boundless, did not enter into the world directly. For it is only by a quasi-biological process that first the elements and then the world itself come to be. Moreover, we saw that the elements, under the description of contraries, come to be out of each other and perish into each other. Thus the hot perishes into the cold in giving rise to it, and vice versa. Now if the elements behave like this, they are coming to be and perishing, that is, changing in the category of substance or being, not merely changing in quality. Hence they do not instantiate MM, nor is there any continuing substratum for the world. Consequently Anaximenes could not have inherited MM from his immediate predecessor.

Could he, then, have inherited the theory or the major concepts of it from Thales? Here we run into a serious problem of evidence. Textual evidence for all the Milesians is extremely exiguous, but in the case of Thales, not even Aristotle possessed any reliable primary sources.[23] The only doxographical basis for a guess comes from Hippolytus:[24]

> Thales of Miletus, one of the Seven Wise Men, is said to have been the first to pursue natural philosophy. He said the beginning and end of the world was water. For from this the world is composed when it is condensed and in turn dissolved, and the world is borne on it. From it come earthquakes, windstorms, and the motions of the stars. (Hippolytus *Refutation* 1.1.1–2)

To say the beginning and end of the world is water is not sufficient to

[22] On the past, present, and future, cp. Homer *Iliad* 1.70; Hesiod *Theogony* 32, 38; Solon fr. 4.15 Bergk; Euripides *Helen* 14; Empedocles B21.9 and *P.Strasb.* a(I) 8; Anaxagoras B12; on archaic style, see Deichgräber 1933, 352–53; Democritus in Ps.-Plutarch *Miscellanies* 7 (Diels *DG* 581.11). Heidel 1913, 692, argues that the phrase is an intrusion; but it is not the sort of phrase a doxographer is likely to have invented. A compression of the original is more likely.

[23] *Metaphysics* 983b20–27, 984a2–3, with KRS 90: Aristotle is tentative in his assertions about Thales.

[24] DK regards this as so suspect that it does not print it (but it is in found in Diels *DG* 555).

show that the world, most especially the earth itself, *is* now water. We would need an inference such as Barnes argues for, based on condensation. But as we have seen, even this argument is not sufficient. In any case, the attribution of condensation in the next sentence must itself be an interpretation, for we are told that Theophrastus found the theory of rarefaction and condensation expressed only in Anaximenes.[25]

Before Thales we have only mythographers. Hesiod explains the world by a birth of cosmic gods, but this account presupposes the nonidentity of begetter and begotten, since they are independent personages. About the same time as Anaximander, Pherecydes provides a cosmogony based apparently on the model of sexual generation.[26]

After Anaximenes we have Xenophanes and Heraclitus. I shall postpone a discussion of the former until later in this chapter, and devote chapter 5 to Heraclitus. But for now note that Xenophanes is often taken as a dualist rather than a monist, which would undermine his value as a parallel for MM.[27] As for Heraclitus, his reputation as a flux theorist at least raises major questions about the possibility of identity through time of an alleged substratum for the world. There is, then, no clear indication that Anaximenes left MM as a legacy to his immediate successors.

Aristotle himself has trouble maintaining MM as an interpretation of the early Ionians. In the first place, he cannot fit Anaximander satisfactorily into his scheme:

> As the natural philosophers maintain, there are two ways [to account for change]. Some make the underlying body one, one of the three elements or something else which is denser than fire but finer than air, and they generate the other things from this by condensation and rarefaction so as to produce a plurality. . . . The others separate out the contrarieties from the one in which they are present, as does Anaximander, and everyone who says there is a one and a many, such as Empedocles and Anaxagoras. For from the mixture they too separate out everything else. (Aristotle *Physics* 187a12–23 = A16)

In the present passage, Aristotle bails out of his standard account and groups Anaximander with the pluralists Empedocles and Anaxagoras. The reason for assimilating Anaximander to the pluralists is that Aristotle thinks he may understand the boundless as a mixture of preexisting elements, which then separate out by a kind of extraction rather than by alteration, as in the theories of the pluralists. But in the second sentence

[25] Simplicius *Phys.* 149.28–150.4.

[26] Damascius *De Principiis* 124b = Pherecydes A8. On the dating of Pherecydes in the sixth century, see KRS 50.

[27] Finkelberg 1997, 9ff. "What sharply distinguished Xenophanes' theory from the doctrines of his Ionian predecessors and contemporaries was the assumption of two *archai* rather than a single one" (ibid., 15).

above, Aristotle mentions an element between air and fire; this Alexander of Aphrodisias identifies with Anaximander's boundless, with some justification.[28] It is not altogether clear if this in-between element really is Aristotle's analysis of the boundless, but there are no better candidates. If that is so, Aristotle seems to have assigned Anaximander to both the Material Monists and the pluralists.[29]

But the problem of how to classify Anaximander is a relatively minor issue. We glimpse a much larger problem when we look at Aristotle's general treatment of change. At the beginning of On Generation and Corruption (Book I, ch. 1, passim), Aristotle discusses early attempts to account for change:

> All those who say the universe is one thing and who derive all things from the one are committed to saying that coming to be is alteration, and what comes to be is really altered. But all those who posit more than one kind of matter, such as Empedocles, Anaxagoras, and Leucippus, are committed to saying they are different. Yet Anaxagoras misconstrues his own theory. For he says that coming to be and perishing are reducible to alteration. (314a8–15)

Aristotle explains how different metaphysical assumptions entail different theories of change: monists must reject coming to be and perishing because their "substratum always remains one and the same" (b3). Pluralists, on the contrary, are committed to accepting coming to be and perishing because "coming to be and perishing result from the compounding and dissolution [of the elements]" (b5–6). Clearly, then, one's ontology entails one's theory of change for Aristotle.

But at the end of the passage cited above, Aristotle points out that Anaxagoras is inconsistent: he denies coming to be and perishing in favor of alteration. Empedocles too denies there is coming to be, as Aristotle recognizes.[30] Both of them, then, are inconsistent with their own theory on this point (b10–12). Here, where Aristotle is most incisive, he reveals the cloven hoof. For he ascribes a theory of change to each Presocratic based on (his reading of) that philosopher's ontology and the logical consequences of the ontology. On Aristotle's interpretation, a pluralist will say that coming to be is a conjunction of elements and perishing is a separation of elements. Anaxagoras and Empedocles, however, say that there is no coming to be and perishing; hence they have misconstrued

---

[28] Metaphysics 60.8–10 = A16a, with Aristotle GC 332a19–25.

[29] Gottschalk 1965 argues that Aristotle really takes Anaximander to be a monist: "He is not making a pluralist of Anaximander, but monists of Empedocles and Anaxagoras" (39).

[30] Empedocles B8–12, Aristotle GC 314b6ff., and 315a3ff., in the latter of which Aristotle focuses on the way the elements emerge from the "One," i.e., from a unified state in the Sphere. This part of Empedocles' cosmic cycle is poorly attested; see Empedocles B30, B31, Aristotle On the Heavens 301a14–20.

their own theories. We shall discuss the pluralists and their motives in more detail in chapters 7–9. For now, we may provisionally adopt the common contemporary understanding that they develop their view of change in response to Parmenides' criticisms of coming to be and perishing. They are, in fact, committed to avoiding these kinds of change. Even though Aristotle is perfectly aware of the Eleatic challenge in his own philosophical program and develops his own ontology at least in partial response to it,[31] he does not see the pluralists in their own context. From their point of view, combination is *not* coming to be and separation is *not* perishing. Aristotle is guilty at least of *ignoratio elenchi*, of failure to appreciate their argument.

What then of the Material Monists? Aristotle says nothing about their specific account of change in this context. But if we look at Hippolytus's exposition of Anaximenes cited above, we notice that it is expressed almost completely in terms of things that *come to be* from the source, in a passage reminiscent of early language, and where "come to exist" is clearly meant. Evidently the ultimate witness to Anaximenes, Theophrastus, had no reason not to use this terminology. Even Simplicius in his account, cited at the beginning of this chapter, uses the terminology of coming to be rather than of alteration. Air comes to be wind; wind comes to be cloud; cloud comes to be water, etc. Is this Simplicius speaking, or does Anaximenes express himself exclusively in terms of coming to be? And, before Aristotle worked out his complex theory of change based on his own ontology, what did *Anaximenes* mean by *gignesthai*, "come to be"? We have echoes of Anaximenes' language showing that everything in this world comes to be from air. But when each thing (element, organism, etc.) exists, is it still air?

The closest we seem to get to Anaximenes is a statement that air is the source:

Anaximenes and Diogenes [of Apollonia] posit air as prior to water as the simple body that is most properly the source. (Aristotle *Metaphysics* 984a5–7 = A4)

Anaximenes, he too being from Miletus, the son of Eurystratus, said the source was boundless air, from which the things that are and were and will be and gods and divinities come to be, the rest from the offspring of these. (Hippolytus *Refutation* 1.7.1 = A7)

They say Anaximenes held the source of the world to be air, and this was boundless in quantity, but determinate in its qualities. All things were generated by a sort of condensation and thinning, respectively, of this. (Ps.-Plutarch *Miscellanies* 3 = A6)

[31] *Physics* I.2–3, 8–9.

Anaximenes, son of Eurystratus, of Miletus, was an associate of Anaximander, who says, like him, that the underlying nature is single and boundless, but not indeterminate as he says, but determinate, calling it air. (Simplicius *Physics* 24.26–28 = A5a)

All of these say that air is the source or *arche* except the last, which says
it is the underlying nature. Clearly the latter phrase is based on Aristotle's
own ontology and is not original. We cannot even be certain that the term
arche is original, but at least it could be. Yet it may be that the only reason
for saying that Anaximenes' air is an underlying nature is that it is an
arche, which at most (before Aristotle) would mean a source or starting
point for change. In any case, we do not find any solid piece of information in the doxographic tradition for MM other than the mere assertion,
in historically anachronistic language, that air is the underlying nature.

From what we can see, Anaximenes allows unqualified coming to be,
though on Aristotle's theory he should not, while Anaxagoras and Empedocles do not allow it, though they should. Aristotle criticizes the latter
two for misconstruing their own theories. Is the tail wagging the dog?
Are Anaximenes, Anaxagoras, and Empedocles so ignorant of their own
theories, or is Aristotle himself guilty of misconstruing the evidence? To
be sure, there is one source and starting point for Anaximenes. But is that
source taken as a substratum that is always present throughout all the
changes and transformations of air? Or could it be that air comes to be
something different from what it was? At least we do not see any sign
that the concept of coming to be was viewed as problematic in the theory
of Anaximenes. We should recognize that Anaximenes and his contemporaries will have no developed technical vocabulary for change. But it is
also likely that they will have had no technical ontological vocabulary.
Under these conditions, a later interpreter will have to reconstruct the
theory, to translate it into his own rich philosophical language, by observing nuances and implications of a relatively impoverished philosophical
language. And that will require a certain amount of patience and respect
for the sources that Aristotle does not always show.

There is another domain where we may see hints of early theoretical
conceptions: meteorology—taken in the broad sense of the study of *mete-
ôra*: things aloft. Aëtius, in a passage describing reasons why Thales might
have chosen water as his source, says:

third, [he may have chosen water because] the very fire of the sun and the
heavenly bodies is fed [*trephetai*] by exhalations of waters, as is the world
itself.[32]

---

[32] Aëtius 1.3.1, *DG* 276.

Here we get a recognition that the heavenly bodies were thought to receive nourishment from moisture, especially in the form of vapors arising by evaporation from bodies of water on earth. Anaximenes derives the heavenly bodies from evaporations:

> The earth is flat riding on air, likewise the sun and moon and the other heavenly bodies, which are all fiery, float on air because of their flatness. The heavenly bodies came to be from earth because of the moisture arising from it, which being thinned came to be fire, and from fire floating aloft the stars were composed. (Hippolytus *Refutation* 1.7.4–5)

In this passage Hippolytus does not say explicitly that vapors nourish the heavenly bodies, but on the assumption that the heavenly fires need fuel, the most obvious assumption is that a constant flow of vapor supplies that fuel.[33] In his report of Anaximenes' cosmogony, Pseudo-Plutarch describes the earth as being formed before the heavenly bodies.[34] The most obvious reason is that the heavenly bodies need a constant supply of fuel that can only be supplied by processes arising from the earth.

Xenophanes has a complex astronomical theory that seems to draw on the same principle:

> Xenophanes [says the stars come to be] from burning clouds; being quenched every day they catch fire again at night, like coals; their risings and settings are kindlings and quenchings. (Aëtius 2.13.14 = Xenophanes A38)

> Xenophanes [says] the sun comes from burning clouds. Theophrastus in the *Physics* has written that the sun is composed of tiny sparks being gathered together from the moist evaporation, and gathering together the sun. (Ibid. 2.20.3 = A40)[35]

There are a number of complexities concerning the stars which we cannot enter into here.[36] Suffice it to say that the doxographers are in agreement that stars are fiery clouds. The moon is a "felted cloud."[37] The sun like the stars is a fiery cloud that is fed by moist exhalations, as Aëtius tells us. Furthermore, eclipses can be accounted for on the present model:

> Xenophanes [says] there are many suns and moons according to the regions, sections, and zones of the earth, and at a certain time the disk falls into a section of the earth not inhabited by us, and just as if it were "walking on nothing" it produces an eclipse. The same said the sun goes on without end but seems to circle around because of its distance. (Aëtius 2.24.9 = A41a)

---

[33] Pace Cherniss 1935, 134, n. 542.
[34] *Miscellanies* 3 = A6, to be quoted and discussed below.
[35] For more texts bearing on this view, see Keyser 1992.
[36] See now Mourelatos 2002.
[37] Aëtius 2.25.4 = A43, with Bicknell 1967c.

The sun is eclipsed probably because it travels over a desert area in which it has no moisture to feed it.[38]

Heraclitus also accounts for the heavenly bodies in terms of evaporation:

> Fire is increased by the bright vapors, the moist by the others. What the stuff surrounding the world is he does not explain; there are, however, in it bowls with the concave side turned toward us, in which the bright vapors are collected to produce flames, which are the heavenly bodies. . . . The sun and moon are eclipsed when the bowls turn up; and the phases of the moon come about as its bowl gradually revolves. Day and night come about, as well as months, seasons, and years, rains, winds, and similar events, as a result of different vapors. (Diogenes Laertius 9.9–10)

Although Heraclitus uses a completely different mechanism from that of Xenophanes to explain eclipses and the like, he sees the fiery bodies as being fed by vapors arising from the earth (cf. Ps.-Aristotle *Problems* 934b33–36). That this pattern is widely influential is seen from a text from the later fifth century preserved in the Hippocratic library:[39]

> The path of the sun, moon, and stars is through the air [*pneuma*]. For air is the fuel [*trophê*] of fire, and fire deprived of air would not be able to survive. So thin air supports the everlasting life of the sun. (Hippocrates *On Breaths* 3)[40]

Quite explicitly in this case, we see that the heavenly bodies live by consuming air, which is then the sine qua non of the heavenly fires. A constant supply of vapor from earth must feed the appetite of the stars. Aristotle knows the view well:

> Thus everyone before me who supposed the sun was nourished by the moist was being ridiculous. And for this reason some claim the sun makes its turnings (solstices), for the same places cannot always supply its nourishment. . . . Just as ordinary fire burns as long as it has fuel, and the moist is the only fuel for fire, so, they assume, the rising moisture reaches as far as the sun, and its rising is like that of ordinary flame, as the basis for their explanation. (*Meteorology* 354b33–354a8)

Aristotle argues against the view, but sees it as a common misconception of his predecessors.[41] (As for his observation that the moist is the fuel for

---

[38] As pointed out by Bicknell 1967b, followed by Lesher 1992, 217, n. 57.

[39] For a general study of nutrition in early Greek thought, see Heidel 1911, 141–70.

[40] Cf. also Philolaus A18 (Aètius 2.5.3), according which "evaporations of [fire and water] nourish the world."

[41] Cherniss 1935, 133–34 with nn. 541–42 argues that this account applies only to the Heracliteans. But this seems to ignore the significance of the passages cited above. See criticisms in Kirk 1954, 265–66; Marcovich 1967, 315–16. See also Alexander *Meteorology* 67.4–8.

fire, it is perhaps helpful to remember that the ordinary form of illumination in a Greek house was a lamp fed by olive oil.) We find an approximation of the view Aristotle criticizes in Antiphon the Sophist:

> Antiphon [says that the sun is] fire feeding on the moist air around the earth, making its risings and settings by always abandoning the air that is burned up and instead adhering to the air that is somewhat moistened. (B26)[42]

Lucretius criticizes this view according to which "the sun's flame grazes in the azure pastures of the sky."[43]

Even Herodotus takes over this view. In his discussion of the Nile floods, he gives his own account of how they come about:

> To explain in more detail, traveling through upper Africa the sun has this effect: because the air in these countries is continually clear and the region is hot and devoid[44] of cool winds, the sun has the same effect it usually does in summer as it travels through the midst of the sky. It draws water to itself and then disperses it to the upper regions, where the winds pick it up and dissolve it by scattering. And appropriately there are winds that blow from this region, the south and the southwest winds, which are by far the most rainy winds of all. I believe that the sun does not disperse all the annual water from the Nile every time, but it retains some of it around itself. (*Histories* 2.25.1–3)

The final sentence is taken by commentators as recognizing the role of the moist vapors in fueling the sun's fires.[45] The same moisture that accounts for evaporation from water sources and which forms the basis for storms and rains feeds the sun.

The widespread—one might say, standard—account of the heavenly fires makes them fires very much like the ordinary fires we build on earth, as Aristotle recognizes. On this account, the heavens are continuous with the atmosphere, and the same general forces that produce winds, rains, and lightning keep the heavenly fires lit. The heavenly bodies require a constant supply of moist vapor, which fuels their conflagrations. What is significant about this for present purposes is the fact that the consumption of fuel is itself a kind of metabolism, as even the term *trophe*, "food, fuel" indicates. Moist vapors are transformed into fire. There is no more striking example in nature of physical destruction than the way a log is consumed in minutes by a flame. The death of the fuel is the life of fire, as we shall see in Heraclitus. It is only with difficulty that post-Parmenidean theorists can account for metabolism on a model of continuing elements.[46]

---

[42] See Pendrick 2002, 295.

[43] *et solis flammam per caeli caerula pasci*, 1.1090; but Lucretius seems to endorse the view at 1.231.

[44] Adding ἄνευ with Madvig.

[45] Weidemann 1971, 110–11; How and Wells 1912, 171; Lloyd 1976, 106–7.

[46] Anaxagoras B10, cf. B17.

Metabolism is itself a kind of paradigm of destruction and transformation of one material into another. Without a strong doctrine of ontological continuity to back it up—such as that found in post-Eleatic systems—it is difficult to see how a tale of cosmic fires kindling in measure and being quenched in measures could be taken as typical of the alterations of a single substance. Metabolism itself offers a kind of paradigm of change which does not accord with Material Monism. To neutralize it, we might expect the advocate of MM to give an account of how metabolism is *really* not a transformation but an alteration of accidents. How one might achieve this without any terminology of accidents and essences, any typology of changes, or any complex ontology, it is difficult to say. But in any case, we get no report of an analysis of metabolism in terms of some preferred kind of change; rather metabolism seems to provide at least a prominent image and model of change.

So far we have not seen any significant historical antecedents or consequents of Anaximenes' alleged Material Monism. We have, on the contrary, seen indications that some kind of coming to be and perishing was recognized by Anaximander and perhaps by Anaximenes himself, and that a kind of dramatic change that seems to exhibit elemental transformations played a significant role in Ionian explanations. Aristotle claims that the Material Monists should reject coming to be and pluralists accept it. What we find is that Material Monists seem to accept coming to be and pluralists to reject it. Aristotle may be right that the Presocratics misunderstood the implications of their own theories; but it is also possible that the misunderstanding is Aristotle's.

3.3.2.2 Material Monism Is Philosophically Incoherent. Scholars often believe that MM is the inevitable interpretation for the early Ionians because it is the only philosophically respectable way to understand what they have to say about the world. Part of this belief is an assumption that MM is philosophically coherent; part is an assumption that there is no other reasonable interpretation. In this section I wish to challenge the former assumption. I will attempt to address the second assumption by developing a viable alternative to MM later in this chapter.

If air is the one substance that is always present in the universe, then when it turns into, say, water, it is still air. Thus water is really just a state of air in which certain sensible properties manifest themselves that are not manifest in the normal uncondensed state of air. Let us return for a moment to the modern scientific analogue to MM, that in which water can manifest itself as a solid (ice) or as a gas (water vapor). The same reality, the chemical compound H_2O, can be present as a solid, a liquid, or a gas. That this conception is philosophically coherent is guaranteed by the fact that the world is actually structured in this way. Since MM structures the world in the same way, one might argue, it too must be philosophically coherent.

There is, however, a fundamental disanalogy between MM and the modern physical/chemical conception of reality. The physical/chemical conception specifies kinds of matter independently of immediate perceptions of it. The chemical formula for water tells me that every molecule of it is composed of two atoms of hydrogen bonded with one atom of oxygen. If then the clear, tasteless liquid in a pot on the stove disappears into the air, the scientist can identify it in the air even though it now has none of the same sensible properties as before, by using tests for the molecule. But with the elements of Anaximenes, their nature seems to be exhausted by the sensible properties they have: water, for instance, may be thought to be cold and wet. Air is problematic in the first place because it has few if any sensible qualities. If now you say that water is *really* air, and I disagree, how can you prove to me that it is? And if the water turns into earth, as Anaximenes holds that it can, how can you prove to me it is still air? Furthermore, how can you prove that the substratum of the water and the earth is the same substratum? In fact, I think this same problem arises for Aristotle in relation to the substratum he posits for his four elements, or simple bodies, namely prime matter.[47] But however that may be, it seems to make no sense to claim that X is really Y, which it does not resemble, if there is no criterion for identifying Y apart from its normal manifestations which it is not then exhibiting. And it is equally irrational to claim that Y continues to exist through a change when there is no criterion for identifying it along the way. What makes the modern physical/chemical theory viable is precisely its use of a formal account of matter independent of sense perception. (This is not, of course, to say that sense perception is not relevant to the theory of matter, only that it is not directly correlated with it.)

No early Ionian, so far as I can see, ever proposes any structural account of matter apart from its perceptual qualities. It appears, then, that the early Ionians do not offer us the minimum theoretical machinery needed for a coherent monistic theory based on a continuing material substratum.

3.3.2.3. Material Monism Is Dialectically Irrelevant. Suppose that MM is the correct interpretation of early Ionian philosophy. Then the Ionians begin the Western philosophical tradition with a rich metaphysical theory, according to which there is a single source of everything which remains throughout all changes, if not as the manifest substance, then as the underlying nature, whose appearance only is changed. Thus Anaximenes' air will be the nature underlying the appearance of water. Water is not an independent substance, but merely a state of air, a temporary appearance.

---

[47] See Graham 1987b.

In Aristotle's terms, air is the mater (*hulê*) of water, and of every other stuff. In terms of his fourfold causal scheme, air is a *material cause* of the change, the continuing factor. The matter gets its phenomenal properties from features that inhere in it, which in the case of MM will be qualities such as hot and dry. Now if Aristotle is right, the Ionians provide striking anticipation of Aristotle's own metaphysics, according to which there is always a substratum for every change, including change from one simple substance or element to another (which I shall call "elemental change"). There is one key difference, namely that for Aristotle the matter for elemental change (prime matter), is itself characterless and not to be identified with any particular elements,[48] whereas the advocates of MM identify the matter with their primary element, such as Anaximenes' air.

Now it appears that Aristotle himself invented the concept of matter, at least as an abstract generality. The record of his argument for matter, and, I contend, of his "discovery" or invention of the concept, can be traced in *Physics* I.[49] He deploys the concept there to answer the Eleatic challenge, that is, Parmenides' challenge to explain how what-is can come to be from what-is-not. Aristotle answers (ch. 8) that "what-is-not" is ambiguous, and that by distinguishing between the substratum (matter) and the privation (negative predicate such as not-hot or not-man), one can say that what-is comes to be from what-is-not in one sense: hot comes from not-hot, man from not-man, but this does not entail it comes from nothing. For the substratum is always present. Thus what-is comes from what-is-not in one sense, but not from nothing, because it comes to be from the substratum in another sense. Thus Parmenides and the Eleatic school are answered and change is vindicated.

Aristotle claims the theory of form and matter as one of his proudest accomplishments. Yet if he is right in his reading of the Ionians, they had *already* invented matter in the metaphysical and not merely physical sense, two and a half centuries earlier. Of course, in *Metaphysics* I he tells the story of how the Presocratics (and Plato) developed the concepts of causes, starting with the material cause, in a halting and primitive manner. The theory was brought to perfection only by Aristotle's comprehensive grasp of the several causes in their interrelations. Yet if Aristotle is right, the Ionians used matter in precisely the same context as he uses it, to provide a continuing foundation for change through time. So, however incomplete was their grasp of the system of causes, they had the principles necessary for a correct explanation of change in relation to being and not-being.

But were the Ionians working on the same problem? Remember that Aristotle is responding to the Eleatic challenge. Parmenides and the Elea-

---

[48] *On Generation and Corruption* 329a24–6. On prime matter, see above, ch. 2, n. 40.
[49] See Graham 1984, 1987a, ch. 5.

tics, on the other hand, did not appear until the early Ionians were gone. Parmenides, for his part, was exercised by the problem of how what-is could come to be.[50] Why did this problem bother him? In particular, why did it bother him *if the Ionians had already solved the problem?* They had already, according to Aristotle's reconstruction, laid the foundation for an adequate account of change, in which some identical reality is always present as the substratum for the change. There is no radical change (something always remains the same), and hence no radical problem of change. Did Parmenides miss this brilliant achievement of his predecessors? Did he have some other worry, some concern about Meinong-like nonbeings?

If we accept MM as an interpretation of early Ionian philosophy, there is, I submit, no reason for Parmenides to be bothered by problems of coming to be.[51] The ultimate problem of change, how what-is can come to be from what-is-not, has already been solved. Furthermore, there is no reason for the post-Parmenidean pluralists to invent a new model of explanation, involving a plurality of beings, given that the matter of MM works just as well as the matter of Aristotelian hylomorphic theory. MM makes no dialectical sense: it cannot tell us why Parmenides was bothered, or why there was a shift in ontology and etiology after him. It is dialectically uninformative, even stultifying. Scholars have, indeed, tended to turn away from the question of Parmenides' motivation in the last forty years, a point we shall return to in chapter 6. But that Parmenides was keenly aware of his historical situation seems assured by the fact that the second half of his argument (the part usually shortchanged by interpreters) is devoted to developing a cosmology. To be sure, Parmenides uses his cosmology for destructive purposes, but it is clear that his philosophical competition, as he sees it, has something to offer that he must counter to win the contest (see B8.60–1). There is, then, an untold story about his relationship to his cosmological predecessors, and no way to tell the story so long as we adhere to MM. To use Parmenides' own road imagery, MM is a dialectical dead end.

Thus we see that MM is historically implausible, philosophically incoherent, and dialectically irrelevant. Are there promising alternatives?

3.3.3 Alternative Theories to Material Monism

There are two alternative theories to account for early Ionian philosophy. Both have been introduced, but neither has been developed extensively nor defended in depth.

---

[50] B8.3, 6–7, 9–10, 12–13, 19–20.
[51] This point has already been made by Cherniss 1935, 368, and Stokes 1971, 33–34, in a more logical than metaphysical context.

(1) The Theory of Powers. According to this interpretation, the ultimate realities of Ionian philosophy are not the primary substances posited by each philosopher, but the *dunameis* or powers which operate in the world. These are, roughly, the qualities like hot and cold in Anaximander, rare and dense in Anaximenes, which are to be understood not as concomitants of the real underlying substance, but the real things (*eonta, onta*) themselves. These "character-powers," "quality-things," "activity-things," or "fluid-qualities"[52] are prominent in the medical treatises of the late fifth century, and indeed in the medical theory of Alcmaeon in the early fifth century. They appear in the Ionians' discussions, where theorists systematically discount their importance, but they are the real explanantia of Ionian philosophy. These characters are genuinely active, hence really active powers, not passive qualities, and account for the changes that take place. It is only after the Eleatics require reality to be self-identical that theorists submerge these powers in continuing substances, that is, elements. But originally they are the real things, and they are subjected to an ontology of substances only later (and gradually).[53]

(2) The theory of a generating substance. On this interpretation, we may take the doxographical tradition at face value for much of what it says about the Ionians. They do recognize an original substance, which at one time was the only thing in the universe. This substance does change so as to produce different substances, which then are ordered into a world system. What is wrong with the doxographers is the claim that the stuffs produced from the original substance are merely states of the original stuff. For the original substance gives rise to other substances in such a way that it is not identical with them, nor does it continue or underlie them. It generates them or comes to be them in such a way that it perishes into its successor substance, and each successor substance perishes into its successor.[54]

---

[52] The terms are from Mourelatos 1973 (he now prefers "characters/powers"); Cornford 1930, 87, on Anaxagoras; Bicknell 1966b, 40; and Zafiropulo 1948, 284, also on Anaxagoras.

[53] Those who attribute a theory like this to at least some Presocratics are Heidel 1906; Cornford 1930; Cherniss 1935; Zafiropulo 1948; Klowski 1966; von Fritz 1971, 20–22; Mourelatos 1973. But Cornford and Zafiropulo see applications only to Anaxagoras, who, to my mind, is too late to be much influenced by these features of what Mourelatos calls the naive metaphysics of things. Cf. also Fränkel 1973, 261: "The notion of substance or matter did not arise until much later [than the Milesians], as an antithesis to form or force. The principal philosophical category of those days was quality, and by 'water,' 'air,' or 'fire' was meant a possessor of certain characteristic properties."

[54] Those who attribute a theory like this to the Milesians are Hölscher 1944, 1953; Cherniss 1935, 359ff; McDiarmid 1953, 91ff; Kahn 1960, ch. 2; Stokes 1971, ch. 2. Notably, Cherniss (ibid., 371—though elsewhere he favors Heidel's reading—see previous note) says, "In the case of the 'air' of Anaximenes we are not justified in assuming with Aristotle a

3.4 Anaximenes and the Generating Substance Theory

3.4.1 Problems for the Theory of Powers

Although the Theory of Powers is attractive, it faces serious challenges. Anaximander seems to have assigned an important role to powers such as hot, cold, wet, and dry. But in his cosmogony, we observe that he accounts for the development of the world in terms of the boundless, fire, air, and earth. Again in his meteorological explanations, he deals with water, wind, cloud, earth, and the like. It is simply not clear whether the powers he discusses have any identity apart from the substances in which they manifest themselves. In Anaximenes we have seven elements or basic substances. They are all conceived of concretely—or as concretely as stuffs such as fire, air, and wind can be. But the point is that Anaximenes appeals not to hot and cold to explain things, but to substances. They are characterized according to the rarity and density, but the rarity and density have no existence apart from the substances that exemplify them. After Anaximenes, Xenophanes deals with water, earth, cloud, and fire. Even Heraclitus, famous as the philosopher of flux, envisages elemental change as taking place between the substances of fire, water, and earth (B31). Later still we see Anaxagoras in developing his cosmology speak indifferently about substances and qualities.[55] It appears that, important as powers are, they are not thought of as realities independent of substances, but as characters in substances.[56]

For these reasons I do not find the Theory of Powers to be promising. It has not, unfortunately, received any lengthy or detailed development. And it has been attacked as implausible.[57] Certainly it would be worthwhile to see its supporters develop the interpretation and attempt to de-

---

persistence in his sense which implies logical identity and physical homogeneity." Kahn (ibid., 155) observes perceptively, "The fundamental difference between the sixth and fifth centuries lies not in the abandonment of monism for plurality, but in the passage from a world of birth and death to one of mixture and separation." Unfortunately, many of these accounts are more critical of MM than constructive of a coherent alternative; sometimes they are difficult to distinguish from the theory of powers. Perhaps for these reason they have gone mostly unnoticed. For the majority of scholars continue to adhere to MM, and typically do not even advert to any controversy about its attribution to the Milesians. See now also Decher 1998, who, however, does not seem to realize that the theory he attributes to Anaximenes and Heraclitus is contradicted by most interpreters, ancient and modern, including Kirk in KRS, which he liberally cites.

[55] B4 seems to conflate the categories of substance and quality; B1–2 stress substantial elements; B8, B12 contraries, while in a cosmological setting B15 speaks of contraries, B16 of substances.

[56] Cf. Lloyd 1964, 92–100; Longrigg 1993, 220–26.

[57] One version of it, at least, by Hölscher 1944.

fend it at length. But in the absence of such a defense, the theory does not seem to me to be viable.

3.4.2 The Generating Substance Theory

The Generating Substance Theory (GST) has not been developed in great detail either, though it has been ably defended.[58] But I shall attempt in these pages to defend it at length as an interpretation of the early Ionians. I begin with some preliminary considerations that count in its favor. First, we have seen that Anaximander portrays contraries perishing into and arising out of each other. We have also seen that he uses everyday elementary substances such as fire, air, and earth to explain the development of the world and its continued operation. These two kinds of accounts can be unified if we view the contraries as embodied in elements, and acting on each other *according to* their powers. The theories of elemental change we pointed out as problems for the Theory of Powers count as support for GST: elemental change seems to be prominent in the theoretical accounts of the Ionians. Finally, so long as we see the original substance in a cosmogony turning into another substance without remainder, we see cosmogony as based on a purely genetic account of one stuff turning into another. This much could be inherited from Hesiod's theogonical account. No subtle or complex metaphysics is needed, no fine distinctions between appearance and reality, patent and latent properties, substrata and characters. We are stuck with stuffs and powers, but both of these entities are, in their own way, manifest, and their importance in Ionian explanations is well documented. We can, in fact, accept a good deal of what Aristotle and his followers say about the Ionians, absent the claim that the original substance remains as a substratum, and the logical consequence that there is no coming to be or perishing.

The advantage, and concomitant disadvantage, of GST is that it overlaps with MM in much of its domain. It can use the elemental changes recounted by doxographers as evidence. But, on the other hand, it is difficult to find decisive texts that exclude MM while supporting GST. This is due in part to the very scarcity of texts available, and also to the slant of the sources toward MM. But there are definitely points at which they clash, and we can adduce indirect evidence from the way that the explanations of the world proceed. One point I would repeat here is the use of the language of coming to be and perishing, in both Anaximander and Anaximenes. According to MM, this language should be ruled out, or at least carefully qualified. As I have noted already, the lack of a technical

---

[58] By Stokes 1971. Stokes's study, however, focuses on the theme of the one and the many, in which the Generating Substance Theory comes in obliquely.

vocabulary in early Greek philosophy makes any kind of evaluation difficult. But it also raises the question of what kind of statements the Ionians could make that would convince Aristotle that they believe in the continuing existence of the original matter even after it has (apparently) changed its nature into something else. The language of coming to be (*gignesthai*) and perishing (*phtheirein, apollusthai*)[59] would not be so loaded as later when Aristotle had distinguished four kinds of change, of which the first was coming to be and perishing, that is, change in substance or being. But it remains true that the terms for coming to be and perishing were also the everyday terms for being born and passing away.[60] The analogy, especially against the backdrop of theogony, is with the generation and destruction of living things. Why do Anaximander and Anaximenes continue to use such language if they are so hostile to its implications?

One way in which the Ionians might signal MM is in distinguishing between appearance and reality. The problem of distinguishing reality from appearance has a long history in literature, including deceptions wrought by the gods themselves,[61] and it has philosophical parallels. We have one clear instance of the theme in Xenophanes:

> She whom they call Iris, this too is in reality cloud,
> purple and red and green to the view. (B32)

Iris is the goddess of the rainbow. Xenophanes is pointing out that the multicolored phenomenon we behold is not a deity, but a certain kind of cloud. It is in reality (*pephuke*) cloud. An Ionian could in principle make a statement like this to prove that one manifestation is really (*eteêi*) or by nature (*pephuke*) an underlying substance.[62] But the statement by Xenophanes is the only one of its kind among the early Ionians. And it is interesting to note that Xenophanes does not identify rainbow with one of the two elements he is alleged to recognize, earth and water, but rather with cloud—which happens to be one of Anaximenes' elements. If the Ionians made statements like this reducing derivative substances to their original substance, no doxographer records it for us. This is, perhaps, an argument from silence, but it has some weight given that the doxographers liked to support their interpretations with textual evidence. But per-

---

[59] See Kahn 1960, 172–74: the term *phthora* which appears in Anaximander B1 is not otherwise attested before the fifth century (although nothing rules out the possibility that he used the term). In any case, synonyms are readily available.

[60] "Since the early philosophers clearly regarded the world as alive, its emergence was a genuine *birth* in their eyes. The two senses of γίγνεσθαι were still one and the same" (Kahn 1960, 158).

[61] Homer *Iliad* 2.5ff., Zeus sends a lying Dream to the Greeks. Hesiod *Theogony* 26–8: the Muses know how to tell true tales as well as lies.

[62] For the former term, see Democritus B9 (and in a skeptical context, B7, B8).

haps this is hasty: there is one statement by Heraclitus saying that the world is fire (B30), about which I shall speak later.

According to GST, the original substance produces a series of derivative substances or elements. The original substance is more fundamental than they for a number of reasons. First, at some time in the primeval past, there was nothing but this substance in existence. Second, it is allegedly the normal or undifferentiated kind of stuff. Third, it has some sort of life force and apparently intelligence, like Anaximander's boundless, by which it controls some things. But when it undergoes change, it ceases to exist and is replaced by a successor substance. This substance can then turn into another substance in the ordered series of substances (such as Anaximenes has) or turn back into the original substance. Since all other substances come out of the one substance, it is indeed the original substance; since it generates all substance it is the generating substance. Inasmuch as its function is to generate all substances rather than to underlie and support them, we may call the theory the Generating Substance Theory.[63]

I want to assert here what I suggested in the first chapter, that I take Xenophanes and Heraclitus to be early Ionians whose views at least approximate those of the Milesians. The fact that their physical theories so closely resemble those of the Milesians speaks for itself.[64] I shall devote a chapter to Heraclitus, whom I see as a critic and transitional figure, but for all that one who intimately understands GST, and indeed perhaps appreciates it better than its orthodox adherents. Hence I shall appeal to them for positive evidence and also see them as providing potential counterevidence to my thesis.

3.4.3 Meeting Objections: Xenophanes' Theory of Matter

Because of the scarcity of evidence, there is no quick argument to establish GST. But a strong case can be made for it by showing how it satisfies the criteria of historical plausibility, philosophical coherence, and dialectical relevance we have already identified. To some extent we have already shown how GST fits with historical texts describing Ionian philosophy better than MM. There are, however, significant objections that can be raised against GST. Here I wish to confront those, and by doing so I hope to show that there is even more historical evidence than has so far been adduced for GST:

Objection 1: Xenophanes is a dualist who composes all things out of two basic substances in a way incompatible with GST.

---

[63] I borrow the name from Pepper 1942, 93, an intriguing account. But his own characterization of the theory is intuitive, and he ascribes it to all or most of the Presocratics, whereas the view as I wish to more narrowly define it holds only (at most) for the early Ionians.

[64] On Xenophanes as closely related to the Milesians, see Heidel 1943; Lesher 1992, 4.

Objection 2: Heraclitus says that "This world . . . is . . . everliving fire" (B30), showing that fire is the material substratum of the world.

Objection 3: There is no ancient authority that attributes anything like GST to the early Ionians. Hence it has no historical support.

I shall argue that the first objection is likely to be false, and that the second and third are certainly false. But I shall postpone discussion of Heraclitus, and Objection 2, to Chapter 5.

Two fragments from Xenophanes seem to support MM:

> All things which come to be and grow are earth and water. (B29)

> For we all come to be from earth and water. (B33)

Barnes observes, "It is plausible to conjoin these lines: B33 supports B29, and Xenophanes makes an explicit inference from originative to constitutive stuff" (Barnes 1982, 42). This analysis, though it does seem to instantiate the inference, is problematic in another way. For in order to have a scheme like MM, we must have only one source, not two. For everything but the original element is a mere phase of the original element, and not an independent reality. If, on the other hand, everything comes from two sources according to Xenophanes, it may be that they are elements in the manner of the later pluralists, and that human beings, for instance, are compounds of earth and water. But in order to understand Xenophanes, we need to look at his theory of matter more closely.

Xenophanes envisages a major alternation between dry and wet periods on earth:

> Xenophanes thinks a mixture of earth with sea occurs and in time earth is dissolved by the moist, claiming to provide as evidence the fact that sea shells are found in the midst of earth and in mountains, and in the quarries of Syracuse impressions of fish and seaweed[65] have been found, and in Paros the impression of coral [or: bay] in the depth of a rock, and in Malta fossils of all sea creatures. He says these things happened when all things were covered with mud long ago and the impressions in the mud dried out. The human race becomes extinct when earth is carried down into the sea and becomes mud, and then the process begins again, and this change occurs in all the world-orders. (Hippolytus *Refutation* 1.14.5–6 = A33)

In this famous passage Hippolytus tell us that Xenophanes uses fossil evidence to argue that the earth is drying out: impressions of sea animals found on land show that these places were once under water. Since, for Xenophanes, the earth reaches down without end below our feet (B28),

---

[65] Reading φυκῶν with Gomperz.

the whole of the infinite plane of earth seems to have been inundated.[66] Where did the water come from? It is possible that it was dissolved in the air; on the other hand, the massive precipitation is difficult to account for unless it results from the transformation of another stuff. One other testimony gives a hint:

> And in fact in some caves water drips down. (B37)

It appears that Xenophanes may see the dripping of water in caves as the transformation of earth into water, or possibly the source of rock formations in caves, or both. It may be, then, that water and earth turn into each other, and that this is the cause of the flood cycles.

Is there, then, the possibility of reducing the two elements of earth and water into one? One fragment supports this interpretation:

> For from earth are all things and into earth do all things die. (B27)

This statement gave rise to a debate among doxographers: Stobaeus, Theodoret, Olympiodorus, and Sabinus saw Xenophanes as saying that earth was the original substance, while Galen objected that this was incompatible with Theophrastus's report.[67] Xenophanes' statement is an intriguing one, which I have already referred to as possibly reflecting an Ionian formula for the original substance which Aristotle follows.[68] The original substance is that which everything comes from and returns to. As Deichgräber has pointed out, B27 provides the one statement we get of a cosmic *arche*, while B29 and B33 deal with the composition of living creatures ("all things that come to be and grow," "for *we* all come to be").[69] But if we now make it plausible that Xenophanes has a single original substance, do we not support Barnes's argument that Xenophanes confirms

[66] I take it that for Xenophanes the earth is an infinite plane separating the earth below from the air above; this seems to be implied in Aëtius 2.24.9 = A41a, which has the sun going on in a straight line *eis apeiron*; this view is defended by Mourelatos 2002, 332–36. Such a cosmology of one substance separated by an infinite plane from another is perhaps to be reconstructed for Thales (Aristotle *On the Heavens* 294a28–33 = A14: water below, air above). Fehling 1994, 137–47, holds that all the early cosmologists held to such a view. However, he ignores the even earlier model of Hesiod, which has the earth surrounded by a (at least a roughly) spherical heaven: *Theogony* 721–5; cf. McKirahan 1994, 12–13, which seems to provide an archetype for Anaximander and Anaximenes.

[67] The texts are collected at DK A36. Stobaeus and Theodoret clearly ground their interpretation on B27, whereas Theophrastus may be following Aristotle, who says that no one chose earth as the original substance (*Metaphysics* 989a8–12). Aristotle, however, did not have a high opinion of Xenophanes as a natural philosopher and may have ignored him: ibid. 986b21–4; Deichgräber 1938, 13. Lucretius also speaks of an anonymous theory that makes earth a single source, one which is transformed into other things: 1.709–10, though some think Lucretius has Xenophanes in mind rather at line 713.

[68] See above, sec. 2.2.

[69] Deichgräber 1938, 13–17.

MM? Not yet. The question, as with Anaximander and Anaximenes, is how he conceives the several basic substances to be related. But the most important point to be drawn from B27 is the recognition that Xenophanes can in principle derive everything from a single substance; his statement is inconsistent with dualism or pluralism, which derive the many substances from a mixture of two or more elements.

The best view we can get of Xenophanes' theory of elemental change comes from his treatment of water:

> Sea is the source of water, the source of wind;
> For neither <would there be wind> without great sea,
> nor currents of rivers nor rain water from the sky,
> but great sea is the begetter of clouds and winds
> and rivers. (B30)

Sea produces wind, cloud, and rain water. Xenophanes prefigures here the water cycle recognized by modern science: evaporation producing vapor, condensation producing cloud, and accumulation of condensed particles producing water drops and precipitation. Sea is the begetter (*genetôr*) of the atmospheric conditions, suggesting that it generates them.[70] We learn further that moist vapors fuel the sun, as though fire were somehow continuous with vapor. Xenophanes takes clouds to be the source of many meteorological manifestations, including rainbows, St. Elmo's fire, lightning, comets, and meteors.[71]

If there is a continuum between earth and water, on the one hand, and water and cloud on the other, and further between cloud and fire, we get a picture like that of Anaximenes. Filling in all the terms, we have a series of earth, water, cloud, wind, and fire, five of Anaximenes' seven basic substances. But Xenophanes does say we are earth and water; what does this mean? Presumably flesh, blood, etc., are most closely connected to earth and water—either mixtures of them or substances intermediate between them in the continuum of stuffs. But there is no clear statement that everything is now earth, or Theophrastus would have exploited this to put Xenophanes in the tradition of MM. Of course there was a tendency from Plato on to interpret Xenophanes as an Eleatic with mainly metaphysical concerns and a nonmaterial monistic ontology.[72] But in any case there is no unimpeachable evidence that he adheres to MM, and in general his theory seems to fit better with GST than with MM.

---

[70] We might need to make a distinction here between sea, the part of the world, and water, the element; but we find Xenophanes' near contemporary Heraclitus interchanging these, e.g., in his B31. In any case, the concept of generation is vividly present. See Deichgräber 1938, 5–7; 15–16.

[71] A39, A45, A44.

[72] Plato *Sophist* 242d, Aristotle *Metaphysics* 986b18–27.

3.4.4 Meeting Objections: Plato as a Witness for GST

According to objection 3 against GST, the interpretation is not credible because it lacks any support from ancient authorities. In this section I will address this problem and argue that there is good early evidence for GST. At the same time I wish to address a problem that has recently been raised to, or rather reiterated against, the whole account of Anaximenes' theory of matter.[73] Georg Wöhrle (1993) is critical of MM, but also of the theory of change ascribed to Anaximander, for instance, by Simplicius in A5.[74] He attempts to cast doubt on MM in part by attacking Anaximenes' theory of change (hereafter TC). His attack draws heavily on the cosmogony attributed to Anaximenes by Pseudo-Plutarch:

> They say Anaximenes held the source of the world to be air, and this was boundless in quantity, but determinate in its qualities. All things were generated by a sort of condensation and thinning, respectively, of this. Motion has existed from everlasting. When air was felted he says the earth was formed first, being completely flat. For this very reason it floats on air. The sun and the moon and the other heavenly bodies have their source of generation from earth. At least he declares the sun is earth, and because of its rapid motion it gains a very considerable amount of heat.[75] (Ps.-Plutarch *Miscellanies* 3 = A6)

What is troubling about this cosmogony, according to Wöhrle, is that it does not exemplify the sequence of changes. If TC were right, we should expect [ii] air to turn first to [iii] wind, then [iv] cloud, then [v] water, and only then to [vi] earth and [vii] stones; and then to go through a similar sequence—or to start from the original air itself—to arrive at the fiery heavenly bodies. Instead the intervening steps are inexplicably omitted, and we have the jump from [ii] air to [vi] earth to earthy heavenly bodies—or perhaps [i] fire. Pseudo-Plutarch should follow the doxographic interpretation by which he is influenced; since he does not, he must be relying on an independent source which turns out to be more reliable. This and some textual considerations in the report of Hippolytus provide the main texts for questioning the theory of change ascribed to Anaximenes. TC, Wöhrle argues, is not original, but was invented by Theophrastus and foisted on Anaximenes.[76] With this theory neutralized, he can argue that there is no basis for MM either.[77]

---

[73] See Graham 2003b.

[74] See esp. pp. 18–23, 57–58.

[75] Reading ἱκανῶς θερμότητα λαβεῖν with Zeller.

[76] "Das gesamte Schema der Verwandlung in zwei Richtungen rührt demnach von Theophrast her" (Wöhrle 1993, 58).

[77] In this argument Wöhrle follows an interpretation of Hölscher 1953 developed in detail by Klowski 1972.

Now I have argued that TC, along with the mechanism of rarefaction and condensation, does not entail MM, and it should be plain from the accompanying argument that TC is logically independent of MM, and hence we see that to deny TC would not undermine MM. We may formulate TC as follows:

TC (Theory of Change)

1. There is a determinate sequence of materials ordered by relative density,
2. which consists of fire, air, wind, cloud, water, earth, and stones.
3. The materials arise from adjacent materials by being compressed or rarefied, respectively.

If we compare TC with MM (above, section 3.2), we see that MM makes no mention of the seven basic stuffs, of rarity and density, or of the causal role of compression or rarefaction; on the other hand, TC makes no mention of the priority of one of the stuffs, nor of the fact that the primary stuff underlies all the others. Again, MM does not entail TC, nor does TC entail MM; they are logically independent theories. Thus TC could hold true regardless of whether there is a single material substratum characterizing the several elements in the series.

Now on the traditional account, MM was invented by Thales, and perhaps followed by Anaximander. Anaximenes' only important innovation is TC.[78] Thus to reject TC would be to destroy the only significant contribution Anaximenes can make to cosmological theory. Of course it may be a true, if disappointing, fact that we cannot avoid, that Anaximenes has no interesting theory of change. But we should seek to save TC if at all possible, for fear of making Anaximenes even more nondescript than he has usually been taken to be.

Part of the problem in rescuing Anaximenes is that no source earlier than Aristotle mentions him by name,[79] and all those after Aristotle are dependent to some (often major) degree on his interpretation. But there is a so-far overlooked source of information in Plato, which I have argued elsewhere should be taken as a testimony.[80] Here we find an interesting correlation between TC and a meditation of Plato's on elemental change:

First, what we have now called [v] water we observe, as we believe, turning into [vii] stones and [vi] earth as it is *compacted*; but then *dissolving* and dispersing, this same thing becomes [iii] wind and [ii] air, and being ignited, air becomes [i] fire, and being *compressed* and quenched in turn, fire departs and turns back into the form of [ii] air, and again air coming together and being *condensed* becomes [iv] cloud and mist, and from these being *felted*

[78] See, e.g., Classen 1977; Hussey 1972, 28.
[79] Wöhrle 1993, 31; cf. n. 7 above.
[80] Graham 2003a.

still more comes [v] flowing water, and from water come [vi] earth and [vii] stones again, these things thus passing on to each other in a circle, as it appears, their generation. (Plato *Timaeus* 49b7–c7)

Above (section 3.1) we noted in the doxography a number of features of Anaximenes' theory:

(1) air is the source of everything by virtue of (2) being the underlying nature; (3)(a) there is a set of elements or basic stuffs: (i) fire, (ii) air, (iii) wind, (iv) cloud, (v) water, (vi) earth, (vii) stones, (b) ordered by their relative density, (4) which arise from each other precisely by their being rarefied or condensed; (5) the condensation process is analogous to the action of felting.

In this passage we find all seven of Anaximenes' basic stuffs, if we make allowance for the synonymy of cloud and mist. The arrangement is not as strict as one would like—on the upward path Plato omits cloud, and on the downward path wind; he inverts the order of earth and stones in their first occurrence, but gets them right in the second—yet clearly all the elementary bodies are present, and the correct order is adumbrated if not rigidly followed. We find then (3a) the seven elements in their correct order, (b) as determined by their relative densities, (4) which are produced by rarefaction or condensation, as Plato tells us using several synonymous terms, (5) including an allusion to the process of felting.

We can, by a process of elimination, verify that no other early philosopher besides Anaximenes recognized just the set of basic substances Plato enumerates. Thus there is every reason to think that Plato is borrowing Anaximenes' theory of change, as commentators on Plato have often recognized.[81] The crucial factor missing here, indeed, is the claim (1) that one basic substance, air, is original and superior to the others. But given that Plato's topic is how elements change into one another and not cosmogony, the priority of any element is not necessarily relevant. The important thing is that Plato has closely reproduced Anaximenes' theory of change, including his list of elemental bodies, his mechanism, and his favorite model (felting). No other early figure has a theory that embodies all these features: it is Anaximenes' theory that Plato has in mind. This point is sufficient to overthrow the claim of Wöhrle and others that Theophrastus invented Anaximenes' theory of change.

Point (2) of the doxographical account is also missing in Plato's account. As the opening lines show, Plato finds this account of change appealing precisely because it seems to be confirmed by our experiences ("we observe . . . "). Yet what it shows us is that no element stays the same,

---

[81] E.g., Taylor 1928, 314–15; Cornford 1937, 180; Vlastos 1975, 80, n. 22. Chalcidius secs. 280, 325 mentions Anaximenes among other ancient theorists without showing any special appreciation of his influence.

but each one is constantly being transformed into its neighbor in the series of changes. Evidently Plato believes that he can accept the sequence of changes, the mechanism, and the model of felting, without any commitment to a single underlying matter that stays the same throughout the change. The lesson is that the elements pass "on to each other . . . their generation" in a never-ending process. Now, it becomes clear as Plato's argument progresses that he himself does not fully accept the account given here. (The theory he is about to elaborate entails that earth cannot interact with nonearthy elements.[82] And instead of Anaximenes' seven-element scheme he endorses the scheme of four elements introduced by Empedocles.[83]) Hence Plato is borrowing this theory only as an initial approximation of what he regards as the true account of change,[84] not developing his own theory at this point. This point helps to confirm the fact that Plato is not the author of the theory he introduces here.

In the following paragraph, Plato draws out what he sees as the implications of this model:

> Since each of these things never appears the same, which thing can we steadfastly maintain is this determinate thing and not something else, without being embarrassed? Nothing, but by far the safest course to take is to say this concerning such things: whatever we observe always changing from one thing to another, for example fire, we should in every case call not "this" but "this sort of thing," or water in every case not "this" but "this sort of thing"— never calling anything else of the sort "this" as though it had some constancy, of all the things we indicate using terms like "this" or "that" with the aim of picking out something in particular. For such an object flees, never abiding to receive the designation of the phrase "this" or "that"[85] or any expression which refers to them as stable objects. But we must not use these terms, but rather call whatever anything is like at its particular stage in the cycle of changes, each and every one, "this sort of thing," and especially fire we should call "the completely such as this," as well as everything that undergoes generation. (*Timaeus* 49c7–e7)

We see Plato here extracting from TC, as he understands it, a flux theory: everything—that is, every element or basic substance—participates in an ongoing process of transformation. If Anaximenes had already been branded as a Material Monist in Plato's time, it is baffling to see how Plato could read a doctrine of radical change into TC. If, on the other

---

[82] *Timaeus* 56d: earth cannot change into other elements insofar as it is composed of incommensurable triangles, 55b–c.

[83] Plato deduces the four elements, *Timaeus* 31b–33b.

[84] Or as reasonable as any account of the sensible world can be: *Timaeus* 29b–d.

[85] Omitting καὶ τὴν τῷδε with Cornford.

hand, Plato did not read Anaximenes as saying that everything is ever and always air and nothing else, he would be free to understand TC as showing how change can proceed without identity. For it is precisely identity through time that Plato is denying in elemental change—that which is allegedly supplied by the underlying matter. The failure of reference reveals in language the lack of continuity in reality.

Plato's interpretation, indeed, looks suspiciously Heraclitean. It follows the pattern of Heraclitean—even radical Heraclitean—interpretation Plato seems to have derived from Cratylus.[86] Nothing stays the same, but is always changing into something else. Plato adds to this doctrine the proviso that it applies in the sensible world, but not in the world of intelligible realities. This Heraclitean connection does not undermine Plato's reading. It shows that his reading may be mediated through another line of interpretation—another interpretive tradition—than Aristotle's. This tradition would apparently go back from Cratylus to Heraclitus. Whatever distortions and appropriations this tradition may impose on Anaximenes, it at least allows us another perspective on what Anaximenes said and meant. And it is perhaps significant that Heraclitus is a near contemporary and neighbor of Anaximenes. So we may have in Plato a point of view illuminating aspects of Anaximenes' thought that would be plain to his fellow Ionians but obscure to fourth-century Athenians in general. Minimally the existence of a competing tradition constrains us to posit an original document, an urtext, that is amenable to the two conflicting traditions of interpretation. We can no longer be satisfied with the doxographical tradition alone.

Once we recognize Plato as a reader of Anaximenes, we have grounds for looking backward to see if his reading can be confirmed. In chapter 5 I shall have much to say about Heraclitus as responding to Anaximenes, and above I have argued that Xenophanes' theory of change may well be similar to Anaximenes'. Here I shall refer to two texts that show other Presocratics were aware of Anaximenes' theory of change. In developing his cosmology Anaxagoras says:

> From these things being separated earth is compacted. For from [iv] clouds [v] water was separated, from water [vi] earth, and from earth [vii] stones are compacted by the cold. These stones move out more than water. (B16)

Here we have four members of Anaximenes' set of basic substances, as well as the mechanism of compacting. Anaxagoras goes out of his way, as it were, to adhere to Anaximenes' scheme. For his own basic theory of matter is one of mixture and separation, in which compression is not a basic process, but at best only a (misleading) way of describing how heavy

---

[86] Aristotle *Metaphysics* 987a32–b1.

substances are extracted from lighter ones.[87] Furthermore, this account seems to occur in a cosmogonical account, in other words, precisely where, on Wöhrle's interpretation, we should not find it.

In his criticism of human beliefs, Melissus reports:

> But it appears to us that the hot becomes cold and the cold hot, the hard soft and the soft hard. . . . And from [v] water [vi] earth and [vii] stone seem to come to be. (B8.3)

The "appears" and "seem" represent misleading human experiences. Like Plato, Melissus seems to think of Anaximenes' theory as appealing to everyday experiences. In a passage in which he alludes to the theories of Heraclitus (first sentence), Empedocles, and Anaxagoras (omitted), he includes the sequence of changes unique to Anaximenes.

Anaximenes' theory of change is in fact so well attested it seems surprising that anyone could question its existence. In light of later responses, it would seem that Anaximenes' theory of change was one of the most influential of early physical theories, and one that struck his successors as especially commonsensical.

What then are we to make of Pseudo-Plutarch's treatment in which several steps of TC are omitted? One obvious possibility is the author of the account is confused. It is likely, indeed, that he has misrepresented Anaximenes' theory, in which the heavenly bodies probably are fiery.[88] The theory of earthy or stony heavenly bodies being heated by friction seems especially to come from later figures such as Anaxagoras and Leucippus.[89] Furthermore, Pseudo-Plutarch contradicts the more self-consistent account of Hippolytus, which seems to be especially complete and well-informed.[90] Wöhrle claims that precisely because he disagrees with the more obvious interpretation of Hippolytus and others, Pseudo-Plutarch should be considered to provide the *lectio difficilior* or more original interpretation.[91] Yet he may simply provide the more ill-informed reading.

Still, there is a more basic problem with using Pseudo-Plutarch as a witness of TC. In fact what we have in his account is a cosmogony, not a theory of change. What the author is focusing on is the order of formation of the *heavenly bodies*, not the elements. Hence he moves immediately

---

[87] B17, cf. B6 et passim.

[88] Aëtius 2.20.2 = A15; Theodoret 4.23; Hippolytus *Refutation* 1.7.5 = A7.

[89] On Anaxagoras Hippolytus *Refutation* 1.8.6 = A42; on Leucippus, Diogenes Laertius 9.33 = A1; cf. KRS 155, Moran 1973. But Bicknell 1969, 68–69, follows Zeller in seeing the stars as having earthy cores and fiery peripheries: Zeller 1919–1920, 325–26; cf. also Longrigg 1965; Schwabl 1966.

[90] *Refutation* 1.7.5 = A7.

[91] The image is from Klowski 1972, 134 (incorrectly cited by Wöhrle 1993, 20, as being from p. 132).

from the air (the original stuff) to the earth, to the sun, moon, and stars. When he reports that "the earth was formed first," he means *first of the heavenly bodies*, not first of the elements.[92] In the context of world formation there is no need to retail the transformations of the elements, especially if that has already been covered. If, then, Pseudo-Plutarch was following the cosmogonical section of a Theophrastean report, we should not expect him to talk about elemental change. He points out that the earth was the first heavenly body to appear, followed by the sun, moon, and stars. The cosmogony is fully compatible with the theory of change attributed to Anaximenes. Thus there is no reason to think that Pseudo-Plutarch undermines TC. And we have Plato's confirmation that Anaximenes did embrace TC.

Wöhrle claims that TC was invented and foisted on Anaximenes by Theophrastus. Because TC does not hold, MM cannot apply to Anaximenes either. At this point we have seen that the traditional Theory of Change can reasonably be attributed to Anaximenes, independently of Material Monism. Since MM is logically independent of TC, the ontological theory must be evaluated on criteria other than whether Anaximenes holds TC. I have raised serious problems for MM independent of TC and tried to show that there is at least a viable alternative ontology to MM in the Generating Substance Theory. And as we have seen in Plato, TC can be interpreted in light of the alternative ontology.

3.4.5 GST and the Criteria

The criteria we raised in connection with MM were as follows. An adequate theory should be (A) historically appropriate, (B) philosophically coherent, and (C) dialectically relevant. It should be evident that GST is (A) more historically appropriate than MM because it does not presuppose a sophisticated ontology. According to GST, one stuff changes into another, and that stuff into yet another. But we do not need to posit some unseen substratum that continues throughout the change. The events that we observe are just the changes that actually occur: one stuff is transformed into another. In the fourth century Plato still sees this account as consistent with our commonsense observations. Part of the beauty of GST as an interpretation of change is that it posits nothing beyond the actual stuffs of the world. X changes into Y changes into Z changes into Y changes into X. The stuffs we experience in everyday life are the realities.

[92] Note also the article with "earth": τὴν γῆν; the article is absent from the two later occurrences of γῆ, where the term denotes the element; and Ps.-Plutarch calls the earth "completely flat," confirming the reference to the heavenly body, not the element. For a different account of Ps.-Plutarch's motivation, see Moran 1975.

The changes we seem to observe are just the real changes. Change between stuffs on this account is what Aristotle calls coming to be and perishing. But the Ionians do not need an advanced typology of changes because they do not have to explain away one apparent kind of change in favor of another. They do not have to say what looks like coming to be and perishing is really alteration. The theory of change is a genetic theory that tells how one original stuff can turn into all others, and how a certain balance of changes can preserve the status quo and hence sustain a cosmos. GST does, to be sure, presuppose certain metaphysical principles, but they are not extravagant or far removed from what everyday experience tells us about the world.

As to (B) philosophical coherence, I have argued that MM cannot be sustained without some account of what it is to be the ultimate reality apart from its sensible attributes—an account which no early Ionian seems to supply. GST does not need such an account because it posits no latent realities. In GST what you see is what you get. And if one kind of stuff seems to change into another kind—for instance, if water seems to turn into air by evaporation—then it really does. Thus on one key point, GST is more defensible than MM. This is not, however, to say that GST does not have philosophical problems. It does. We shall examine them in some detail in the next chapter. How then can we say GST is superior to MM? In light of the problems to be discussed later, I do not think we can claim GST is superior to MM philosophically. But I shall try to show that the problems of GST make sense with reference to the third criterion, whereas the issues of MM never get raised at all. In other words, the problems of GST are dialectically relevant, while those of MM are not.

There is, however, one area in which it may appear that my defense of Anaximenes fails to supply a viable alternative to MM. We saw in the previous section that both Melissus and Plato viewed the account of elemental transformations as at least consistent with common sense. If that is the case, perhaps what Anaximenes gives us is not a *theory* but merely a *description* of elemental change. In other words, he tells us the sequence of changes we experience for basic stuffs, and he may even tell us how they happen (by condensation and rarefaction), but he does not tell us why they happen. I do believe that part of the enormous appeal TC seemed to have for later thinkers (as pointed out in the previous section) was its transparency: it seemed to give an immediate account of how changes happen. We can observe, as it were, water boiling off in a heated pot, turning into air; we can feel the convection currents rising off the pot (wind) and observe steam (cloud) forming over the pot, then dissolving into air. We can observe silt (earth) settling out of water at a river mouth, see sand and earth being compacted into successively harder and denser materials, and sedimentary rocks crumbling into sand or clay in their turn. We can, then, almost see the basic stuffs in reciprocal motion.

Is this account of change merely a low-level empirical generalization Anaximenes has produced? I think not. In the first place, the fact that the account seems commonsensical in retrospect should not conceal from us the conceptual advance it may represent. By way of comparison, we might notice how commonsensical the law of gravity appears to us, now that we all understand it. In the second place, we should notice that the mechanism of condensation-rarefaction represents an enormous advance. A change in quality or in nature is derived from a change of quantity (as expressed in Aristotelian categories). This is one of the greatest explanatory insights of all intellectual history. The fact that the account may subsequently appear to be superficial and empirical says more about the expectations of thinkers in the Eleatic and Socratic traditions than it does about Anaximenes. Surely TC is a powerful *theory* of change that entails a set of standard changes that can then serve as a *description* of change.

(C) The dialectical relevance of GST and MM we have hardly discussed so far. But in the next three chapters I shall try to show that the major debates of the late sixth and early fifth centuries revolve around the tensions inherent in GST, while they have nothing at all to do with MM. Indeed, were MM the theory in force, there would be no reason for Parmenides to oppose coming to be, for generation would already in effect be excluded from scientific explanation. Parmenides would be beating a dead horse, and his perceived contributions to the logic of explanation would be completely specious. But if GST, or something like it, provides the dominant model of explanation, generation and destruction are fundamental kinds of change for cosmologists. There is at least a theoretical target for Parmenides to take aim at, and the game is afoot.

All of this is merely a preliminary assessment of GST. But it shows promise as a more fruitful account of matter and change in sixth-century scientific philosophy than MM. There is one large question that remains unasked: how could Aristotle possibly get Anaximenes and his contemporaries so wrong that he completely mistook the kind of theory they were proposing? I shall attempt to answer that question later, to give a diagnosis of Aristotle's error, in chapter 10. For now, suffice it to say that Material Monism did appear, in the fifth century, in a form sufficiently reminiscent of sixth-century theories to be easily confused with them.

3.5 Anaximenes' Achievement

Freed from Aristotle's false perspective, Anaximenes appears as much more than just another Material Monist who happened to choose air as

his matter. Like Anaximander he sees the world as emerging from a primeval unitary state. Unlike Anaximander, he identifies the character of the
primal stuff and allows it to continue in the resulting world. Because his
stuff has a definite character, he can identify certain properties that make
it fit to be the original and ruling stuff. Like Anaximander he envisages a
stepwise articulation of the original stuff into successor substances, and
like him he recognizes an oscillating balance between contrary states. But
unlike Anaximander, Anaximenes identifies a specific means by which
changes take place: rarefaction and condensation. And because he identifies a mechanism, he can at least adumbrate the laws that operate on
physical objects and ultimately maintain the cosmic scales in balance.

Furthermore, by virtue of the fact that Anaximenes' source is a determinate stuff existing in the present world, he eliminates an embarrassing
principle from his theory. Whereas Anaximander's world begins with a
something-I-know-not-what that is forever beyond our ken, a mysterious
and inaccessible stuff with divine characters but unknown and in principle
unknowable properties, an inscrutable source, Anaximenes begins with
something that surrounds us all. Anaximenes' source is, to be sure, imperceptible in a certain way, as air is invisible, and of itself without sound,
feel, taste, or smell.[93] In a sense, then, it is unknowable—inaccessible to
perception at least. Yet in another sense it is the medium for all perception,
such that even the slightest perceptible sound, scent, or tinge of color is
immediately conveyed by it, and thus indirectly known in every perception.[94] More important, it is as familiar to us as our life's breath: if we are
deprived of it but for a few moments, we sink into unconsciousness and
then death.[95] Air is inside of us as well as all about us, penetrating our
deepest recesses and constituting our very consciousness and intelligence.
Air is eminently knowable as a concomitant of all perception, thought,
and action, and is the principle of life.

Anaximenes' great achievement is to fill in the gaps of Anaximander's
grand vision with details that allow application and expansion. In Anaximenes' world there is no principle or existent, explanans or explanandum,
that is in principle unknowable. The world has become knowable through
and through. A finite set of everyday substances plus a perspicuous process of connection between them in principle allow the theorist to explain
every phenomenon from the birth of the world itself to its most peculiar
meteorological occurrence. Anaximenes adopts the explanatory project

[93] "This is the character of air: when it is very uniform it is imperceptible to sight, but it is
discerned by being cold or hot or damp or in motion" (Hippolytus *Refutation* 1.7.2 = A7).

[94] "Although it is invisible to sight, it is manifest to the understanding" ([Hippocrates]
On Breath 3).

[95] Ibid. 4: "If one cuts off the passages of breath, in a brief portion of a day the patient
will expire."

of Anaximander, but he replaces an inscrutable starting point with a familiar source belonging to a comprehensive set of basic substances; he enriches a credo in a lawlike order with a comprehensible physical process. He replaces biological procession with mechanical process. In short he writes the first fully natural ontology and the first fully natural etiology. With him Anaximander's scientific program gets the scientific principles it needs for further elaboration.

4

THE GENERATING SUBSTANCE THEORY

AS AN EXPLANATORY HYPOTHESIS

WE HAVE ISOLATED the Generating Substance Theory as a set of assumptions governing the construction of theories in Ionian science or philosophy. If GST has any value for understanding the development of philosophy and science, it will be as a model for understanding the world. What restrictions does it place on explanation? What are its advantages and disadvantages? In order to understand GST, we must look at it as a kind of guide to theorizing. In this chapter we shall examine, first the standards by which we must judge GST, then its advantages, and finally its disadvantages.

4.1 GST Formalized

Thus far we have operated with an informal understanding of GST. I now offer a formal characterization:

The Generating Substance Theory (GST)

  1. There is a set of substances $\{S_1, \ldots, S_n\}$ of which the world is comprised.

     Def.: $\{S_i\}$ are the basic substances.

      a. $S_i \neq S_j$ if $i \neq j$.

  2. There is a substance S_g belonging to $\{S_i\}$ such that at some time t_0 prior to the present only S_g existed.

     Def.: S_g is the original substance.

  3. The S_i arise from S_g by transformation relation T.

     Def.: S_g is the generating substance.

  4. There is some mechanism M which governs T.

  5. S_g governs M.

  6. (a) The world comes to be through the ordered transformations of S_g, and (b) continues to exist by a balance of transformations.

In (1) we identify a set of "basic substances," what will later be called elements, of which everything in the world is composed. In (1a) we make explicit the fact that this is a pluralistic system, in which no basic substance is identical with or reducible to any other. (2) posits an original substance which is itself one of the basic substances but which at some time in the past was the only substance in existence, and from which all the other basic substances arise. According to (3) the generating substance turns into the other basic substances by the operation of a rule of transformation, which by (4) is governed by a mechanism. Point (5) makes the generating substance the controlling factor or governor of the world. And (6) identifies the ultimate product of the transformations as the world or the cosmos. Points (1) and (2) roughly prescribe the system of basic entities or the ontology of the system, points (3)–(5) the causal relations or etiology, and point (6) the domain of application, the cosmology and cosmogony.

In the points listed here, I attempt to formalize the features I have argued characterize early Ionian explanations. GST is, strictly speaking, not a theory at all, but a schema for explaining the world, which can be instantiated in any number of theories. But there is perhaps no need to be fastidious about the terminology. Any theory which satisfies the conditions laid down can be called a generating substance theory. It is my contention that the formal conditions given here are in fact instantiated, at least once, namely by Anaximenes, and that the scheme is approximated by Thales (for all we know), by Anaximander, and even by Xenophanes.

In the formalization of GST I have used one loaded but undefined term, "substance." This term corresponds to the notion of *ousia*, which became a technical term of great importance in Aristotle, though he did not invent the term.[1] Aristotle's theory is complex and controversial. His substance is determinate and definable, independently existing, and (sometimes) particular.[2] For him the elements and material stuffs, which he calls "homoeomeries," are not full-blown substances because they are too indeterminate. Here I use the term "substance" in a looser sense than Aristotle's, but I do wish to stress the concrete physical character of substance (which is sometimes present in Aristotle) and its independent existence. In fact, the sense of "substance" I intend here is fairly close to Aristotle's sense (if not the reference) of *prôtê ousia*, "primary substance" or "ultimate reality," whatever that may be.[3] As I indicated in the last chapter, I think

---

[1] Plato first uses the term in a philosophical sense at *Euthyprho* 11a; but it had a nonphilosophical use before this. See von Fritz 1938, 53ff.

[2] Aristotle *Metaphysics* VII.3, 4, 13.

[3] Aristotle views the question of being as central to all philosophy: "[T]he question that is and always has been asked and always has puzzled philosophers is this: What is being? [τί τὸ ὄν;] and this question is just: What is substance? For some say this is one while some

the Presocratics, while lacking a clear-cut distinction between properties and substances, do tend to think of their primary entities as physical stuffs first and as characterized by properties second. Aristotle would agree in one sense: he sees the Presocratics as searching (in vain) for an understanding of substance, and so as committed to identifying the true substances of the world.[4]

The picture that emerges from this scheme is that of a sequence of substances produced in a lawlike way from the generating substance. In Anaximenes we have a detailed sequence of basic substances:

fire $\leftrightarrow$ AIR $\leftrightarrow$ wind $\leftrightarrow$ cloud $\leftrightarrow$ water $\leftrightarrow$ earth $\leftrightarrow$ stones.

At some time prior to the present there was only air, here marked by capitals to indicate its special role. Then a process of rarefaction produced fire and a process of condensation produced wind, cloud, water, earth, and stones. But rarefaction and condensation are not unconnected: they must be different aspects of a single process. And indeed the substances are transformed into one another only to be transformed back again.[5] Thus we have one sequence of transformations whereby one basic substance is mapped onto another in order, for instance $S_2 = T(S_1)$, $S_3 = T(S_2)$, and in general $S_{n+1} = T(S_n)$. This sequence is balanced by the inverse relation T^1 such that $S_1 = T^1(S_2)$, etc. The relationship between the basic substances is linear and bidirectional rather than cyclical.[6] There must be constant interchanges such as that found in the transformation of water to air and cloud and back again to water to avoid a rapid exhaustion of the elements. Hence the need for a two-way process. The simple relationship of rarer-denser makes a linear sequence obligatory.

One curious feature of GST is the presence of (5). As we have seen, there are hints in Anaximander and Anaximenes, echoed later in Heraclitus, that the generating substance has divine powers and controls the world.[7] Point (5) then anchors a kind of theology in GST, and also provides a source of teleological reflection in which things may happen for a

---

say it is more than one; and some say it is of a limited number while others of an unlimited number. Therefore we too must consider especially and first and only, so to speak, what is the nature of this kind of being" (*Metaphysics* 1028b2–7). On Aristotle's account, all philosophy is an ontological quest for the ultimate reality.

[4] Aristotle *Metaphysics* 1028b2ff.

[5] In a sequence we will observe in Heraclitus in the next chapter.

[6] Interpreters often call the transformation of elements in the Ionians cyclical. But there is no circle because, e.g., Anaximenes' stones do not turn into fire and vice versa; there are end points and reversals in the scheme of transformations, but no circle. The cyclical view comes from Aristotle *On Generation and Corruption* II.4, esp. 331b2–4, cf. Plato *Timaeus* 49c6–7.

[7] Aristotle *Physics* 203b10–15; Cicero *On the Nature of the Gods* 1.10.26, Aëtius 1.7.13 = Anaximenes A10; Heraclitus B67, B64.

purpose. From what we can gather, however, these hints were never developed in detail, but rather the burden of explanation was carried by (3)—(4), the scheme of transformation and the causal mechanism driving it.[8] There is an apparent tension between GST-5 and GST-3–4, between teleological and mechanistic explanation, between natural theology and reductive science. For now it will be enough to notice the broad outlines of this kind of theory and glimpse its potential for elaboration. Point (5) is axiomatic for GST; were it to be developed in detail, it could give rise to a creation story much like that of Plato's *Timaeus*; yet it seems to remain idle and unexplored in the actual instances of the theory that have come down to us.

4.2 A Compromise View?

At this point it might be well to consider how GST differs from MM. The crucial difference between the two views is to be found in GST-1a. By changing that to say that all the other substances are modifications of the generating substance, we would have MM—or almost.

$$\text{MM-1a} \qquad \text{For all } i, S_i = S_g.$$

For a proponent of MM, every other basic substance is merely an accidental modification of the original substance. Hence, it is one in essence with the original substance. Thus (assuming Thales is a Material Monist) we must say air and earth are just forms of water, and everything we encounter in the world *really is* water. This may seem paradoxical in light of some principle such as the identity of indiscernibles—which in its strongest form says that two items are identical just in case they have all the same properties. By this principle, if air is light or rarefied water, then it is discernible from water and hence not identical to it. Here, however, we are concerned not with the identity or nonidentity of individuals (token identity), but with the identity of types of things (type identity). If "identity" seems to refer to too narrow a concept, we may use Aristotle's broader notion of sameness. What we are concerned with in GST and MM is sameness or difference of kinds of reality: is air the same reality as water? According to Aristotle's account of MM, there is only one stuff, and every other apparently elemental stuff really has the same essence as the original substance, for instance, Thales' water. Air, earth, etc., are all the same in *essence*, namely they have the essence or nature of fire.[9]

---

[8] Aristotle *Metaphysics* I.4, 1085a10ff. and I.7.
[9] Aristotle *Metaphysics* 983b6–13; on essential sameness, ibid., 1018a5–7.

Formally, we might define MM as follows:

Material Monism (MM)

1. There is a set of stuffs $\{S_1, \ldots, S_n\}$ of which the world is comprised.
   Def.: $\{S_i\}$ are the basic stuffs.
   a. There is a stuff S^* belonging to $\{S_i\}$ such that
      i. $S_i = S^*$
      ii. S^* is the substratum (material cause) of S_i.
   b. S^* is the only substance (independent reality).
2. At some time t_0 prior to the present only S^* existed.
   Def.: S^* is the original substance.
3. The S_i arise from S^* by alteration relation A.
   Def.: S^* is the source (*arche*) of the world.
4. There is some mechanism M which governs A.
5. S^* governs M.
6. (a) The world comes to be through the ordered alterations of S^*, and
   (b) continues to exist by a balance of alterations.

Points (MM-1a-i) and (-1a-ii) above may seem difficult. The point is that, on the assumption that the substance of a thing peculiar to it, and is the substratum of a thing, the ultimate substratum is the ultimate substance.[10] Hence (MM-1b) there is only one substance, the original matter of the world, and we have a monistic scheme.

One who held MM might well hesitate to call the original substance a "generating substance" because the theory rules out generation; it is, however a "source" (*arche*) in a strong sense—not only a point of origination but also the continuing substratum of everything, and hence a principle for Aristotle. Similarly, an advocate of MM might hesitate to view the other basic substances as independent substances, as suggested by GST-1. And the concept of transformation found in GST-3 is a loaded one. But with a change of name for the generating substance, a suitably weak sense of "substance," and an interpretation of "transformation" as alteration, or more generally as qualified coming to be, a proponent could accept the remaining principles of the formal scheme. Hence we see that GST-1a is the real bone of contention.[11]

If we look back at the conflicting testimonies of Ionian philosophy suggested by Plato and Aristotle, as noted in the last chapter, we might see some common ground between GST and MM. Suppose that Anaximenes

---

[10] Aristotle *Categories* 2a11–14, 2b15–17, 2b37–3a1; *Metaphysics* 1028b33–36, 1029a7–27, 1038b9–10, 15 with Graham 1987a, ch. 9.

[11] I have considered giving this point a number of its own. But it seems to me that it comes in as a kind of implication or presupposition rather than as a primary claim of the Ionians. If it were more explicit there would be less confusion about it in the tradition.

does not articulate either (1a) or (1b) above: he is silent on the ontological status of the substance or substances undergoing change. At this point we would have to attribute to Anaximenes a scheme something like GST-1–6 without (1a). Let us call that "weak GST." Could it be that we are justified in attributing to Anaximenes only weak GST, involving terms so vague that they would be compatible with either MM or GST? In other words, Anaximenes might not explain what the precise relation is between the original stuff and the stuffs that arise from it; he might not explain the nature of the change between them—whether it is unqualified coming to be or qualified coming to be (a distinction which is not made explicitly until Aristotle in any case). Thus his account of cosmogony might not commit him either to MM or to GST. Anaximenes' failure to commit himself either way would allow us to explain how Plato and Aristotle could read Anaximenes so differently: Plato reads him as committed to (1a), Aristotle to (1b); since Anaximenes is silent on the issue, both may be said to have a valid interpretation of Anaximenes. But both would be going beyond what he said to define a relationship left indeterminate in weak GST.

This move appears attractive precisely because it seems appropriate for the early stage of development we might expect in the mid-sixth century, and it would account for the radically different readings found in philosophers of the fourth century. But note that accepting weak GST provides little comfort for MM. For on the latter view there should be a monolithic tradition from the early sixth century at least until the late fifth (with Diogenes of Apollonia), which strongly prefigures Aristotle. If the early Ionians accept only weak GST, we are left with the possibility that there is only one advocate of MM, namely Diogenes himself. That still leaves historians of philosophy with a century of debate to reconstruct on lines very different from those suggested by the traditional interpretation.

Furthermore, considerations of the language of coming to be and models such as Anaximander's of invasion and cosmic justice, and both Anaximander's and Anaximenes' of biological generation—models, that is, in which the beings do not continue, but give rise to different beings—suggest that strong GST is the more natural reading. In other words, the route from weak GST to strong GST is shorter and easier to follow than that from weak GST to MM. The standard mode of explanation in the archaic period is genetic explanation, as seen already in Hesiod's theogony. The thing generated need not be, and generally *is not* identical to the thing generating it. We have seen talk of sea as a generator of clouds still in Xenophanes: the mental habit is deep-seated and automatic. The leap from seeing water as a successor-substance of air, to seeing it as a modification of air seems to be a very great one that needs to be motivated by some philosophical necessity and defended by some philosophical argument. Thus even if Anaximenes explicitly embraces only weak GST, the most obvious articulation of it would be strong GST. The indications of his language, theoretical approach, and intellectual background seem to me to point in the direction of strong GST. In their historical contexts, not all extensions of a theory are created equal. Strong GST is historically easier to motivate than MM in the sixth century, and more natural as an expression of insights contained in the stated theory.

4.3 GST as a Paradigm of Explanation

4.3.1 Paradigms in the History of Science

Historian and philosopher of science Thomas Kuhn has called attention to the role of paradigms in the development of science. By providing striking solutions to a set of scientific problems, paradigms have "served for a time implicitly to define the legitimate problems and methods of a research field" (Kuhn 1996 [1962], 10). A paradigm attracts followers from another scientific approach, and is open-ended enough to suggest possible further applications of the method to new problems. "In the absence of a paradigm or some candidate for paradigm," Kuhn notes, "all of the facts that could possibly pertain to the development of a given science are likely to seem equally relevant" (15). But "[w]hen the individual scientist can take a paradigm for granted, he need no longer, in his major works, attempt to build his field anew, starting from first principles and justifying the use of each concept introduced" (19–20). Now a number of questions can be raised as to how reliable Kuhn's account of scientific development is and as to how helpful it is in the explication of a development so early in the history of thought as the Ionian tradition. Here I merely want to make some basic observations about the relevance of Kuhn's conceptual tools.

Clearly at some level the very possibility of scientific research depends upon some kind of consensus operating among the researchers concerning core areas (and perhaps some lack of agreement on certain peripheral issues as well, spurring on research efforts). Without some set of background assumptions, each researcher would be called upon to reinvent the wheel. When a basic level of agreement is reached, shared inquiry of some sort is possible. Much controversy surrounding Kuhn's theory has centered on possible antirealist implications of a kind of social process of determining a paradigm.[12] But this controversy need not detain us here, for at the beginnings of scientific inquiry we are far from a direct confron-

[12] See Kuhn 1996, 174ff.

tation between theory and reality (if this is possible).[13] The social construction of reality will be the first step in theoretical articulation, whether that construction will later be constrained by real laws of nature or not. And without some shared assumptions, no inquiry is possible.

What we seem to see in early Ionian theory is the pursuit of a common type of research based on a set of common assumptions. These assumptions, or some of them, I have attempted to articulate in GST. They seem to operate as a background theory against which basic questions such as, What is the real *arche*? and more specific questions such as, What is lightning? What is an earthquake? make sense. I take it that it would not be proper to call GST itself a paradigm, since it is merely a formal abstraction. In Kuhn's thought a paradigm must be something of a concrete exemplar which can then serve as a model for further research.[14] I presume that something like the writings of Anaximander or Anaximenes would have served such a function in the sixth century BC.[15] I see Anaximenes especially as providing the best exemplar simply because, at least as far as the scanty record allows us to see, he has worked out the details of ontology and etiology to a greater degree than Anaximander. He is probably not the most original or the most insightful Milesian, but he may be the most systematic and articulate. The role of his paradigm would have been, if anything, more important than any modern paradigm. For in the absence of formal logic, a well-developed mathematics, or any methodological sophistication, a concrete example of explanation would have to carry the whole weight of scientific or prescientific method. There would be no other way of learning how to inquire about the natural world than to assimilate and then to imitate a model.

We must admit from the outset that the Ionians can hardly be called scientists with well-defined methods and procedures. Yet they do at least pursue some broad research program within a framework of shared assumptions about the world. And these assumptions make the world an autonomous realm with its own natures and laws. If there was no preceding scientific system to win researchers away from, at least there was an unscientific way of looking at the world as a realm ruled over by gods, guardian spirits, nymphs, and satyrs, a world in which human affairs were at the mercy of capricious beings endowed with immortality and superhuman power. That picture of the world, with its superstitious practices

---

[13] On the Greeks' inability to provide decisive tests, see Lloyd 1966, 78–79; Lloyd 1979, 139–46; Vlastos 1975, 86ff. On the modern philosophical debate, see e.g., Quine 1960, ch. 2, for a classic statement.

[14] Kuhn 1996, 10–11. Kuhn calls the paradigm a "concrete scientific achievement" which is "prior to the various concepts, laws, theories, and points of view that may be abstracted from it" (11). He further articulates the notion at 181ff.

[15] Since Thales does not seem to have left a written record.

recommended in Hesiod's *Works and Days*, was still dominant in the archaic period.[16] The paradigm of the Ionians offered a radically different way to view the world, with concomitant implications about how to behave in relation to the world. It was a prescientific paradigm. Yet it was the paradigm that prepared the way for all other paradigms prescientific and scientific. For it first defined the world as an object of knowledge in principle accessible to human inquiry. And it allowed its adherents the luxury of identifying problems to be clarified and questions to be examined. It suggested a program of research: to specify the S_i, the S_g, the T, and the M of the model. Without some such model there would not be a research program or any common quest for early cosmologists to debate about. With it, there is a framework of shared assumptions that can found a common inquiry into the nature of things.

4.3.2 Anaximenes' Paradigm

In Kuhn's analysis of scientific development we begin with a concrete instance of scientific explanation that serves as a guide to future research by offering a concrete instance of problem solving which then becomes definitive of the correct method. The instance is not itself an abstract exercise in definition, but a case study from which one might abstract principles and assumptions. It is, in Kuhn's terms, a paradigm. The concept of a paradigm seems useful in the present case precisely because at the earliest stages of scientific theorizing, we find no reflections on how to theorize or what method to use. We find simply concrete examples of explanation. Among these, as I have already suggested, that of Anaximenes is the most developed as to a systematic approach with implications for theory and method. Accordingly, I shall attempt to articulate features implicit in Ionian explanation, using Anaximenes' theory as the chief example and those of other Ionians secondarily.

4.3.3 The Competition

In the sixth century BC the competition for Ionian theory was not some other scientific or philosophical theory, but mythological explanation. Homer's world embodies mythological cosmology, but it does not develop it. It is Hesiod's account of the world, chiefly in the *Theogony*, but also in the *Works and Days*, that defines the standard view of the world. Heraclitus testifies directly of his influence in the sixth century:

[16] As Heraclitus testifies, B57, cf. B106, B40. Homer is reputed wisest, but he is a fool: B56, B42.

The teacher of the multitude is Hesiod; they believe he has the greatest knowledge—who did not understand day and night: for they are one. (B57)[17]

Early in the fifth century Parmenides recognizes the influence of Hesiod by beginning his poem with a proemium influenced by Hesiod, and by echoing his powers of Day and Night in his own cosmology.[18] We have considered the differences between mythological explanation and scientific explanation briefly in the first chapter. Let us now take a closer look at the two kinds of explanation.

In the first chapter, we saw that Hesiod tells the story of how the world came to be by providing a genealogy of the gods. Here again is the beginning of that genealogy:

Indeed, first was Chaos born, but then
broad-bosomed Earth, steadfast seat ever of all
[the immortals, who hold the peaks of snowy Olympus],
and misty Tartarus in a recess of the wide-wayed earth,
and Eros, who fairest among the immortal gods,
looser of limbs, of all gods and all men
overcomes the thought in their breast and their wise counsel.
From Chaos Erebus and black Night were born,
And from Night Aether and Day were born,
whom she bore being with child after mingling in love with Erebus.
And Earth first bore equal to herself
starry Heaven, that he might cover her all around,
that he might be a secure seat for the blessed gods always.
And she bore long Hills, lovely haunts of the divine
Nymphs, who dwell on the woody hills.
And she bore the fruitless deep, with raging swell
Sea, without desirable love. But then
lying with Heaven she bore deep-swirling Ocean,
Coeus, Crius, Hyperion, Iapetus,
Theia, Rhea, Themis, Mnemosyne,
golden-crowned Phoebe and lovely Tethys.
After them was born the youngest, wily Cronus,
most terrible of her children. And he hated his flourishing sire.
Next she bore the Cyclopes with overweening heart,
and Thunderer, Lightener, and mighty-minded Flash
who gave Zeus thunder and wrought for him lightning. (*Theogony* 116–41)

---

[17] Cf. Heraclitus B40, B106.

[18] Parmenides' proemium, B1, is widely recognized as an allusion to Hesiod, especially its use of the house of Night (*Theogony* 744–57, cf. Parmenides B1.9, 11, and perhaps echoed in B8.53–59). For references see below, ch. 6, n. 12.

Chaos, signifying not some sort of primeval disorder, but a yawning gap,[19] was born first—providing a place for the world to appear—a womb of creation. The Earth is born, then Tartarus, the underworld. Eros appears as an agent of sexual generation, which then appears in many (but not all) of the subsequent births. Many of the first generation of deities are cosmological figures—they not only rule, but *are* the corresponding features: Earth, Tartarus, Heaven, Hills, (the inland) Sea, and Ocean (the stream that flows around the circumference of the disk-shaped earth). In addition atmospheric and meteorological features appear: Night, Day, Aether, and the gods of lightning and thunder.

The genealogy follows a roughly cosmic blueprint: Chaos, Earth, Tartarus, Heaven (conceived as a dome covering flat earth), Hills, then, in union with Heaven, Ocean, which flows around the earth where heaven and earth meet. The story invites a reading in which we demythologize it and read it as a straight cosmogony. But of course the story is undeniably a genealogy of the gods, and intermixed with cosmic features are minor deities, legendary figures, and monsters such as the nymphs, Iapetus, and the Cyclopes. What Hesiod gives us is not primarily a cosmogony but a theogony. And his story does not end with the formation of the physical cosmos. It continues on with dynastic struggles between generations of gods: Heaven is replaced by Cronus, and Cronus by Zeus and the Olympian gods, and continues with more genealogies of gods. For Hesiod, cosmogony is implicit in theogony, and cosmology in the orderly rule of the gods when the civil war between gods was ended. Thus for Hesiod:

theogony (cosmogony) → theonomy (cosmology),

where by "theonomy" I signify the apportioning of power and dominion among the gods.[20]

For Anaximander, by contrast, cosmogony is paramount. The world comes to be when a kind of generative power or seed separates off from the boundless and produces a concentration of dense, earthy matter, surrounded by a sphere of flame. The flame is broken up into circles constituting the heavenly bodies, and the present world emerges. Apparently a cosmogonical account leads to a cosmological account of how the world functions now. Anaximenes adds an additional stage: he provides a theory of elemental change which probably supports both the cosmogony and the cosmology. The transformations of air produce an earth and a heaven

[19] On the etymology and interpretations, see West 1966, 192–93; two competing interpretations identify the gap with space in the underworld (West et al.) or with space above the earth Cornford 1950, 98–104; Cornford 1952, 194–95; KRS 37–9. I think both are wrong: Chaos denotes the unified space in which the earth and its heavens come to be; but I cannot argue the point here.

[20] Homer *Iliad* 15.189–93; Hesiod *Theogony* 881–5; Cornford 1912, 15–17.

around it, which then continue to exist by a balance of elementary changes. Thus for Anaximenes:

$$\text{ontogony} \rightarrow \text{cosmogony} \rightarrow \text{cosmology},$$

where by "ontogony" I signify the coming to be of the *onta*, the basic substances that provide the structural elements of the world.

In the Milesian account, (1) things, specifically stuffs or kinds of matter, have replaced persons as the ultimate players in the drama. Instead of a story about how divine persons with supernatural powers were born, came into conflict, and certain deities won the contest and set themselves up as rulers of the world, we have an account of how stuffs were differentiated and came into a harmonious balance. (2) Physical processes such as condensation and rarefaction have replaced birth and human agency as the significant kinds of action portrayed. What matters, consequently, is not particular events or behaviors but general types of action. (3) Physical explanation has replaced historical explanation—drawing on social interactions and psychological motivations—as the dominant kind of causal account. Instead of a struggle between would-be rulers over who should rule, the Milesians tell how physical forces ordered and now maintain the world. The account does not depend on historical accidents, alliances, and personal attitudes, but on general patterns of physical action—something like laws of nature.

In chapter 1, I enumerated scientific features of Anaximander's theory in contrast to its mythological antecedents: (A) the explanantia were natural events; (B) the explananda were natural events, (C) resulting in a closed system of explanation (D) based on material cause explanation (E) in a reasoned exposition in prose, (F) drawing on everyday experience. We might sum up all these points by saying that GST offers a *physical explanation* of phenomena. This new style of explanation displaces, and aims to replace, the quasi-historical style of explanation found in mythology.

Furthermore, GST offers a *mechanical explanation* of events. Not only are human or quasi-human relationships not relevant to explanation, human or quasi-human agency is ruled out as well. The causal factors of an explanation are properties of physical stuffs, which behave in nonhuman ways. The point may seem obvious and unworthy of attention; but if we recall Hesiod's mythology we can note that his *Theogony* is inhabited by many quasi-cosmic or physical entities such as Heaven, Earth, Ocean, and Sea. His entities are personified and act out of recognizably human motives for recognizably human ends—for instance, power and sexual gratification. In GST the players are not only physical stuffs, they are demythologized and dehumanized stuffs which tend to act in a fixed way in response to purely physical conditions and forces. Of course divine qualities are assigned to the generating substance, and we have in GST-5 a reference to the power of the generating substance to control the mechanism; there is, then, the potentiality for GST to turn into a story of how a divine being creates the world. Yet we never see the Ionian philosophers pursuing this line. The mechanism of change plays the lead role.

GST further appeals to *empirical evidence* to substantiate its explanations. Mythological explanations require adjudication by individuals with special prophetic gifts, or with inspiration of the Muses to give an accurate account of why and in what order events happen. Hesiod reports having a experience in which the Muses appeared to him in person and gave him a branch of bay, which presumably he used as his poet's staff.[21] Homer begins the *Iliad* by invoking the Muses, and later calls on them again to tell him the catalog of ships which sailed to Troy—a feat beyond human ability.[22] By contrast, GST appeals only to everyday occurrences, or analogies with them, to account for phenomena. The earth is like a raft floating on a vast sea, or the heavenly bodies are like chariot wheels, or like a felt cap. The philosopher's source of knowledge is his own ability to reason and make connections between events. We should not, however, expect too rigorous an appeal to experience. According to feature (F) of Ionian theories, the philosophers appeal to everyday experience. What we see in models such as the chariot wheel and the felt cap are, by comparison to the refined experiments sought by contemporary science, very homey pieces of evidence. Nonetheless, the Ionians do support their theories by making analogies between exotic phenomena such as the motions of heavenly bodies and everyday events such as the motion of a wheel or a cap. They expect us to understand and ultimately to be persuaded of their theories by transferring our everyday experience to the phenomena of the world.

Not content with particular explanations of individual events, GST aims at *comprehensive explanation*—accounting for everything in the world, at least in principle. Not only do Ionian thinkers account for the coming to be of the cosmos, they also try to account for its present structure, for the rise of living things and of human beings, and for human culture. To judge by explanations later in the fifth century, even the gods themselves are fair game to become objects of explanation. To the extent that the Ionians are successful, there will be no phenomenon or event left out of account.

One feature that has drawn much attention from ancient students of the early Ionians is the emphasis on a single *arche*. Although their views

---

[21] *Theogony* 1–34, with a reflection on their ability to lie as well as tell the truth (27–8). Hesiod invites the Muses to inspire him in telling the story of the gods' birth, 104–14.

[22] *Iliad* 2.484–92: the Muses were present, but we know these things only by hearsay; furthermore, the narrator could not tell the ships and their leaders if he had ten tongues, ten mouths, and a heart of brass, without the help of the Muses.

do not necessarily commit them to one underlying substance of all things, as I have argued, they clearly are committed to *a single explanatory principle*: water or the boundless or air or fire. In some sense, they seem to agree, explanation has a single starting point.

4.4 Advantages of GST

In this section we shall consider the advantages of GST, relative both to mythological explanation and to MM. What does it have to offer as an explanation of the phenomena?

4.4.1 Use of Mechanical Explanation

About physical explanation I shall have little to say. This is the single most miraculous advance made by GST. But it is, on the one hand, extremely difficult to track the step or steps by which the Ionians moved from mythological to philosophical explanation.[23] And on the other hand, GST shares with later models of explanation this feature, so that it does not differentiate GST from other Presocratic models. In the case of *mechanical explanation*, however, there is much to be learned by how this and succeeding theory types account for phenomena. In GST-4 we have a mechanism M which governs transformation relation T of the basic substances. What is crucial is that M operates in a wholly predictable way: if some set of conditions C is met, resulting state R is produced. Most clearly this occurs in Anaximenes' account when a basic substance is either condensed or rarefied. If, for instance, air is condensed, it becomes wind. If wind is condensed, it becomes cloud. If cloud is condensed, it becomes water. If, in general S_n satisfies the condition for condensation, it changes into S_{n+1}. There is no decision to be made, no act of will to intercede between the input and the output.

Such an account stands in stark contrast to mythological explanation, in which a social situation leads to a reaction on the part of a divine being. At the beginning of the *Iliad* the priest of Apollo is insulted; he prays to Apollo, who is indignant with the treatment the Greeks have given to his priest, and, by extension, to himself. He attacks the Greeks until, suitably chastised, they relent. There is something lawlike in social interactions. But the actions involved depend upon social duties being observed, and any agent by a willful or rash action can upset the social balance. Midway between the purely mechanistic and the purely social interaction is the

---

[23] Stokes 1962–1963 gives a good account of the continuity between Hesiodic and Milesian cosmologies.

situation portrayed in the fragment of Anaximander. The several substances must make retribution to each other for their transgression according to the assessment of Time. The seasonal or cyclical excesses of one contrary are seen as transgressions, the retreat of the contrary as requitals. Governing the process is an authority figure, Time, who dispenses justice impartially.[24] The players are themselves impersonal forces, but they are personified in the imagery. An interplay of mechanical forces is represented as a human social interaction. With Anaximander we are at a crossroads between mechanical and social explanation. After Anaximander it becomes increasingly clear that outcomes are determined by initial conditions alone, not by cosmic arbitration or intervention.

4.4.2 Appeal to Empirical Evidence

It has been pointed out, quite rightly, that such empirical evidence as the Presocratics (including the later Presocratics) are able to appeal to is not adequate to establish—or, more significantly, to overthrow—their theories or any particular point of them.[25] By modern standards, there is no test that early Greek theories suggest that would be sufficient to confirm or disconfirm them as against competing theories. If, however, we examine the theories in light of their mythological alternatives, we at least see that they make extensive appeals to experience in areas in which it would not occur to a mythographer to seek confirmation. To explain earthquakes, Anaximenes explains that as the moist earth dries, it cracks and crumbles:

> Anaximenes says when the earth dries out after having been wet, it cracks, and when the broken crags fall it is shaken by them. That is why earthquakes happen during droughts and again during times of heavy rain. For in times of drought, as has been said, it cracks as it dries out, and being soaked by the waters [in times of heavy rain] it crumbles. (Aristotle *Meteorology* 365b6–12 = A21)

Anaximenes can appeal to the everyday behavior of clay to provide a small-scale model of this large-scale event. He also seems to claim that his theory is confirmed by the fact that earthquakes occur regularly in times of drought or heavy rain, that is, when the earth is apt to crack and when, after it has been cracked, it is apt to crumble.[26] To a mythographer,

---

[24] At least on the most robust interpretation of the imagery. See above, sec. 2.2.

[25] "[W]e do not know of a single fact observable by the *physiologoi* with the means at their disposal which could 'upset' any major theory of theirs in the domain of terrestrial physics" (Vlastos 1975, 87).

[26] This seems to be Anaximenes' view, not Aristotle's interpretation, since in the following lines Aristotle uses the inference to criticize the theory.

all that one needs to know is that Poseidon shakes the earth—and why the god is angry.[27] So at least the appeal to empirical evidence sets Presocratic explanation apart from earlier modes of explanation. We find in the Ionians simple analogies like that of Anaximenes which offer everyday events as models of astronomical, meteorological, or geological phenomena. The analogies themselves prove nothing, but they do reveal the assumption that exotic phenomena are not in principle different from everyday events: the regularities we observe can in principle account for strange and seemingly prodigious or ominous occurrences.

Furthermore, the new regularities we identify can be confirmed by pointing out how they account for other regularities: earthquakes occur (allegedly) in times of drought or heavy rain. In principle the method is like that of modern science: find a hypothesis that accounts for the primary phenomena, then see if it makes sense of secondary phenomena as well. When we find, for instance, that a formula for the attraction of bodies can explain not only the elliptical orbits of planets, but also the parabolic course of projectiles, we think we have a genuine law of gravity. Now in Ionian theory the characterization of the phenomena to be explained is often crude, the secondary regularities contrived, and in general the standards of rigor much too low to confirm anything. But if we keep in mind the state of research in the sixth century BC, we shall perhaps not be wrong in emphasizing the merits rather than the demerits of the theory.

Thus far we have not looked specifically at the advantages peculiar to GST (in contrast to other Presocratic theories). The salient feature of GST is the concept of transformation between one substance and another. In this we find a good deal of everyday empirical support. We may observe, for instance, that boiling water in a pot causes the water to gradually disappear; it steams off and then is seen no more. According to GST, it has turned into air—or, to follow Anaximenes precisely, to wind, cloud (steam), and then air. Sense experience seems to confirm the sequence: an updraft is created, steam rises off the pot, and the water gradually disappears. Moreover, we may posit a cycle of changes: as air is cooled and compressed, it forms clouds, which generate winds, from which rainwater falls. We have, then, a series of reversible changes which corresponds to the modern concept of the water cycle. Let us return to Xenophanes for the earliest statement of the cyclical relationship:

> Sea is the source of water, the source of wind:
> for neither <would there be wind> without great sea,
> nor currents of rivers nor rain water from the sky,

[27] Thus at the beginning of the *Iliad*, the Greeks seek to avert a plague by finding out the cause. By consulting a seer, they determine that Apollo is sending the plague because Agamemnon insulted his priest; consequently they seek to propitiate the god (1.57ff.).

but great sea is the begetter of clouds, winds,
and rivers. (B30)

There is ample empirical evidence that water turns into clouds, which produce winds, which dissolve into air; and the reverse steps lead to precipitation, which restores water to the surface of the earth. What experience does *not* teach us, of course, is exactly how water "turns into" air. We now know that water molecules are suspended in air in a supersaturated state in steam or cloud, and dissolved in air as a solution. But it is perhaps significant that what *appears* to happen is taken by GST as what *actually* happens: one substance turns into another.

Not only is there a water cycle in GST; air turns into fire or aether, so that air may be taken as the fuel of the heavenly bodies. The heavenly bodies would need fuel only if they were understood to be burning by analogy with everyday objects, for instance, wood in a fireplace. But of course the fuel to wood relationship can be seen precisely as a transformation of the sort posited by the theory in the first place.[28] If we cover up a fire, for instance, put a snuffer on a candle, the fire goes out. Thus air or some kind of vapor can be seen as the nourishment of fire by an intellectual of the late fifth century:[29]

Air [*pneuma*] is the fuel [*trophê*] of fire, and fire deprived of air would not be able to survive. ([Hippocrates] *On Breaths* 3)

Water, as we have noted, also turns into earth according to Anaximenes and perhaps Xenophanes, and certainly Heraclitus. We find empirical evidence for such a relationship in the mud deposited by streams in deltas. Earth absorbs moisture from rains, but sometimes seems to give up moisture as it dries out. That earth turns to stone we can infer from fossils embedded in stone, as did Xenophanes. That stone turns to earth can be seen from the detritus of sandstone or the heating of limestone to make lime. And indeed there is a cycle of earthy materials to stone: sand to sandstone, clay to shale to slate, carbon to peat to coal to diamond, calcite to limestone to quartz. Erosion turns sandstone back into sand, and acid reactions dissolve limestone. There is further a larger carbon cycle involving the creation of complex carbohydrates by plants, the destruction of carbohydrates to water and carbon dioxide, the capture of carbon dioxide by minerals, etc. All of this is at least prefigured in GST.

Most basic to Anaximenes' version of GST is the correlation between hot and cold, on the one hand, and rare and dense, on the other. This connection Anaximenes supported by an appeal to experience:

[28] See discussion in sec. 3.3.2.1.
[29] The author is likely a sophist: Jones 1923, 221–22. See previous chapter for a discussion of this passage in the context of cosmology.

[Anaximenes] says that what is compressed and condensed is cold, what is rare and "relaxed" (using just this word) is hot. Hence it is not unreasonable to say that man releases both hot and cold from this mouth. For when breath is pressed and condensed by the lips it is cold, but when the mouth is relaxed it becomes hot as it escapes, because of being rarefied. (B1, Plutarch *On the Principle of Cold* 947F)

This everyday experiment purports to show that what is compressed is made cold and what is rarefied is made warm. Of course the physical theory is wrong. Indeed, the converse is true: compression gives off heat, while rarefaction absorbs heat—the principle on which refrigerators and air conditioners operate. But Anaximenes does invoke the evidence of the senses to establish the connection between compression and temperature. One would like to raise objections: what do you make of hot earth and cold air? Why does water expand when it freezes? We have no record that Anaximenes worried about such problems. At least, however, he does recognize that his mechanism of change must meet empirical standards, and he tries, however crudely, to supply empirical evidence to support it.

If we look at the whole pattern of explanation in strong GST, the most striking feature is the fact that every phenomenon is manifest. Water seems to change into air—and it does. Air seems to change into fire— and it does. Water seems to change into earth—and it does. There are no invisible factors posited to account for changes. Or rather, the only invisible factor is air itself, but we can adduce at least tactile evidence to show that air is something.[30] What makes GST so marvelously commonsensical is that in the scheme what you see is what you get. The phenomena themselves turn out to be transparent in that the disappearance of some substances and the appearance of others is taken to be a basic fact. Two centuries after Anaximenes we find Plato expressing a similar sentiment (in a passage discussed earlier in a different context):

what we have now called water we observe [*horômen*],[31] as we believe, turning into stones and earth as it is compacted; but then dissolving and dispersing, this same thing becomes wind and air, and being ignited, air becomes fire, and being compressed and quenched in turn, fire departs and turns back into the form of air, and again air coming together and being condensed becomes cloud and mist, and from these being felted still more comes flowing water, and from water come earth and stones again, these things thus passing on to each other in a circle, as it appears, their generation. (*Timaeus* 49b7–c7)

---

[30] See [Hippocrates] *On Breaths* 3–4.

[31] Compare Aristotle commenting on Anaxagoras (*Physics* 187b1–2): "Accordingly, they say everything is mixed in everything, because they observe [*heôrôn*] everything coming to be from everything."

Plato carefully qualifies the pictures with phrases such as "we believe," and "as it appears." Nevertheless, he regards the picture of physical change represented as so attractive as to pass for the result of sensory observations. Notice that not a single entity in the series is a merely theoretical entity: there are no atoms, powers, Platonic Forms, not even any Aristotelian substrates. GST offers us as the ultimate realities just a set of actually observable stuffs such as we encounter in everyday life. And the changes it posits between them are changes such as we observe—if not in detail, at least in their overall effects—such as evaporation, condensation, silting, and oxidation. We do not need, then, to make a great leap of faith to accept GST—at least once we are committed to some kind of physical explanation—but only a willingness to generalize certain types of processes as definitive of all change.

One other feature to notice in GST is the fact that the basic substances of the theory more or less correspond to the major masses of the world, what Lucretius calls the *maxima membra mundi*, the ultimate limbs or parts of the world.[32] Later, in Empedocles, these world masses will be standardized as earth, water, air, and fire: earth on the lowest stratum, water in the seas that rest on earth or fill its hollows, air all around us, and fire manifest in the heavenly bodies and various fiery or luminous celestial phenomena from sunsets to lightning to comets. Presumably Empedocles' typology owes much of its appeal to the fact that his elements (or "roots," *rhizômata*, as he called them), just are the world masses. Because they are the most plentiful substances of all, it seems evident that the world is made up of such things. Anaximenes' list of basic substances is not so compelling. Yet it contains Empedocles' four elements, as well as wind, cloud, and stones. Wind perhaps has the least claim to being a basic component of the world; Anaxagoras omits it in his retailing of meteorology (B16). Other early thinkers identify wind as air in motion.[33] Clouds, however, occupy vast stretches of the sky and sometimes cover it completely. Stones, or better here, stone as designated by a mass term, forms the foundation of the world, the bedrock of continents, on which earth occupies only a limited area, as is plain to see in many parts of Greece. Thus the framework of Anaximenes offers something like a theoretical articulation, not only of the basic stuffs of the world, but also of the most common and plentiful stuffs. And these stuffs seem to form the framework of the world, the strata and structures that define what it is to be a cosmos: the mountains, fields, seas, atmosphere, and heavens. In other words, although the emphasis is always on materials, the articula-

---

[32] See Kahn 1960, ch. 2.

[33] [Hippocrates] *On Breaths* 3: "Wind is a moving current [*rheuma kai cheuma*] of air." Seneca *Natural Questions* 5.2 = Democritus A93a.

tion, *diakrinein,* in Anaxagoras' vocabulary, of the materials is at the same time an arrangement of matter into a structured unity.

In summary, GST draws on familiar materials and everyday processes; it seems to invoke no mysterious or invisible realities or actions, but rather describes a system in which the reference points just are the components of the world as we know it.

4.4.3 Comprehensive Explanation

The one important point about the domain of explanation is that the Ionians seem to leave nothing out of consideration. Not only do they explain the earth and the heavens, they also explain the origins of life and the development of human beings. As Gregory Vlastos puts it, contrasting their type of explanation with the mythological:

> To attack the supernatural head-on would turn [the Presocratics] into out-
> laws. So they do the next best thing. They proceed by indirection. They so
> fill up the universe with *physis* as to leave no room for anything else. (1975,
> 19–20)

It is not obvious that the Ionians are consciously setting out to subvert the mythological worldview.[34] But whether by design or not, they do manage to displace mythological explanation so completely as to leave no room for it. On the other hand, so far as we know, they did not yet carry theory into all the realms that it would later extend into. Ethical issues are notably absent from the early Ionians, with the exception of Heraclitus, who, as we shall see, is a maverick in a number of ways. Thus there is room for expansion in some domains, particularly matters of human culture and norms of human behavior, for Ionian theory. But there is in principle no reason why naturalistic explanation cannot be applied to all domains—and it would be in the fifth century.

4.4.4 Unity of Explanation

From the time of Aristotle it has been commonplace to identify the early Ionians as advocates of a single *arche*: water, the boundless, air, or fire.

---

[34] "Es ist merkwürdig, dass die griechische Philosophie nicht mit der Bekampfung des Mythos beginnt, sondern zunachst eine Reihe positiver rationaler Welterklärungsversuche unternimmt, ehe sie sich ihres Gegensatzes zum Mythos bewusst wird" (Nestle 1942, 82); the conscious attack on myth begins only with Xenophanes (86ff.). "However justly [the dethroning of Zeus] might be said [to be the consequence] of the natural philosophy inaugurated by the Milesians as they appeared after a lapse of more than a century, there is nothing to show that Anaximander and Anaximenes were conscious of such implications of the attitude they adopted toward nature" (Heidel 1943, 257).

Construed as in the light of MM, that principle was viewed as a changeless reality present everywhere, either in its common form, or as a substratum underlying whatever other appearance it took on. For instance, if water is the original principle, then everything is either manifestly water, or water modified by some quality; for instance, air might be warm, light water that takes on the appearance of a gas. I have rejected MM and the notion of a continuing substrate. What then is left of the one principle of the Ionians?

We are left with a single principle of explanation. This principle still plays a large role in the theory. At one time everything was S, for some S, namely S_g. As S_g was differentiated into other substances, it lost its complete dominance. But it remains somehow the ruling power in the system, and the one which has the capacity to create a world in the first place. To be sure, the generating substance ceases to be everywhere and everything. But at one time it was everywhere and everything, and outside our little world order, it remains everything. MM offers an ontological monism, a type monism[35] in which the only kind of reality there is, is the generating substance. GST offers less: recognizing a type pluralism, it posits a single origin of change and a single principle of explanation. GST is clearly a genetic theory in which to explain something is to tell where it came from. Within that framework, there is only one source of everything in GST, and hence there is an explanatory monism. It might be possible to urge the superiority of an ontological monism to an explanatory monism, although I am not certain on what grounds. But even if that were the case, it would not settle the historical question of whether the Ionians were in fact ontological monists. Hesiod's mythological explanation is predominantly genetic insofar as he explains the origin of the world as a genealogy of cosmic, and then Olympian, deities. GST is closer to the Hesiodic model of explanation than is MM. But, while that may suggest that GST offers a more primitive and hence less sophisticated reading of Ionian explanation than does MM, it may equally suggest that GST is more historically plausible than its rival. GST is, I would claim, just as much a rational reconstruction as MM, in the sense of offering a philosophically meaningful analysis of Ionian philosophy and science. The real question is, Which is historically accurate? not: Which is more philosophically sophisticated? To answer the former question we must appeal to historical evidence, not to ahistorical philosophical analysis.

In what follows I hope to be able to tell a philosophically coherent, indeed enlightening, story about how Ionian thought develops, based on GST. I believe MM will come to seem ever less plausible in light of the dialectic interactions of Ionian theory. But I think that within the frame-

---

[35] See ch. 6 below for typologies of monism.

work of GST we can take a certain kind of monism seriously: an explanatory monism.

Given a single principle of explanation, we can be impressed with the accomplishments of Ionian theory. Starting with a single stuff that allegedly is present in a primeval chaos, the Ionians show how an initial differentiation took place into basic substances; how the basic substances separated out into their own places, into strata that were to become the *maxima membra mundi*; how the process (or, more precisely, the original process and its inverse) that differentiated the substances produced a stable environment from which, in a secondary development, plants and animals could emerge to create, in modern terms, a biosphere. Genetic in conception, the theory goes on from explaining the origins of the world to accounting for its long-term stability—i.e., it goes on from giving genetic explanations to giving structural or systematic ones. Surely Ionian theories embody one of the grandest, and most unanticipated, intellectual leaps of all time. From a single kind of matter and a single process of change, they derive the whole world in all its structural and dynamic complexity and fill it with the variegated forms of life. We should never underestimate the power of Ionian explanation—even without the, as I think, unnecessary prop of Material Monism.

4.5 Disadvantages of GST

Thus far I have enumerated and briefly developed some advantages of GST. But that is not the whole story. For the theory has serious demerits as well as merits. To examine these in a more or less abstract and systematic way will prepare us for the historical conflicts GST faced at the end of the sixth century and the beginning of the fifth. Here I shall deal exclusively with strong GST except as otherwise noted.

4.5.1 The Problem of Primacy

The most striking, and most troublesome, feature of GST may well be its elevation of one substance to the status of generating substance. In GST-2 we see that any of the basic substances can be generated out of any of the others by the appropriate process of change. Of course some substances are more remote from each other than others, but at least by a mediated process of transformations, any member of the set of basic set of substances plus the generating substance can give rise to any other. Why, then, is one member of this set prior to the others—ontologically prior? Why, to be more concrete, should, e.g., Anaximenes make air prior to fire, water, or earth? Let us call this the Problem of Primacy. The problem becomes acute when we see that different basic substances appeal to different philosophers as generating substances: water to Thales, air to Anaximenes, water and earth to Xenophanes, fire to Heraclitus. There seem to be three reasons ascribed to the Ionians by their early interpreters.

(a) At time t_0 before the creation of the cosmos, the whole universe consisted only of that substance. This of course makes the generating substance a kind of matrix out of which things arise, or, more to the point, a reservoir of being.[36] Anaximander seems to envisage such a relationship between his *apeiron* and the cosmos.[37] His generating substance differs from that of other early Ionians by the fact that the boundless does not exist in the cosmos, but only before and outside it. According to Aristotle, part of Anaximander's motivation was to provide a neutral source of becoming which could account for the presence of both of any pair of opposites.[38] Because the boundless universe consists of the generating substance, the generating substance can be called "boundless."[39] As another possible manifestation of this feature the generating substance is called "deathless": not that it never perishes into anything else in the cosmos, but that the universe is always filled with it.[40]

(b) The ultimate substance of the universe will be the one which is most easily adaptable—the most changeable of all things. Anaximenes' air seems to have this potency, since it is the most indistinct to the senses and hence, presumably, exhibits the greatest equilibrium of contrary qualities; Thales' water and Anaximander's *apeiron* can also be understood in this way.[41]

(c) The generating substance is associated with vitality and intelligence, and hence with autonomy or ruling power; thus Anaximander's boundless and Anaximenes' air are endowed with understanding.[42] The generating substance thus becomes not only the original substance but the ruling substance of the natural order.

There is a certain tension in this set of grounds. Whereas (a) would make the generating substance the most stable, most enduring state of reality, (b) would make it the most unstable and ephemeral. Moreover, (a) would work best as an account of an original matrix which gives rise to the contrasting qualities of the cosmos without itself participating in the cosmos;

[36] Aristotle *Physics* 203b18–20, Simplicius *Physics* 465.5–10.
[37] Ps.-Plutarch *Stromateis* 2.
[38] *Physics* 204b22–9.
[39] See Dancy 1989, 151, 163–65.
[40] Aristotle *Physics* 203b13–15, Hippolytus *Refutation* 1.6.1, and Anaximander B2.
[41] Aristotle *Physics* 189b1–8, 205a25–8; Simplicius *Physics* 36.8ff., *On the Heavens* 614.13ff.
[42] Aristotle *Physics* 203b7–15 (Anaximander et al.), Aëtius 1.3.4 (Anaximenes).

for once it becomes a member of the cosmos with its cycles or paths of transformation, it seems to have a place in the scheme and to be no different from the other players.[43] On the other hand, if the generating substance is found within the cosmos, it would seem that (b) would at least provide a potential justification for its superiority. Thus (a) would seem most appropriate for Anaximander's *apeiron*, (b) for Anaximenes' air.

But there is a more general problem with all the proposed explanations. They remain extrinsic to the system of explanation embodied in GST. If X turns into Y during one part of a meteorological cycle, and Y turns into X in another, it does not matter which of them is designated the generating substance: X generates Y and Y generates X. There is no *formal* reason for preferring one member of a series of basic substances. Furthermore, the identification of one substance as the original substance remains beyond the pale of any conceivable empirical justification—contrary to the empirical spirit of Ionian explanation. Of course, (a), (b), and (c) are meant to precisely fill the explanatory gap and to provide philosophical reasons for preferring some substance. But among them, only (a) seems to offer some noncontentious reason for the superiority of one substance, and that only in the case in which the substance does not enter into the cosmos. By its very aloofness from the processes of the cosmos it becomes different. But in this special case, the generating substance also becomes problematic in a new way: it is in principle beyond our experience. It is a pure postulate, forever inaccessible to verification.

As to (b), it does not even succeed in singling out a candidate. For it remains unclear on what grounds we can say that one substance is more adaptable than another. We see water freezing to ice, evaporating to fog; air allegedly changes to fire on the one hand, to wind and cloud on the other. But if we accept the general scheme of GST, the case is even worse: every substance is, technically, just as changeable as every other![44] Whenever the conditions are right for a portion of X to change to Y (whatever those conditions are), X changes to Y.

The connection between (c) life, vitality, intelligence, and different substances might seem to be a more empirical question than either (a) or (b) can present. But here again different possible relations can be suggested: water is associated with living things, birth, reproduction;[45] air is associated with the breath of life and the possession of soul;[46] fire is associated with vital heat.[47] Earth is associated with the generation of plant life and

[43] As Aristotle recognizes in a criticism: *Physics* 189b8–10.

[44] A point which Plato seems to exploit at *Timaeus* 49c–d.

[45] Aristotle sees this as the chief justification for Thales' original substance: *Metaphysics* 983b20–7.

[46] Aëtius 1.3.4, Diogenes of Apollonia, B4, B5 [Hippocrates], *On Breaths* 3–4.

[47] Aristotle *On Respiration* 474a25–31, *On the Soul* 416b27–30.

indeed with spontaneous generation.[48] There is, then, no clear-cut winner in a contest based on (c). But even if there were, it is doubtful that this criterion could settle anything. For all of these claims remain extrinsic to the explanations produced by a scheme such as GST: we could indifferently assign to any basic substance primacy and still have the same physical explanation of the cosmic processes. The real explanation of the events and processes—the origins of life, the meteorological cycles, astronomical phenomena—are performed by T and M of the scheme, not by the identification of S_g.

I have already mentioned the tension between the several criteria for picking out the generating substance. Let me refer briefly to a quite different tension that seems to arise. It has been argued with some plausibility that the background picture of cosmic justice and hence cosmic order is a democratic one: Anaximander envisages a world in which each set of opposites rules in turn, just as in a radical democracy different citizens have their turn to rule.[49] Yet the political model which GST invokes is a monarchical one: there is a single substance whose right it is to rule. Even when it interacts with other substances it is somehow superior to them. How are we to reconcile the egalitarian system of justice presupposed by cosmic models with the monarchical and autocratic principle of the generating substance? The political background is as confusing as the theoretical foreground.

4.5.2 The Problem of Origination

Other problems arise from an appeal to (a) as the ground for choosing one substance as the generating substance. If at t_0 everything was S_g, why did something other than S_g arise at t_1?[50] Let us call this the Problem of Origination. Why would the most excellent of the basic substances change into something else, something inferior, particularly when it was completely uniform and omnipresent? At this point the only possible answer seems to be that something like (c) obtains, and the generating substance itself initiated a different state of affairs.[51] Yet (c) seems to be problematic, given that the mode of explanation of changing phenomena in GST is M, the mechanism of change. There is, in other words, a built-in tension between the mechanistic accounts that seem to be indicated by the pre-

[48] Aristotle *History of Animals* 539a22–4, 547b18–19; *Generation of Animals* 714a24–5, 715b25–30, 721a7–9.

[49] Thus Vlastos 1947, cp. 1953a.

[50] This problem seems to have been first made explicit by Parmenides, B8.9f.

[51] That the early Ionians subscribed to a theory of eternal motion, as Aristotle asserts, *Physics* 250b15–23 (and the doxographical tradition echoes, e.g., Hippolytus *Refutation* 1.6.2), I take to be merely Aristotle's inference of what they are committed to.

ferred form of explanation, and the teleological or psychological account supported by (c). We can perfectly well imagine Anaximander saying that the boundless decided to produce a world. But we cannot imagine that the resulting picture would be the same kind of picture which he and his successors wanted to portray. The Milesian picture is a picture of a self-regulating cosmos without any significant interventions by any god or any personified power or substance. At best feature (c) will have to remain in the background if the cosmology and the cosmogony are not to degenerate into a theogony.

But we should notice that, while the Ionian scheme seems to rule out theogony, it does parallel theogonic accounts in offering a cosmogony. Indeed, it appears that cosmogony is the overall goal of explanation. GST-6 expresses the use of the generating substance to explain how the cosmos arises. And cosmogony seems to characterize most Ionian accounts. Do we really need cosmogony, or would cosmology by itself do? We might be perfectly happy with cosmology alone. But the early Ionians seem not to envisage an explanation of the cosmos without cosmogony. This suggests a presumption that to explain the cosmos is to tell how it arose: it is to give a specifically genetic explanation. But why this? Here we could note how natural it is to explain events by appealing to their origins. We could also note that it is the pattern established by theogony itself. Thus there are both natural and traditional reasons for the first philosophers to give a genetic account of the cosmos. But of course a reflective critic could challenge the presumption itself.

4.5.3. The Problem of Being

In addition to the two problems noted, a deeper problem looms. If X turns into Y and Y into Z, and then, in the inverse process of change Z changes into Y and then into X, what is it that has changed? If, as GST would have it, X is no longer present in Y or Z, how are we to analyze the change at all? Y replaces X without remainder; Z replaces Y, and then in the return journey Y replaces Z and X Y. What is the thing that changes? How can we analyze the change, that is, make philosophical sense of it? One of the alleged properties of the generating substance is that it is everlasting. But it is not—at least not in the cosmos. We have seen this in chapter 2 in the case of Anaximander's boundless, which produces a seed-like intermediary, which produces the contraries, which in turn seem to perish into each other. We have a succession of substances, not a single everlasting substance. And if the original substance is changeable and indeed interchangeable with all other basic substances, what is it? Here we have what we may call the Problem of Being. It grows out of some of the same puzzles as the Problem of Primacy, but here we are dealing with a problem not of ranking, but of fundamental definition. What ontological properties does the generating substance have that make it superior to all substances? Indeed, what is it in its own right? When there is a change, what entity changes, and how does it change? The Ionians have, so far as we know, no sophisticated philosophical analysis with which to back up their account of change. But if GST fairly represents Ionian theory, it must at some point undergo scrutiny as a general theory of change as well as a theory of the natural world. For it proposes its basic substances as the basic entities of the world and their transformations as the fundamental changes of the world.

Conclusion

Ionian theory provides a powerful and brilliant method of explaining the unknown phenomena of the world in terms of events and processes familiar in everyday life. As we have seen, it constructs mechanical explanations of these phenomena, drawing on empirical evidence in an effort to give a comprehensive account of the world, based on a single explanatory principle, the generating substance. It succeeds to a remarkable degree in grounding all explanations on a single principle, while appealing mainly to knowable entities and processes.

On the other hand, we may reasonably ask why one substance should be promoted to the place of the original and dominant substance when GST itself treats every substance as on a par with every other. Furthermore, it remains obscure why the initial undifferentiated state should give way to a differentiated cosmos. Finally, what sort of ontological status does the generating substance have, and how are we to understand its relationship to other substances into which it changes?

The problems as I have sketched them here are heavily informed by the model of GST. In the major traditional account of Ionian philosophy, Material Monism, there is no Problem of Primacy, and similarly no Problem of Being. For the *arche* just is the continuing substrate of change in an ontological analysis, and by virtue of that fact it is primary: every other substance is a mere manifestation of qualifications to the *arche*. If the *arche* is water, air is qualified water, and earth is differently qualified water. Water is always present and always basic. It is ontologically determinate, stable through all changes, and primary to every other entity. Now there could be a problem concerning why one should recognize one stuff as the primary reality in contrast to others; but this would not be a problem about why we should favor one stuff among its peers—for according to MM, the real stuff has no peers. It is ontologically unique and fundamental, and hence primary. We need only to identify some trait of being that is itself fundamental and hence will characterize the primary substance. Only the Problem of Origination still confronts MM, for its primary stuff is ontologically secure and axiologically beyond challenge.

In the following chapters I shall attempt to show that the future of Ionian theory can be plotted as a series of criticisms based on the three problems outlined above. Heraclitus, I shall claim, reacts to the Problem of Primacy and the Problem of Being. Parmenides, responding to Ionian philosophy as characterized and modified by Heraclitus, responds chiefly to the Problem of Being (in a way diametrically opposed to Heraclitus) and secondarily to the Problem of Origination. To the degree that weaknesses of GST and not MM can be seen to motivate Heraclitus and Parmenides, the case for GST will be strengthened and that for MM weakened.

5

HERACLITUS'S CRITICISM OF IONIAN PHILOSOPHY

O F THE EARLY IONIAN philosophers Heraclitus is the most difficult to categorize and evaluate. He is seen either as just another Ionian philosopher with a new *arche* to promote, or as a radical critic of the establishment who defies not only the conventions of his Ionian forebears but also the laws of logic as well—that is, as either very conventional or very radical. Although he makes radical pronouncements, some of his views seem strictly in keeping with Ionian physics; although aspects of his theory seem conventionally Ionian, he seems to draw radical conclusions from that theory. In this chapter I shall attempt to show that by assuming GST as our interpretation of Ionian theory, we can see both what is conventional and what is radical in Heraclitus's theory. Moreover, we can see him as an insightful—and constructive—critic of the Ionian tradition.[1]

5.1 Extreme Interpretations

In the ancient tradition of interpretation, Heraclitus is known for five doctrines which are of importance for an overall interpretation: (1) Fire is the *arche* or original material cause of the world;[2] (2) there are periodic episodes of world conflagration (*ekpurôsis*) when the cosmos is consumed, to be regenerated later;[3] (3) everything is in flux;[4] (4) the opposites are identical;[5] and (in consequence of [3] and [4]), (5) Heraclitus violates the Law of Non-contradiction.[6] The period of modern interpretation be-

---

[1] This chapter is an expansion and revision of Graham 1997.

[2] Aristotle *Metaphysics* 984a7–8, *Physics* 205a3f., *On the Heavens* 298b25–33, Diogenes Laertius 9.8. That fire is in some sense the primary reality for Heraclitus is not controversial; but that it is primary by way of being a material cause in the full Aristotelian sense is.

[3] Aristotle *On the Heavens* 279b12–17, Diogenes Laertius 9.8.

[4] Plato *Cratylus* 401d, 402a, *Theaetetus* 152d–e, 160d, 179d–e; Aristotle *Metaphysics* 987a32–4, *On the Heavens* 298b25–33, *Topics* 104b21–2.

[5] Aristotle *Eudemian Ethics* 1235a25–8, *Topics* 159b30–2.

[6] Aristotle *Topics* 159b30–3, *Physics* 185b19–25, *Metaphysics* 1012a33ff., 1012a24–6, cf. Plato *Theaetetus* 183a.

gins with a concerted attack on all of these doctrines.[7] John Burnet[8] argued against (2), (4), and (5); Karl Reinhardt[9] argued against all five; his general position was continued by G. S. Kirk[10] and Miroslav Marcovich.[11] Here are some statements of the view:

> Finally, need it be repeated that the flux doctrine too as a Heraclitean doctrine is only a misunderstanding, spun out from the constantly returning comparison of the river which remains the same while the water in it continually flows in and out? Not a single fragment expresses the thought that all things are in flux, that in general only transition and change, never continuity and persistence are found—which will show us where the *panta rhei* is truly at home. The basic idea of Heraclitus is much rather the most precise opposite imaginable to the flux doctrine: persistence in change, constancy in alteration, *t'auto* in *metapiptein*, *metron* in *metaballein*, unity in division, eternity in destruction. (Reinhardt 1916, 206–7) The conclusion that emerges from these facts is clear: Heraclitean natural philosophy promises likewise to provide a physical resolution to the problem of contradiction. (ibid., 204)[12]

> Plato and all later ancient critics took the river-analogy to apply to changes in every individual thing, and to illustrate the continuity of those changes; actually it illustrates the measure which must inhere in large-scale changes taken as a whole. Heraclitus did not believe, any more than any of his predecessors, that everything was changing all the time, though many things are so changing and everything must eventually change. (Kirk 1954, 366)

> Heraclitus' logical-metaphysical world-order remains, due to the universally valid equality of opposites in the Logos, in continuous balance, very similar to the resting but taut bow . . . with a minimum of change . . . and motion. (Marcovich 1965, col. 293)

Let us call the Heraclitus that emerges from this approach "Heraclitus the Constancy Theorist" or "Heraclitus-C" for short. For it is a central tenet of the present interpretation that Heraclitus' real theme, his contribution to the philosophical debate, is his recognition of constancy in the

---

[7] Earlier Schleiermacher 1807, reprinted in Schleiermacher 1838, 2, at 86–104, and Lassalle 1858, reprint Lassalle 1920, 11ff., 192ff., had rejected both (1) and (2); but their position remained a minority view. But even earlier Hegel had rejected (2): Hegel 1971, 333. For a defense of (2) see Zeller 1919–1920, 863–78.

[8] Burnet 1930, ch. 4.

[9] Reinhardt 1916, 1942.

[10] Kirk 1951, 1954, KRS, ch. 6.

[11] Marcovich 1965, 1967.

[12] Reinhardt's analysis of the problem of contradiction is colored by his view that Heraclitus comes after Parmenides and responds to him. Reinhardt's historical sequence has been almost universally rejected; but it is defended by Holscher 1968, 161–65.

change that goes on at all times. Heraclitus-C may recognize the reality and universality of change, but he stresses the fact of constancy.[13]

The construction of Heraclitus-C secured the general rejection of (1) and (2).[14] But it provoked a backlash by critics who argued in favor of (3), (4), and (5). Gregory Vlastos, Karl Popper, Rudolfo Mondolfo, W.K.C. Guthrie, C. J. Emlyn-Jones, M. C. Stokes, and Jonathan Barnes[15] have all argued in favor of the flux theory (3), usually with (4) the identity of opposites,[16] and have accepted the consequence (5) that Heraclitus's philosophical theory was formally incoherent. Here are some representative quotations:

> What is only occasional and intermittent, though recurrent, in Anaximander [namely, transgression of the elements], becomes universal and invariant in Heraclitus. Why this difference? We cannot answer this question without granting the obvious implication of the river-fragments which Kirk formally denies ([1954] p. 367 et passim), the universality of change. (Vlastos 1955, 356–57)

> Heraclitus' speculative imagination transforms this straightforward cosmological theorem [Anaximenes' theorem that all things are differentiations of air] into an assertion of the unity of all differences whatever, including moral ones, and pursues its consequences to that reckless and bewildering conclusion that "for the god all things are fair and good and just" (B102), which, if true, would be fatal for all morality, not excepting his own. (ibid., 367)

> We need not expect Heraclitus's thought to be by our standards completely logical or self-consistent. (Guthrie 1962–1981, 1.461) He ignores the law of contradiction, he insists that opposites are identical. (ibid., 463)[17]

[13] Kirk presents Heraclitus as legitimately, though not exclusively, a cosmologist (1954, xii–xiii); Marcovich sees him as chiefly a metaphysician and religious thinker (1965, 295). Rabinowitz and Matson 1956, 256–57, criticize Kirk's emphasis on cosmology.

[14] The sense in which (1) is rejected is the sense in which fire is a material substratum for change. Note that this Aristotelian style of interpreting Ionian *archai* came under independent attack, esp. by Heidel 1906, 333–79, and Cherniss 1935, during the period when the new interpretation of Heraclitus was emerging. Cherniss 1935, 29, n. 108, gives important arguments against (2). Barnes 1982, ch. 3, has revived Material Monism and applied it to Heraclitus (ch. 4), as we shall see. Kahn 1979, 134–55, has revived (2), and Robinson 1987, 186, supports the view. See also Mondolfo 1958, 75–82, and the rejoinder by Kirk 1959, 73–76, and earlier Gigon 1935, 48–49, who reasserted the view thirty years later: Gigon 1968, 222–23. Both (1) and (2), however, are very much minority views now; see, e.g., Reeve 1982. But there is a recent defense of (2) in Finkelberg 1998b.

[15] Vlastos 1955a, review of Kirk 1954, 310–13; Popper 1958, 1963; Mondolfo 1960; Guthrie 1962–1981, 1, ch. 7; Emlyn-Jones 1976; Stokes 1971, ch. 4; Barnes 1982, ch. 4. In a review article of Kirk 1954, Rabinowitz and Matson 1956 also reject Heraclitus-C. See now Tarán 1999 for Heraclitus-F.

[16] Sometimes featuring it, as in the case of Emlyn-Jones 1976, and also Popper 1968, 145.

[17] Cf. also pp. 468f., where Guthrie discovers another contradiction "from our point of view," namely the conflation of matter and soul. This alleged contradiction is outside the

[T]hese opposites combine in ways which defy logical analysis. The element of paradox which is everywhere apparent cannot be attributed to stylistic or rhetorical idiosyncrasy, but must be closely associated with a mode of thought whose linguistic origins may well constitute, for Heraclitus, the ultimate origin of his belief in the identity of opposites. (Emlyn-Jones 1976, 111)[18]

[Heraclitus] did not attempt, even while declaring the identity of opposites, to avoid admitting some opposition between them. . . . So Heraclitus did not give vent to his paradoxes in total naivete, without knowing their paradoxicality at all. But, if he knew their paradoxicality, then he was only a step from knowing that there was something wrong somewhere with the argument; only he could not lay his finger on the flaw and continued to proclaim the paradoxes with his unique vigour. (Stokes 1971, 100)

Now [the] logical notion of contrariety was certainly not available to Heraclitus. . . . [W]ith such resources, Heraclitus might well have failed to see the necessary falsity of his position. (Barnes 1982, 80)

This school of interpretation usually stresses, as against Heraclitus-C, the important role of flux theory in understanding Heraclitus. According to them a central and essential theme of Heraclitus is universal change. I shall, then, call the figure they construct "Heraclitus the Flux Theorist," or "Heraclitus-F." Despite the disagreement over flux, there is some measure of agreement between the two sides: most partisans of Heraclitus-C are willing to concede that Heraclitus acknowledges change as an important phenomenon, while partisans of Heraclitus-F usually dismiss the extreme Cratylean flux of all things changing in all respects, as un-Heraclitean. This may raise the suggestion that the debate is a mere verbal disagreement about what is to count as flux.[19] There are, however, substantive areas of disagreement. Minimally, the two sides disagree about the coherence of Heraclitus's system and about his fundamental contributions to the history of philosophy. More specifically, the two sides disagree over whether one can step into the same river twice.[20]

---

domain of contradictions based on phenomenal opposites or flux which I wish to examine in this paper.

[18] Unlike most interpreters, Emlyn-Jones (113) holds that the unity and opposites and cosmological flux are not closely related doctrines.

[19] Thus Barnes 1982, 65, n. 19, on Reinhardt 1916, 207.

[20] According to Reinhardt, Kirk, and Marcovich, you *can* step into the same river twice. There is only one river fragment, B12, which presumes that the river remains constant (cf. on this point Kahn 1979, 167–68—with his excellent comments on B12). B49a and B91a are only echoes of B12 (Reinhardt 1916, 207, n. 1; Reinhardt 1942, 18, n. 24 on B91a; Kirk 1954, 382–84, argues that B91b should be attached to B12, but Marcovich 1967, 206–11, rejects this). Advocates of Heraclitus-F, on the other hand, would at least accept the message of B91a, and usually claim it has a source in Heraclitus. On the question of how

Heraclitus-C has the virtue of consistency, but he tends to be a rather pedestrian philosopher-physicist, following in the steps of his Ionian predecessors, while occasionally making radical, but rather misleading, noises to punctuate his originality. Heraclitus-F, on the other hand, is genuinely original, but unfortunately in a bad cause, for his ill-considered innovations lead into the slough of inconsistency. On neither of these interpretations, then, does Heraclitus get high marks. Interpreters end up offering patronizing apologies for Heraclitus such as some cited above.

Recent studies have opened the possibility of more interesting and complex interpretations,[21] but at least the two sides help to define some of the most fundamental and persistent issues in interpreting Heraclitus. I believe it is possible to offer an interpretation that draws on both the major modern alternatives while being superior to each, one which makes Heraclitus both a consistent philosopher and a radical innovator. Moreover, this interpretation will not merely make Heraclitus more innovative than Heraclitus-C and more philosophically coherent than Heraclitus-F— which would be no great victory—but would make him a better philosopher than Heraclitus-C and more innovative than Heraclitus-F. Initially what I wish to show is that Heraclitus holds both positions (3) and (4), i.e., flux and the unity of opposites, in some important sense, but that these positions do not, either individually or in conjunction, entail (5) inconsistency. This much constitutes a negative argument. I wish also to show that a proper appreciation of the historical context of Heraclitus's utterances will reveal a certain logical elegance to his position. Heraclitus's paradoxes, seen from the right perspective, constitute a penetrating criticism of his predecessors and offer a powerful and sophisticated theoretical alternative. I should add here that my final position will be much

---

many river fragments there are, the Reinhardt-Kirk-Marcovich view must surely be right: B49a, B91a, Plato *Crat.* 402a, etc., are all echoes of B12 which can be explained as exemplifying a certain reading of B12 (Marcovich 1967, 206, pace Kahn 1979, 168–69, on B91; but Reinhardt accepts B49a). A glance at Plutarch *Natural Questions* 912a (Marcovich testimonium 40c[5]), to be cited below, shows that the adverb *dis* was simply grafted onto B12 (Reinhardt, ibid.). The solution to the problem in source criticism does not automatically solve the larger debate over whether in general you can step into the same river twice (sufficient ambiguity remains in B12: Kahn 1979, 167–68; Stokes 1971, 291–92), but it does make the debate focus on different textual evidence. Below the debate will focus on a so far less vexed fragment, B88.

[21] Kahn 1979; Wiggins 1982; Mackenzie 1988 all make significant attempts to break out of the limits of previous interpretations. Moravcsic 1983, 1989 offers an interpretation of Heraclitus from the standpoint of evolving methods of philosophical analysis—an approach that has much in common with my own reading. I have certain problems with all of these interpretations, some of which I shall point out in notes; but they make advances in our understanding of Heraclitus. One interpretation which has particularly influenced my thought on Heraclitus, though I have come to dissent from key points of it, is Mourelatos 1973. Thus far no clear alternative to Heraclitus-C and Heraclitus-F seems to have emerged.

closer to Heraclitus-C than to Heraclitus-F, for in many ways the advocates of the former interpretation are on the right track. The error in Heraclitus-C is not a failure of conception but of emphasis: the flux is quite real and Heraclitus is radical. The present argument will, however, vindicate many key points of the Heraclitus-C interpretation. Yet my interpretation will go beyond Heraclitus-C the conservative physicist to Heraclitus-R the radical or revolutionary metaphysician.

Jonathan Barnes has provided the most recent and also most philosophically sophisticated and historically informed argument for Heraclitus-F, one which is, more than most others, a revival of the ancient view. I wish to examine his interpretation, which I believe is wrong, but wrong in a suggestive way that will point us toward a more satisfying interpretation. The criticism of Barnes will lead us to revise the principles Barnes ascribes to Heraclitus, which will, I hope, lead us to see more perspicuously what views Heraclitus shares with his Ionian predecessors and also how he departs from them in other respects.

5.2 Barnes's Argument for Heraclitus-F

According to Barnes 1982, chapter 4, Heraclitus accepts some version of what I shall call the Flux Thesis, that all things change. Inclined to accept Plato's dictum *panta chôrei* as a quotation (*Cratylus* 402a, Barnes 65),[22] he interprets the phrase loosely to mean "everything *always* changes" (65, Barnes's italics), and later clarifies this to mean, not with Cratylus that "everything is always flowing in *all* respects," but that "everything is always flowing in *some* respects" (69, Barnes's italics). This view I shall designate as Universal Flux (UF), understanding it in the sense specified by Barnes. Secondly, Barnes accepts what I shall call the Identity of Opposites doctrine (IO), which involves the claims that (a) for all properties, there is some subject that instantiates both the property and its opposite, and (b) for all subjects, there is some property such that the subject instantiates both it and its opposite (70). But if this is so, then Aristotle was right to claim that Heraclitus transgresses the Law of Non-contradiction (LNC),

---

[22] Barnes (ibid.) says, "It is true that the particular phrase '*panta rhei*' first occurs in Simplicius [*In Ph*. 1313.8]." This is at best misleading. Plato *Cratylus* 439c2–3: ὡς ἰόντων ἁπάντων ἀεὶ καὶ ῥεόντων; ibid. d4: δοκεῖ ταῦτα πάντα ῥεῖν; 440c7–8: πάντα ὥσπερ κεράμια ῥεῖ; *Tht.* 182c3–4: κινεῖται καὶ ῥεῖ ... τὰ πάντα; *Philebus* 43a3: ἅπαντα ἄνω τε καὶ κάτω ῥεῖ; Aristotle *On the Heavens* 298b30: πάντα γίνεσθαί τέ φασι καὶ ῥεῖν; *Metaphysics* 987a33–34: ἁπάντων τῶν αἰσθητῶν ἀεὶ ῥεόντων; cf. ibid. 1078b14–15. Surely the essential phrase was current—and well on its way to becoming a slogan—some nine centuries before Simplicius; that other words occur with the phrase does not weaken its value as a conventional expression of doctrine.

for if a subject x instantiates both a property Φ and its opposite Φ', since Φ' entails not-Φ, it is true at the same time that x is Φ and x is not-Φ (79f., cf. 69f., 73). Furthermore, on the basis of B88 (to be discussed below), Barnes claims that Heraclitus supports the Identity of Opposites from Universal Flux. Hence Barnes's whole interpretation of Heraclitus's metaphysics can be summed up as follows:

UF entails IO entails not-LNC.

Now since LNC is a condition of meaningful discourse in general (Aristotle *Metaphysics* IV), any result which rejects it constitutes a reductio ad absurdum. Hence Heraclitus's theory is incoherent, untenable. And although Barnes wants to maintain that "[e]ven at his most paradoxical, Heraclitus remained a rational thinker," for "his extraordinary thesis of [the identity of opposites], no less than his traditional monism, was based on evidence and arguments" (75), he must in the end concede that "his thesis lapses into inconsistency" (80).

Barnes's interpretation takes into account the relevant fragments and testimonies and provides a sophisticated reading in which philosophical theses are neatly and helpfully arrayed to show what, on this view, Heraclitus is doing and why. Realizing the difficulty of attributing to Heraclitus an untenable position, Barnes observes that LNC was not yet articulated, nor was the conception of contraries as entailing contradictories; hence "Heraclitus might well have failed to see the necessary falsity of his position" (80). Of course we are dealing with an early period of Greek philosophy, a period prior to the development of logic and even of dialectic as we know it from the early Eleatic tradition. Thus it is not unthinkable that Heraclitus should make logical errors. On the other hand, the principle of charity requires that we attribute major logical errors to a philosopher only as a last resort. What we can do in a positive way is to demand that Heraclitus be held only to views he is clearly committed to on the basis of the most favorable interpretation we can make of the texts. There is no need to attribute such an unfavorable view to Heraclitus.[23] The pivotal claim of Barnes's interpretation is that the Identity of Oppo-

[23] Triplett 1986 has argued effectively that there is no need to attribute such an unfavorable view to Heraclitus. Here I shall develop a line of criticism that in part follows Triplett and in part strengthens his case. Marino 1984 also criticizes Barnes's reading of Heraclitus, especially concentrating on hermeneutic assumptions. He is right to fault Barnes for not taking Heraclitus's style seriously enough (Barnes 1983, 106, n. 10 confesses his sin); however, it does not follow from this that Barnes is wrong in supposing that Heraclitus's utterances "are the expression of arguments and/or theories" (Marino, 81), however indirect and complex the expressions may be. See Barnes 1983 for a defense of Heraclitus as providing arguments even in the framework of gnomic utterances.

sites entails not-LNC. His main evidence comes from B88, which we must now consider:[24]

> B88. [i] As the same thing [in us (?)] are living and dead, waking and sleeping, young and old; [ii] for these things having changed around are those, and those again having changed around are these.

According to Barnes, "[S]entence [i] states three instances of the Unity Thesis [IO] and sentence [ii] grounds these instances, as its introductory particle shows, on the Theory of Flux [UF]" (72). Thus B88 provides Barnes with the crucial link between the Identity of Opposites and Universal Flux. But what does it really show? For Barnes, the Identity of Opposites is the claim that every property is coinstantiated with its opposite in some subject, and every subject instantiates some property together with its opposite at the same time (70). But instead of taking the Identity of Opposites according to some dogmatic reading, we might use B88 to show us what Heraclitus actually does mean by statements of the unity of opposites. In Wittgenstein's words, perhaps we should "look and see." If [i] provides the ground and rationale—or at least a general type of rationale—for the unity of opposites, we must expect that the sense of [i] will be illuminated by [ii]. But what does [ii] say? Is it a statement of a mysterious metaphysical truth, or something else? All it really says is that things (in a most generic sense of "things") change into their opposites. Compare the following fragment:

> B126. Cold things warm up, what is hot cools down, what is wet dries out, what is dry grows moist.

This is, at one level, a truism; but it does notice the "fact" that change is into opposites. It does not assert, or presuppose, that all things are changing all the time. But it does presuppose the reality of change.[25] [ii] says no more than B126, and presupposes no more. Now if given properties are the same because of mutual interchanges of qualities that go on in nature, then the Identity of Opposites does not consist in our exemplifying opposite properties *at the same time*. At time t_1 we are F, at time t_2 we are the contrary of F, F'. What Heraclitus is saying is that, since the preceding case holds true, F and F' are the same.

Now precisely what it means to say that F and F' are the same, remains unclear. But nothing he has said clearly commits Heraclitus to holding

---

[24] B88 is on any account essential for an understanding of Heraclitus's theory of opposites. As Gigon 1935, 90, says, "Das Stück ist in seinem logischen Inhalt von der allergrössten Wichtigkeit. Es ist das einzige Fragment, wo die Gegensatzlehre in paradigmatischer Deutlichkeit erscheint."

[25] Cf. Aristotle *On Generation and Corruption* 322b12–18.

that we are F and F' at the same time and in the same respect, so that LNC would be violated. Heraclitus does give other accounts of opposites than that provided in B88. For instance, B61 deals with a different kind of opposition.

> B61. Sea is purest and foulest water: to fish drinkable and life-giving, to men
>   undrinkable and destructive.

Here sea supports opposite qualities, "purest" and "foulest." Presumably it is purest at the same time—but not in the same respect. It is purest to subjects $\{S_i\}$ and foulest to subjects $\{S_j\}$, where the respective sets of subjects are disjoint. "Here," Barnes remarks, "at least it is clear that Heraclitus committed the fallacy of the dropped qualification," a fallacy he commits—though less clearly—in other contexts, including the use of temporal qualifiers as in B88 (74, cf. 72–73).[26] But Heraclitus has only committed a fallacy if he is now willing to assert that sea is purest and foulest at the same time and in the same respect—which he shows no propensity to do. It is one thing to make paradoxical pronouncements, another to reject LNC. The paradox has heuristic pedagogical value; it need not express theoretical commitment to contradiction. Heraclitus need be no more committed to contradiction than Plato's Socrates, who causes Meno to contradict all his previous opinions about virtue. Contradiction may be part of a philosophical enterprise without being part of the theory of that enterprise.[27]

[26] Scolnicov 1983, 1: 108 gives a similar account.

[27] Emlyn-Jones, who like Barnes attributes to Heraclitus a version of the Identity of Opposites which entails not-LNC, wants to purge Heraclitus of clause [ii] in B88. Noting that explanations are rare in Heraclitus, and that the word *metapiptein* occurs only in one earlier passage, he suggests that the term is a later one read back into Heraclitus from a later gloss, mediated by Melissus B8 (3–4) and Plato *Cratylus* 440a–b (93–94). Against this, we must note that such a reconstruction is highly conjectural, bordering on the fanciful. That Heraclitus does not provide explanatory clauses as often as other philosophers is true. But Barnes notes that after purging the fragments of doubtful connective particles (that may be supplied by the quoter), he finds the particle *gar* occurring nine times (including the occurrence at B88). On balance, Barnes 1983 concludes that Heraclitus did not rely mainly on an aphoristic style but employed a modest amount of argument. As to the word *metapiptein*, it is true that the term is not common in fifth-century prose, but that fact does not allow us to immediately assess its place in philosophical discourse. In some few cases, there are reasons for suspecting the earliest quoted use of a philosophical term may come long after its introduction, as with the term *apokrinesthai*. In any case, the fact that the term appears in passages inspired by Heraclitus in both Melissus and Plato can reasonably be used as evidence that they are echoing Heraclitus, rather than rewording him. Moreover, the term does appear in other Presocratic writers after Heraclitus. Finally, the parallel between [ii] and B126 indicates that even if we did not have the second clause of B88, we might be able to reconstruct the reasoning with a little astute observation of Heraclitus's view of change. In general, B88 undermines the case for Heraclitus transgressing LNC, and there seems to be no independent ground for rejecting the second clause of B88.

But what does it mean to say that opposites are the same? If F and F' were really identical, the transition from one to another would not be a change, and hence Universal Flux would not be true. Accordingly, it would appear that the Identity of Opposites entails not-UF. Perhaps that objection will not seem compelling to one who accepts the Identity of Opposites, for that principle seems to be self-contradictory in its own right. But if we are not already sold on the Identity of Opposites, its incompatibility with Universal Flux may warn us that we are on the wrong track. Change presupposes different end points for change, with reference to which we can identify the change. This may indicate that whatever Heraclitus intends by the unity of opposites, he does not intend the identity of opposites. The opposites are one in some more extended sense.

Our results at this point are largely negative: we have seen that the version of the unity of opposites presented in B88 is not necessarily the Identity of Opposites and does not entail not-LNC; we have observed briefly that the reasoning given for the unity of opposites in that fragment does not commit us to Universal Flux. We are left, then, with an argument for the unity of opposites that has an uncertain place in Heraclitus's thought. What Heraclitus is committed to in B88 does not seem radically different from what other Ionian philosophers were saying: there is change, and that change involves interaction among opposites—indeed a mutual interchange of opposites. But Heraclitus presents his insights in a paradoxical, and apparently intentionally perplexing, manner. What is his point, and what role does paradox play in it?

5.3 The Unity of Opposites

There is no question that Heraclitus embraces the unity of opposites in some sense, recognizing that opposites are "the same." But we find even a century and a half later, in Aristotle, that sameness is a broad concept that can have many meanings, most of which are not consistent with the narrow concept of identity.[28] Can we find some sense that (a) fits the texts of Heraclitus, (b) is historically appropriate, and (c) is philosophically coherent? I believe we can. To begin, let us consider one of the problems raised by GST: the Problem of Primacy. How can any substance pretend to be superior to others when they all turn into one another? If we look again at B88 we see that this is precisely the situation Heraclitus envisages: every opposite turns into every other; presumably every substance with one set of characters (for instance, water, being cold and wet) turns into

---

[28] *Metaphysics* V.9, *Topics* I.7. Among the different kinds of sameness are sameness in genus, species, number, and matter, and being accidents of the same subject.

a substance with opposite characters (fire, being hot and dry). And B126 makes the same point. Let us consider a formal relationship two substances have to one another if they satisfy the conditions described in these passages: they are transformationally equivalent (TE) insofar as the first is transformed into the second while the second is transformed into the first. More formally:

> X is transformationally equivalent to Y just in case X can turn into Y and Y can turn into X.

Initially, we may notice that this relation is (i) symmetrical, (ii) transitive, and (iii) reflexive. For, abbreviating "X is transformationally equivalent to Y" by "$TE(X,Y)$," we see that (i) $TE(X,Y)$ if and only if $TE(Y,X)$. For instance, air turns into water if and only if water turns into air. (ii) Furthermore, if $TE(X,Y)$ and $TE(Y,Z)$, then $TE(X,Z)$. For example, if water turns into air and air turns into fire, then water turns into fire—and fire turns into water. (iii) Finally, $TE(X,X)$, for by substitution in (ii) if $TE(X,Y)$ and $TE(Y,X)$, then $TE(X,X)$. For instance, if air is transformationally equivalent to water and water to air, then air is transformationally equivalent to air.

We have already seen that, according to GST, different basic substances turn into each other in a way that fulfils the relation of transformational equivalence.[29] We find the relation also in Heraclitus:

> B31a. The turnings of fire: first sea, of sea half is earth, half fire-wind
>   [*prêstêr*].[30]

Heraclitus characterizes the changes as "turnings," *tropai*, using the same term used for turning points of the sun, the solstices[31]—suggesting a regular and reversible process. Fire turns into water ("sea"), and half of water turns back into fire ("fire-wind"), while the other half turns into earth, in a regular and proportionate process. (We shall return to B31a presently.)

The identity relation itself is symmetrical, transitive, and reflexive. If, then, we think of identity as a paradigmatic case of sameness, there is a good deal of sameness in transformational equivalence. What we must reject is the claim that things that are transformationally equivalent are necessarily identical. They *may* be identical, as in the case of X being identical to X in the relation of reflexivity. But they need not be, and indeed cannot be in all cases. For, assuming that the action of turning into

[29] Expressed in GST-3; see ch. 4, sec. 1.
[30] A fiery meteorological event associated with storms: Aristotle *Meteorology* 371a15–17, Pliny *Natural History* 2.133. Modern commentators have connected it with lightning, tornadoes, and sheet lightning. The important thing is its fiery nature.
[31] Cf. Snell 1926, 359 and n. 1. Cf. Anaximenes A15.

is a kind of change, one thing could not turn into another if there were no others. For a transition from X to Y is a *change* only if X and Y are not identical. It may seem a problem that X can turn into X on this account, but the case is not really problematic because this change always involves an intermediary that is nonidentical to the subject. Since transformational equivalence is a second-order relation, it can have properties the first-order relation does not; in this case, X is transformationally equivalent to X even though X cannot immediately change into X.

Now if Material Monism is correct, there cannot be transformational equivalence between any basic substances. Suppose, for instance, that water is the principle and source of change. Then water is always present: it does not *turn into* anything because it is always *present as* a substratum or foundation for the phenomenal properties that are acquired in a change. Fire is not transformationally equivalent to water because the two substances are not on an equal footing. Fire is rare water, but water is not some kind of fire. In Aristotle's vocabulary, water *alters* in its accidents, but it does not change in its essence, while fire is merely a certain state of water.

Earlier we examined the concept of change in Ionian philosophy. I have argued that the change between basic substances is transformation, not alteration. But the evidence was limited and the conclusions, accordingly, provisional. What is Heraclitus's view of change? Consider the following passages:

> B36. For souls it is death to become water, for water death to become earth; but from earth water is born, and from water soul.

> B76. Fire lives the death of earth, and air lives the death of fire; water lives the death of air, earth the death of water.[32]

Here Heraclitus juxtaposes life and death, coming to be (the same word as that for being born in Greek), and perishing. When X becomes Y, X dies or ceases to be; Y, by contrast, owes its existence to the death of X. There is a reciprocity, for Y dies so that X may come to be. We hear echoes of Anaximander's interchanges between opposites, in which one power flourishes with the death of another in the cosmic war. It seems to me that Heraclitus here endorses *radical change* in which there is no identity between the subjects of the change.[33] Moreover, there seems to be no room

---

[32] There are several versions of this fragment, which present problems for interpretation. Most notably, they seem to suggest a four-element ontology which is inconsistent with Heraclitus's three-element ontology; the statement may be infected by a Stoic interpretation. See Kahn 1979, 153–55. Robinson 1987, 98–101, 130, 186 supports four elements.

[33] Nussbaum 1972, 155 on B36: "The fire simply ceases to exist, and is replaced by water."

for any continuity of anything between them. Certainly he never points to a substratum that continues—not even fire. And he seems to embrace the notion that the changeless (immortal) arises out of what changes:

B62. Immortal mortals, mortal immortals: living the death of these, dying the life of those.

Somehow life and death confer a kind of immortality on things that participate in them, while immortality depends on life and death. One is reminded of Plato's and Aristotle's doctrine that although individual living things cannot be eternal, they partake in immortality by reproducing their own kind, and the kind is eternal.[34] The fundamental sense of B62 may refer to the interrelation of mortal men and heroes, but its more general sense must apply to the elements which prove to be immortal precisely by the recurrence, as Empedocles saw.[35]

Lucretius criticizes Heraclitus's view of fire:

But if by chance they think in any other way [sc. than by rarefaction and condensation]
fires by coming together are able to be extinguished and to change body,
surely in no way will they hesitate to do it:
no doubt the heat will perish completely into nothing
and all that is created will arise out of nothing.
For whatever by being changed leaves its own boundaries,
this is the immediate death of that which was before. (1.665–71)

The only alternative to rarefaction-condensation (which the doxographical tradition ascribes to Heraclitus)[36] as a mechanism for the change of fire, is coming to be and perishing, which picks up the language of Heraclitus himself. And this of course violates the principle *ex nihilo nihil fit*.

Heraclitus does, indeed, acknowledge an order of change (present in the above passages also):

B31a. The turnings of fire: first sea, of sea half is earth, half fire-wind [*prêstêr*].

B31b. Sea is poured out, and measured into the same quantity it had before it became earth.[37]

---

[34] Plato, *Symposium* 206e–208b; Aristotle *Generation of Animals* 731b24–2b1.
[35] For the first sense, see Marcovich 1967, 240–41; for the extended sense, see Kahn 1979, 216–20. Empedocles seems to be echoing Heraclitus in B17.11–13, B26.10–12; cf. Bollack 1965–1969, 3.118–19 on another passage of Empedocles.
[36] Simplicius *Physics* 23.33–24.4.
[37] The source of this fragment, Clement of Alexandria (*Miscellanies* 5.104.3), cites it in two parts. It is appropriately divided into two halves in Marcovich, and Jones 1972 argues for a difference in sense in the two halves.

Fire first turns into water, then into earth. But half of what turns into fire turns back into water. Here we seem to glimpse an equilibrium reaction, where fire never diminishes in quantity because of the reciprocal change. What fire is lost is made up by the reverse change. Secondly, when water turns into earth and back again to water, it resumes the same quantity it had before it changed. Now Heraclitus does not say that a certain quantity of water produces the same quantity (whether weight or volume) of earth; but he does say it recovers the quantity it used to have. Schematically, then:

$$q_1 \, S_1 \leftrightarrow q_2 \, S_2,$$

where q designates a quantity of a given substance. Indeed, this law, as we may call it, helps to strengthen the case for transformational equivalence. Not only is there a qualitative relationship between S_1 and S_2, but there is a quantitative relationship between them, which can be expressed as a ratio: $q_1 : q_2$. Presumably, then, a certain amount of a given substance can be replaced by a proportionate amount of another substance in a series.

There is one place where Heraclitus seems to say that everything is fire:

B30. This world-order, the same of all,[38] no god nor man did create, but it ever was and is and will be: everliving fire, kindling in measures and being quenched in measures.

If measures or portions of the world are alternately being kindled and being quenched, the world is not fire in the sense of having the properties of fire. What happens of course is that the world is constantly changing, and in that sense it is firelike. But Heraclitus calls the world fire, as if to identify the nature of the world with its leading component; and in B31a he talks of the turnings of fire, as though it were fundamental in a way the other stuffs in the series are not. The crucial question that remains is to determine *how* fire turns into other things: is it an alteration or an unqualified coming to be, in Aristotle's terms? The final answer will need to await a discussion of flux in Heraclitus.

But what Heraclitus says of the relation of fire to everything else is instructive:

B90. All things are an exchange for fire, and fire for all things, as goods for gold and gold for goods.

---

[38] This phrase, τὸν αὐτὸν ἁπάντων, is found in Clement, but not Simplicius or Plutarch. I follow Vlastos 1955a, 344–47; Kerschensteiner 1962, 99ff; Marcovich 1967, 268–70; Kahn 1979 in retaining the phrase.

The relationship he claims is not one of identity or composition (part-whole or substratum-substance), but of weak equivalence. Gold does not turn into merchandise or vice versa, but it is given in exchange for them. It is equivalent but not identical. Gold is a universal standard of value, but it cannot be eaten or worn or lived in. When a purchase is made, one party gives up gold for merchandise, the other takes gold and gives up merchandise. The two items are never confounded. B90 is perfectly harmonious with elemental change as happening by generation, and makes no claims beyond those that are already present in GST.

The crucial piece of evidence for understanding B30, however, is one that we have already looked at: according to B36 and B76, supported by B62, we see that the coming to be of one elemental body is the perishing of another. This is the one kind of change MM cannot allow: the change between elemental bodies must be construed as an alteration of the basic stuff. Heraclitus explicitly rejects continuity between basic substances. His view is incompatible with MM. Moreover, as he is the only philosopher who seems to have explicitly analyzed the change between basic substances, his must stand as the best-informed and most reflective view of elemental change of any Ionian philosopher. In all likelihood he is not proposing a new theory of elemental change, but drawing out the implications of the common Ionian theory of change for physical explanation in general.

In summary, the evidence we have seen so far allows us to view Heraclitus as construing the transition from one substance to another in the series of basic substances as instances of generation. On this view there is no continuing element or substratum, but a change of identity in which one substance dies as another is born. The relationship between the two is not a strict identity but an equivalence of value in which a given quantity of the first substance corresponds to a different, but proportionate, quantity of the second. As a quantity of the first substance turns into the second, a proportionate quantity of the second turns into the first. On the other hand, B30 and B31a give pride of place to fire, and seem to assert that it remains when all else changes. Is there indeed an underlying reality that is preserved through all changes? Heraclitus's analysis of elemental changes tells us that there is not. Fire is fundamental just by being symbolic of the constant change that the elements undergo.

Let us return for a moment to the unity of opposites. Before finishing this exposition we must clarify some points. When we talk about opposites changing to opposites, we run the risk of confusing what Aristotle would call particulars and universals. When we say the hot changes to the cold, do we mean Hotness Itself, as Plato would say, becomes Coldness Itself, or do we mean something hot turns into something cold? The Presocratics did not, of course, have the linguistic resources to talk about these differences perspicuously, nor did they make the kind of categorial distinctions that would clarify the problem. Plato himself seems to barely remark the distinction. In the *Phaedo* he argues that opposites come from opposites—generalizing inductively from examples in the natural world (70dff.). He notes a potential ambiguity only later when an anonymous character in the dialogue thinks that Socrates is ruling out what he had advocated before:

> Inclining his head and listening, Socrates said, "You do well to remind us of what was said, but you do not understand the difference between what we are saying now and what we said then. For then we were saying that the contrary *thing* [*pragma*] comes from the contrary thing, but now we are saying that the contrary *itself* can never become contrary to itself, neither the property in us nor the reality in nature. Then, my friend, we were talking about things that *have* the contraries, calling them after the name of those others; now we are talking about those very things which, being present, carry the name we apply to them." (103a–c)

More than a century after Heraclitus, Plato brings in the ontological distinction as an afterthought—thirty-three Stephanus pages after introducing his argument from opposites! Only with Aristotle do we get a confident categorizing of entities and a lucid application of them to the problem of change.[39] So we cannot expect Heraclitus or anyone before him to make Plato's or Aristotle's points in a perspicuous manner. Are they hopelessly confused, so that Heraclitus naively confuses particular and universal, subject and property?

I think not. For the issues that arise for the Ionians surely have to do not with high-level metaphysics, but with the physics of change: how does this stuff turn into that? Specifically, how do particular stuffs change from one quality to its opposite? That is just the background we find, for instance in Heraclitus B31a. If we have any question about the level of generality, we need only remark Heraclitus's use of the terms "sea" and "fire-wind" to designate his subjects of change: Ionian physics is firmly rooted in meteorology. There is no question of Heraclitus thinking that Hotness Itself turns into Coldness Itself in B88. His point is the more mundane, but for his generation especially puzzling, fact that something that is cold and wet (water), for instance, can turn into something hot and dry (fire). It is an apparent fact that such transitions do occur. At least we see liquid olive oil providing fuel for the lamp, and cosmic fires being fed by moist evaporations.[40] The task is to account for them within an acceptable explanatory framework. And to understand what framework

---

[39] Especially in *Physics* I.5–8, drawing on the distinctions made in the *Categories* and expanding on Plato's insights in the *Phaedo*.

[40] See above, ch. 3, sec. 3.2.1.

is acceptable, we must at least appreciate the kind of explanation familiar to sixth-century Ionia. (I shall have more to say about the historical dimension presently.)

What we have seen so far indicates that opposites are the same just in the sense that opposite *things* or *stuffs* turn into one another. One disappears while another appears; we find some causal connection there, but no continuing identity. Rather there is a succession such as happens when one thing dies and another is born. Two opposite things in a cosmic series of transformation are equivalent, then, but not identical. They are, moreover, quantitatively equivalent in the sense described, by bearing a determinate ratio to one another. To say that opposites are the same is simply to say that they are transformationally equivalent.

5.4 The Flux Thesis

As we have seen, Barnes rejects the strong reading of the flux thesis suggested by Cratylus: "everything is always flowing in all respects," for a weaker but still strong interpretation: "everything is always flowing in some respects," which we have designated Universal Flux. This proposition, together with his Identity of Opposites which follows from it, entailed the denial of the Law of Non-contradiction. Here I shall argue that Heraclitus does not hold Universal Flux, but some yet weaker thesis—one that still qualifies as a flux thesis, but one that has very different implications than Barnes's Universal Flux. I shall focus here on the so-called River Fragments, which are generally accepted as the basis for the flux thesis. They are as follows:

B12. On those stepping into rivers staying the same, other and other waters flow.

B49a. Into the same rivers we step and do not step, we are and are not.

B91. [a] Into the same river it is not possible to step twice . . . [b] it scatters things and again gathers them . . . and approaches and recedes.

A6. Heraclitus says, I believe, that all things pass and nothing abides, and comparing existing things to the flow of a river, he says that you couldn't step twice into the same river. (Plato *Cratylus* 402a8–10)

Which passage or passages tell us what Heraclitus's doctrine was?

Recent studies have demonstrated beyond reasonable doubt that B12 is genuine.[41] For it exhibits the concentrated expressiveness of Heraclitus's

---

[41] See Kahn 1979, 166–68, to whose interpretation I am heavily indebted. In light of Kahn's explication, Kirk's defense (KRS 196) seems unnecessarily weak: "[I]n sum we feel

best utterances. In a paraphrase of his meaning, his sayings will lose significance, as by a law of entropy: they are so well put together that any change will damage them. Here is the Greek:

ποταμοῖσι τοῖσιν αὐτοῖσιν ἐμβαίνουσιν ἕτερα καὶ ἕτερα ὕδατα ἐπιρρεῖ

potamoisi toisin autoisin embainousin hetera kai hetera hudata epirrei.

We observe the following features: (1) the second and third words, underlined in the Greek ("the same") appear between the terms for rivers (first word) and for those stepping in (fourth word); the words agree in case and number with both terms, producing syntactic ambiguity: should they be construed with "rivers" or "those stepping in"? Both interpretations make sense, and both, as we shall see, are philosophically meaningful. The positioning of an element that can be taken as common between two disparate elements reflects features of both chiasmus and zeugma, but its particular form is unique to Heraclitus and found in several other texts. (2) There is a contrasting pair of terms, "the same" and "other" (*hetera,* two times). (3) The word endings of the first four words are identical in sound as well as case (which need not happen for words in different declensions), and produce a kind of word painting in which the sound of the river is reproduced. (4) The second half of the expression contains alliteration (*hetera kai hetera hudata*). Clearly the nine words of the fragment are charged with significance, such that the removal or change of any one element would destroy the sense of the whole. The features mentioned are all part of Heraclitus's repertoire (more will be said about several below), and the density of expression and intensity of thought are not to be found in any other philosopher—indeed in any other prose writer or poet. Fragment 12 is a genuine verbatim quotation.

What, then, is the relationship between B12 and the other texts? None shows the same kind of linguistic density. Note, however, that both B49a and B91[a] begin with the word for "river(s)"; both in fact are in the dative case—as is B12—and not in natural Greek word order, in which, as in English, the subject comes first. Both, moreover, use the same word for those who step in—as does B12—and Plato, too, uses this word. Since the other texts contain the same words rearranged, the obvious hypothesis is that they are mere paraphrases in less complex diction, and perhaps more quotable form, of B12.

5.4.1 *The Sameness of Rivers*

If we disambiguate, B12 contains two messages:

---

that fr. 12 has every appearance of belonging to Heraclitus, being in natural and unforced Ionic and having the characteristic rhythm of archaic prose." It is not what he has in com-

i. On those who step into the same rivers, different waters flow.

ii. On the same persons who step into rivers, different waters flow.

For now, we shall concentrate on (i), the interpretation that is most influential in the tradition. What is crucial here is the contrast (remarked as (2) above) between what is the *same* and what is *different*. The waters are other and other, repetitively different. The rivers, on the other hand, are the same. What we find in B91[a] and A6 is the denial that the rivers are the same. B49a, on the other hand, allows that in some sense we do step into the same rivers, in another sense not. Why do the texts not agree? Could it be that we have really two fragments that say different things? I think not.[42] Another paraphrase shows what is happening:

ποταμοῖς δὶς τοῖς αὐτοῖς οὐκ ἂν ἐμβαίης· ἕτερα γὰρ ἐπιρρεῖ ὕδατα.

Potamois dis tois autois ouk an embaies: hetera gar epirrei hudata.

You could not step twice into the same rivers; for other waters flow on. (Plutarch *Natural Questions* 912a = 40c[5] Marcovich)

Plutarch gives the first three words of B12, with Attic endings, adding *dis* "twice," and retaining the participle *embainousin* as a finite verb; he gives a close paraphrase of the second half of B12. Clearly he has in mind B12, and he regards the claim that you cannot step twice in the same river as a paraphrase of the first half. This is as close as we can get to an admission by an ancient source who knew Heraclitus's text better than we can, that Plato's expression in A6 is a paraphrase of B12.

We can trace Plato's mistake in his description: compare *things* (*onta*) to the flow of a river. Plato gets the comparison—broadly at least—and he sees that there is constant change. But he confuses the river with its contents. He affirms what Heraclitus is at pains to deny: that the river just is the water in it. B91[a] and A6 are intended as paraphrases of B12, but they get the point wrong. We must reject them as testimonies of what Heraclitus meant. As for B49a, it sounds Heraclitean with its opposing theses. But it, too, while imitating the word order of B12, contradicts it: one of the opposing theses is that we do *not* step into the same rivers—a point B12 does not concede. Thus B49a, too, fails as a commentary on Heraclitus's thought.[43]

---

mon with Ionian prose writers but what is uniquely Heraclitean that confirms the fragment. See also the insightful study of Petit 1988.

[42] Pace scholars such as Vlastos 1955a, who rejects B12, and even Kahn 1979, 168–69, who vindicates B12, sees it as perhaps countered by another text (see next note).

[43] Marcovich 1967, 206, points out that the interpretation of Craylus and Plato can be gotten by reading *embainousin* in the iterative sense: "repeatedly step in." Persuaded by Cratylus's use of the flux doctrine, Kahn (169) suggests reconstructing a fragment as follows: "One can never bathe twice in the same river. For as one steps into [what is supposed

B12, version (i), contains a striking message. It says that while different waters flow, the rivers remain the same. But why does Heraclitus want to make that point? At one level, he is observing a mundane fact about rivers: the waters constantly change in the same river. Indeed, if we look behind the observation, we shall note that, if there were no waters, there would be no river, but only a dry riverbed. And if the waters did not *constantly* change, we would have not a river, but a lake or pond—a body of standing water. In a certain sense, then, without the different waters, there would be no river—no continuing reality. Thus it turns out that there is a causal connection between the different waters and the same river: the former provide the ground for the latter. In Aristotelian terms, the waters furnish the material cause, the geographical features the formal cause, of the river. Both the formal and the material factors provide necessary conditions for there being a river. On the other hand, to say that the river is changing because the waters are changing is to miss the complex relationships between the waters and the river. Paradoxically, the changing waters preserve the river, while the perennial river concentrates and conducts the changing waters. More generally, local change begets global stability, while global stability focuses local change.

What then of the general interpretation of Heraclitus's physics? As Plato expresses it, the moral of the river image is that all things move (*panta chôrei*) or all things flow (*panta rhei*),[44] the latter using the liquid imagery of the river.[45] If by "all things" we mean the contents of the river, the saying is certainly true. But if we mean to apply the expression to all things without restriction, it is false: there is at least one thing that does not flow as the waters flow, namely the river itself. The point turns out to be surprisingly Aristotelian: there are high-level realities that persevere through changes in their matter. For the high-level realities—the *ousiai* or substances— are not identical with the matter, but are the result of structures supervening on matter, ordering it into permanent patterns.[46] Heraclitus says nothing in B12 about structures and supervenience directly. The general point, however, is reflected in his concept of Logos, the unseen

---

to be] the same rivers, new waters are flowing on" (bracketed passage is from Kahn). The second sentence is a rendering of B12, the first the conjectured original claim that you cannot step twice into the same rivers. But without the bracketed words, the two sentences are incompatible (cf. Robinson 1987, 140 n.), and we know the words are not found in B12: they constitute a desperate attempt to reconcile two irreconcilable statements.

[44] Barnes (65) supports πάντα χωρεῖ as a quotation, unconvincingly. Both ῥεῖ and χωρεῖ are often cited. They presuppose different metaphors and should not be confused: Lebedev 1985, 143–44.

[45] It seems to me likely that the statement of the flux doctrine derives from B12, which uses the compound verb ἐπιρρεῖ and presents the image of flux.

[46] A point congenial to Aristotle, e.g., *Metaphysics* VII.17.

but ever-present structure of nature.[47] As Kahn describes it, "[T]he *logos* of Heraclitus is not merely his statement: it is the eternal structure of the world as it manifests itself in discourse" (1973, 94).[48] The moral seems to be, not that all things change, but that because the parts change, the whole stays the same. But does the river itself change over time? Heraclitus says, or hints, nothing about this. Of course it is a fact recognized by modern geology that rivers appear, change their course, and disappear over time, and the fact of rivers silting up harbors was known in antiquity. But B12 has nothing to say about the long-term changes of rivers. So whatever we may wish to infer about this topic will take the form at best of an extrapolation from the evidence; it cannot be based directly on the text.

We may, however, make one preliminary observation about flux. If Heraclitus subscribes to MM, he must hold that flux is merely an illusion, since fundamentally everything is one and unchanging. But B12 gives us fundamental change leading to higher-level constancy. The message is the reverse of what we expect. Of course it is possible that Heraclitus holds that there is yet a more fundamental unity than the unity he alludes to in B12. Indeed there is a more fundamental unity; but is it a unity of matter or of process?

5.4.2 The Sameness of Men

As Charles Kahn (167) rightly noted, B12 has a second interpretation in which the phrase "the same" attaches to "those stepping into" the rivers. "[T]he point," T. M. Robinson (1987, 84) complains, "seems trivial, and hardly part of [Heraclitus's] intention." On the contrary, Heraclitus has gone to a good deal of trouble to produce a text with syntactic ambiguity. Inverting word order, pluralizing terms, using juxtaposition, omitting the substantive to which the participle refers, he achieves just the right balance of sound and sense. Without any of several devices Heraclitus has employed, there would be no ambiguity. The first four words embody a scheme that is uniquely Heraclitean: a structure in which ABC can be read either AB, C or A, BC, and in which, at least in certain contexts, the common term B becomes a link: A = C. In its simplest form, the structure can be exemplified in three words: *êthos anthropôi daimôn* (B119): char-

---

[47] B1, B2, B50.

[48] Literally, *logos* is something said (cf. *legein*, "say"). But the word is rare in early vocabulary, where *epea*, "words" is usually found and *logos* typically denotes a story. See Boeder 1959; Nussbaum 1972, 3–5, 9–11; Lesher 1983. For a general treatment of *logos* in Heraclitus, see Aall 1896–1899, 1.7–56. He is wrong to identify *logos* too closely with reason, but right to resist the identification of *logos* with fire. See Minar 1939, whose general conclusion I agree with, and Verdenius 1966–1967, whose stress on *logos* as argumentation I disagree with.

acter[49] of-man destiny, or character = destiny. Peculiar to Heraclitus, the form is a kind of Heraclitean trademark.[50] What then is the point of introducing the form here? Clearly to create the same kind of equation we find in other passages. As one fords the river (an everyday practice in archaic Greece, which boasted few bridges), other and other waters flow on to one. But one remains the same. Neither the river nor the traveler changes in the process, though the river sends different waters to the traveler.

So far the interpretation is useful to rule out of court one ploy of the flux theorist: if the river stays the same, the traveler may change, and hence cannot step twice into the same river.[51] On interpretation (ii), the traveler does not change any more than does the river. But more is at stake than just the flux theory itself. By equating the traveler and the river, Heraclitus seems to be offering us an immediate extension of the insight of interpretation (i). As the river is, in a certain sense, constituted precisely by the changing waters, so the traveler is the same precisely in virtue of encountering the changing streams. The obvious suggestion is that in confronting a changing environment—a concept familiar to Greek philosophers as *to periechon*[52]—the traveler is being constituted as the same subject. It is our unitary reaction to outside stimuli that makes us who we are: our character (or customary reaction: *ethos*) is our destiny. To put it otherwise, our soul has its own Logos that governs its interactions with the world. It is a Logos that increases itself:[53] we build our soul by our interactions with the world. But without interactions we would be nothing, just as the bow and the lyre without the tension of opposing limbs would be nothing.[54] Interpreters are perfectly correct in seeing that the river imagery applies to humans. But the message is not, as they usually take it, that humans are always changing, but the contrary: that only by confronting a changing environment do they become stable beings.

The key to human integrity is unified experience:

B1. Of this Word's being forever do men prove to be uncomprehending, both before they hear and once they have heard it. For although all things happen according to this Word they are like the unexperienced experiencing words and deeds such as I explain when I distinguish each thing according to its nature and show how it is.

---

[49] Or, more literally: the guardian spirit.

[50] See below, sec. 6.1.

[51] This view may be behind the extreme Heracliteanism of Cratylus, though Aristotle does not attribute the extreme view to a belief that the subject is always changing, but that objects are always changing: Aristotle *Metaphysics* 1010a7–15. A view of the changing subject is hinted at but not fully developed in Plato *Symposium* 207d–e, *Theaetetus* 179d–80a.

[52] See Anaxagoras B2, B14, Anaximenes B2, Aristotle *Physics* 259b7–12.

[53] B115.

[54] B51.

What experience does is to allow the subject to react appropriately to the stimuli of the environment. The inexperienced is helpless before a new situation because he does not know what is happening and accordingly cannot respond appropriately. The expert grasps the true nature of things, and by showing us the constitution of each thing, teaches us how to react appropriately. Ever-changing conditions make perception possible; experience makes appropriate reaction possible. We are how we react to the world around us. And knowledge is the most powerful tool in making us a stable subject in a changing world.

But even if we are relatively stable structures, do we not change over time, just as, it may seem, rivers change? And if we are ever changing, even if more slowly than our environment as a whole, is it not true that all things flow?

5.4.3 The Sameness of the World

Let us return to Heraclitus's statement on cosmology:

> B30. This world-order, the same of all, no god nor man did create, but it ever was and is and will be: everliving fire, kindling in measures and being quenched in measures.

Heraclitus's cosmos is remarkable for being ungenerated.[55] According to his testimony, and contrary to some doxographers, it does not come to be or perish. It always is as an everliving fire. The fire is constantly changing—being transformed into other substances in the sequence of changes. But the cosmos itself remains through the changes. Like the river, its existence depends on the flow of its components. Like man, it outlives changing circumstances. Its permanence is predicated on the exchange of substances, on ongoing transformations. Without everliving alterations, there would be no cosmos. But because of elemental changes, there is an everlasting cosmos. At least one thing in the world continues unchanging: the world itself.

It turns out that, contrary to Plato's testimony, not all things are changing. Some things are relatively stable: rivers and men. One thing, at least, is everlasting: the world itself. Heraclitus does indeed believe in flux, at an elemental level. But arising out of the flux is a higher-level ordering that is stable. For there is a Logos present in the transformations of matter

---

[55] In almost all other cosmologies the cosmos comes to be. The only deviation is found in Empedocles' cyclical generation—but his cosmos does come to be, repeatedly. Xenophanes does not have a cosmogony per se, but he does seem to allow changing epochs, which he may designate as *kosmoi*: Hippolytus *Refutation* 1.14.6 = A33 with Kerschensteiner 1962, 94. Some still ascribe to Heraclitus a cyclical generation of the world punctuated by a cosmic conflagration (*ekpurôsis*); see n. 14.

that manifests itself as a structural continuity, as, in Aristotle's terms, a primary substance. Flux is one part of Heraclitus's doctrine, and an important one, playing a foundational role. But it is only half of the story. The other half of the story is enduring order based upon elemental change.

5.4.4 The Domain of Flux

According to Barnes's Universal Flux, everything is always changing in some respect. But we see that this is false: many things are stable at least over long periods of time (rivers, men), and one thing forever: the world.[56] Or better, Barnes's Universal Flux is not so much false as irrelevant. For the kind of change that rivers and men undergo does not keep those objects from remaining identical through time. The river is, as virtually all commentators have taken it since ancient times, a concrete example of a general relationship. It exemplifies how things change—and how they stay the same. What changes, in Heraclitus's view—as B30 makes explicit—is the basic stuffs of the universe, what in later terms would be called the elements. What stays the same is the higher-order structures that supervene on the stuffs: rivers, men, the world. The stuffs are always changing in the sense that some portion of them is always turning into its successor, while some of that successor substance is turning back into the first kind of stuff. Barnes's Universal Flux entails nonidentity only so long as the domain is limited to elemental stuffs: the component stuffs of organized bodies are always changing so as not to be identical. But for organized bodies, there is identity through time in Heraclitus just as much as in Aristotle. Thus Universal Flux, insofar as it is true, is not strong enough to entail that opposites are exemplified in organized bodies; hence UF does not entail IO.

The following thesis, which I shall call Material Flux (MF), seems to be true for Heraclitus:

> MF: The basic substances of the world are constantly undergoing transformation.

But we can make an even stronger claim, for Reciprocal Material Flux (RMF):

> RMF. The basic substances of the world are constantly undergoing reciprocal transformation.

---

[56] One could object that even the world is changing in some respect, namely in its elemental constituents. But I take it that this is too weak an interpretation to satisfy Universal Flux. The thesis is that the subject of change is becoming different in its definitive characteristics.

The reciprocal transformations referred to are the changes in which portions of two substances adjacent in the series of transformations turn into each other; for instance, a portion of water turns into fire, while a portion of fire turns into water (assuming a three-member series: earth, water, fire). Moreover, we have seen evidence for an even stronger claim, in B31, for Lawlike Material Flux (LMF):

LMF. The basic substances of the world are constantly undergoing reciprocal transformation in a lawlike way:

i. Each portion of a given basic substance that turns into another substance is replaced by an equivalent portion from another basic substance which turns into the first substance.

ii. Hence the total amount of each basic substance in the world remains constant.

We have the beginnings of a law of conservation, not precisely of mass or of matter, but of material proportions. Now we cannot guarantee constant proportions in any creature of the world. For instance, people grow up, become fat or thin, moist or dry (in the classic accounts of sickness). But creatures do achieve a kind of stability or homeostasis such that they can survive for long periods of time, much longer than their constituent elements. Thus people and rivers are continuing if not everlasting features of the world. Their continued existence results from their ability to channel or metabolize portions of their environment to preserve a stable condition.

5.5 Heraclitus and GST

It is time now to pull together the results of our reconstruction. Barnes, following Plato and Aristotle, claims that Heraclitus adheres to MM, and that in consequence of accepting the Identity of Opposites and Universal Flux (on empirical grounds), he is committed to denying LNC. I have already challenged MM in a general way, offering as its replacement GST. In the present chapter, I have argued that Heraclitus does not advocate the Identity of Opposites, but a more general account of the unity of opposites whose content is exhausted by Transformational Equivalence (TE). Furthermore, I have argued that Heraclitus does not embrace Universal Flux but a more limited (but in another way much more constructive) thesis, Lawlike Material Flux.[57] In this section I wish to show (a) that Transformational Equivalence and Lawlike Material Flux are closely

[57] Or alternately, he holds Universal Flux only in a limited form that does not have the entailments Barnes claims for it.

related, (b) that they provide the foundations of a coherent philosophical theory, and (c) that they are motivated by GST (but not MM). In general, then, I hope to show that the present interpretation rescues Heraclitus from philosophical incompetence and promotes him to a place of honor. To the extent that this point is made, GST is confirmed as the best interpretation of Ionian philosophy.

According to Lawlike Material Flux, transformations between basic substances occur in such a way that a portion of one substance that turns into another substance in the series is replaced by an equivalent amount of the second substance in a reciprocal transformation. Hence, Transformational Equivalence follows from Lawlike Material Flux, or to put it differently, Lawlike Material Flux entails Transformational Equivalence. We might observe that Lawlike Material Flux is a dynamic description of how changes occur in the substances of the world. Transformational Equivalence embodies the static relations between different basic substances that are instantiated in dynamic changes. The static relations would best be expressed in ratios of measurable units such as volumes or weights if Heraclitus wished to give a rigorous or scientific characterization of his system—which I take it he does not.

On the present theory there is no need to reject or undermine LNC. Change is orderly, balanced, reciprocal, and lawlike. And the flux that occurs at the elemental level even generates stable structures at a higher level, which is the level of middle-sized objects that we experience. Indeed, experience itself seems to help to constitute men as stable, psychical substances. Although all middle-sized objects may ultimately be fated to suffer destruction, the cosmos itself is everlasting. The world itself, then, has a permanent cognizable structure, and the middle-sized objects in it have long-lasting cognizable structures. The world is coherent and knowable, precisely because the middle-sized things of the world we encounter are enduring structures like ourselves. The traveler crosses a river; while fleeting currents make up the river and impinge upon the traveler, there is a continuing river and an enduring traveler. Moreover the flow of the current and the course of metabolism make up the river and the traveler; from another perspective a higher principle of order organizes the flux of the world into complex beings.

Opposites are distinct. Particular instances of opposites may turn into each other, but they do so only by ceasing to exist in their own character and nature; hence there is radical change but no radical incoherence. Opposites may be transformationally equivalent, but they are never identical or indiscernible. The present interpretation parallels that of Barnes in a certain way; for Barnes Universal Flux entails the Identity of Opposites; on the present theory Lawlike Material Flux entails Transformational Equivalence. On both, in other words, a flux principle entails a relationship between opposites. But on Barnes's interpretation not-LNC follows, whereas on my interpretation LNC is not challenged. My interpretation seems to fit all the texts, and to fit them better than MM and its supporting theses do. And it avoids logical chaos, as Barnes's view does not.

Furthermore, there seems to be a deep incoherence in Barnes's position. If we take MM seriously, as Barnes is committed to do, we must say that the flux we encounter in the world is, in a sense, trivial. For the changes we experience are only changes of accidents, whereas at the foundation of all change is the unassailable unity of the underlying substance. Fools may get excited by the changing circumstances around them (B87), but wisdom dictates that we dismiss changes as insignificant and focus on the unity (B50). How curious that we should in our analysis of the world focus on flux as the major theme. And how strange that Material Monism, a theory designed to render experience rationally explicable, should lead us to deny the Law of Non-contradiction.

Finally, what historical evidence can we find to support the present interpretation? I have already given a preliminary argument for GST as a characterization of Ionian philosophy. Now I wish to argue that the present interpretation follows directly from GST. For GST-3, according to which the basic substances arise from transformation relation T, entails Lawlike Material Flux—at least so long as T is understood to include both a one-direction relation mapping one basic substance onto its successor, and the inverse relation mapping the higher substance in the series to the lower. But this is precisely the kind of transformation Anaximenes envisages in the paradigm case. The order of the world just consists of an orderly exchange of substances in a series: fire becomes air becomes wind becomes cloud becomes water becomes earth becomes stones, and vice versa. Thus,

GST entails LMF entails TE.

In other words, Lawlike Material Flux is *already* implicit in GST. There is no need to claim, as does Barnes, that Heraclitus's theory is based on empirical observation.[58] Of course it is based on empirical observation just as much (or as little) as any other Presocratic theory—but that does not motivate it. What motivates it is the whole Ionian theoretical apparatus that Heraclitus inherits, as we shall see presently. That the dual transformation is a theme of Heraclitus can be glimpsed from a famous fragment:

[58] Barnes 1982, 67: "[T]he Theory of Flux was no a priori intuition or piece of fanciful imagery; it was a general thesis about the nature of reality, founded upon and supported by a series of empirical observations." More picturesquely: "Flux and the Unity of Opposites are twin horses, bred and nourished on wholesome empirical food, possessed of deep

B60. The way up and down [= the course out and back]⁵⁹ is one and
the same.

At one level this is just an observation about a road. The same road that
leads out of town leads back to town, or that leads uphill in one direction,
leads downhill in another. A road proves to be another instance of the
unity of opposites: one road (*hodos*) is the physical medium over which
two routes (*hodoi*) pass. The same road that carries one out of town in
the morning to work the fields leads one back to town at night. In the
larger context of the city, its existence depends on the flow of commerce
in and out, as well as the exchange of gold for goods and goods for gold.
But there is a further application for the symbolism. For the basic substances follow a path upward as they are rarefied, and downward as they
are condensed. As the literal road up and down is one and the same, so
the figurative road of transformation is one and the same. What goes up
must come down, what comes down must go up. B60 expresses, inter
alia, Lawlike Material Flux.⁶⁰

In GST, Lawlike Material Flux is only implicit, and by making it explicit, Heraclitus hopes to say something more than has been said by earlier Ionians. But what is his point, if it is not (as traditional interpreters
from Plato and Aristotle to Barnes would suppose) to show the world is
a physical and logical chaos? Precisely, it is to show that nothing else is
needed in the explanation of the cosmos than the interchange of opposites. Specifically, what is *not* needed is a generating substance. This, if it
is Heraclitus's point, seems to run counter to all the tradition teaches
about Heraclitus, that he makes fire his principle. But we should notice
here something anomalous about fire: it is the *least* substantial of all the
basic substances, the least permanent, concrete, stable existent. Indeed, it
seems most to partake of energy and change and instability and flux. As
Aristotle would later express it:

> Flame comes to be through the continuous interchange of moist and dry, and
> it is not really fed—for it does not abide as the same thing for any time at all,
> so to speak. (*Meteorology* 355a9–11) Fire ever continues coming to be and

---

strength, and harnessed to the old monistic chariot which Heraclitus inherited from this
predecessors" (78). See criticisms of Barnes's interpretation in Moyal 1990.

⁵⁹ On the terms *anô katô*, see following note.

⁶⁰ Furthermore, the language of the fragment, *anô katô*: out/back, up/down is the language of the racecourse: what makes a race is the turning post around which the runners
change courses back and forth. To complete a race, one approaches the finish line by running
away from it. In a long race, the many turnings are all part of a single motion toward the
goal. (Cf. Aristotle on circular and two-way linear motion, *Physics* VIII.8.) On the racecourse imagery, see Lebedev 1985.

flowing like a river, but this escapes us because of its rapid changes. (*Parva Naturalia* 470a3–5)[61]

Fire and flux are almost synonymous, at least after Heraclitus. If fire is a mere symbol rather than a generating substance, then the whole classical interpretation crumbles. And we may recall that the relation Heraclitus identifies between fire and everything else is a surprisingly weak one: gold for goods and goods for gold—an exchange value rather than an identity (B90). It is at least possible, then, to see fire as a kind of antisubstance that takes the place of the generating substance only to show the absurdity of the whole concept. But fire is indeed more of a process than a substance. If he were an advocate of MM, why did Heraclitus pick something so unstable, so ephemeral as the ultimate reality? Would not his choice undermine the whole notion of a material substrate: something that remains as a permanent foundation for change? Here we have the other side of the paradox Barnes's position poses for us: if Heraclitus is a Material Monist, why does he make such a fuss about flux—since the real story is that everything is absolutely permanent? On the other hand, if the material substratum is an unstable nonsubstance, Heraclitus pulls out the foundation from beneath his own feet. If the even the substratum is unstable, why bother with one at all? These are problems advocates of MM have not really addressed.

On the present view, the paradoxes are precisely the point of the exercise. By singling out one basic substance as special on the basis of its adaptability, and at the same time its instability, we point to the impermanence of substance. Fire, the star substance, is constantly consuming, transforming, appropriating other substances into itself, and constantly threatening to disappear when it ceases to devour its neighbors. It is the one substance that is most like a process of change. As a symbol, it reveals that in truth all other substances that are linked with it in a chain of transformations are no less permanent than it. As a material reality, it cannot subsist in its own right for an instant, without consuming something else, without playing its role in the serial process of transformation. Here we, as modern interpreters, are presented with a choice, like Parmenides' *krisis* between "it is" and "it is not": is Heraclitus hapless in his identification of fire as a substratum, or is he brilliant in his dissection of the Ionian tradition? Is he an ignorant imitator of MM, who cannot see what sort of stuff MM requires to make it work, who bungles the theory and falls afoul of the Law of Non-contradiction? Or is he an incisive critic who locates the Achilles' heel of Ionian theory in its claim that everything

---

[61] Cf. also Plato *Timaeus* 49d4ff. Theophrastus would later develop the concept of fire as a process in *De Igne*.

comes from one stuff, while maintaining that everything turns into everything else? On the latter view, he exploits the weakness inherent in Ionian theory to go beyond it. Flux is integral to GST, and, it turns out, fire is an embodiment of flux. Yet even the constant flux of elements is regular, and its presence reveals a deeper order, an order not of material persistence but of dynamic equilibrium.[62] The latter theory is bold, innovative, and even revolutionary. The former is inept and incoherent. At least, no one has yet defended the traditional reading without convicting Heraclitus of logical incoherence.

Heraclitus's theory, as interpreted through Lawlike Material Flux and Transformational Equivalence, embodies the criticism we have called the Problem of Primacy. If every basic substance changes into every other basic substance, then every substance is equivalent to every other substance; or to be more precise, a certain amount of any basic substance is equivalent to a proportionate amount of any other. On this account there is no primacy, for no basic substance is superior to any other. Some may serve as a better measure than others (as gold does for goods), but it is not on that account better than they. Hence it is false that one substance generates all the others in a way that they do not generate it. In other words, GST-2 is false. But to deny that there is a generating substance is to deny also that there is a state in which everything in the world is the generating substance. And this is precisely what we have seen in B30: the cosmos was and is and will be everliving fire. But everything is not fire at the same time; rather, everything turns into fire by turns. There is no conflagration, contrary to one ancient interpretation (that of the Stoics, who saw Heraclitus as a forerunner of their own cataclysmic theory).[63] Consistent with a denial of GST-2, then, we see evidence of a denial of GST-3 which uniquely generates everything out of a primitive state in which only the generating substance exists.

We have, then, the following surprising deduction:

GST entails LMF entails TE entails not-GST.

Has Heraclitus fallen into the trap of denying LNC? No. For he adopts GST only as a dialectical move to show its failures, as the hypothesis for an indirect proof. He is not committed to GST-2 or -6a, but only to -1, -

[62] Reeve 1982, 303–4, concludes that "the monism which results [from Heraclitus's physical theory] is vastly more successful than any of the incipient . . . monisms of his Milesian predecessors. . . . It emphasizes that Heraclitus' one reality is not a static stuff to which change is extrinsic, but a measured process." This seems to be close to the right answer, but confuses things by calling Heraclitus's theory a monism. At this point the one stuff has been superseded by a process.

[63] See n. 14 above.

3 (in a limited sense),[64] -4, -5, and -6b; he does not accept the theory as a whole. His invocation of GST can be seen as an attempt to show that the theory, as held by his predecessors, is untenable. If we assume GST, we deduce not-GST: a reductio ad absurdum. Although much of GST can be rescued, GST-2 cannot. The claim that there is a primary generating substance and that the world comes to be out of that substance must be abandoned. The Problem of Primacy reveals a structural flaw in the theory, which must be modified accordingly. When the theory is modified, the salient points of the theory will be abandoned, and the result will be a new theory preserving what can be salvaged from the old. The new theory is a successor of the old, but it is not logically compatible with it.

But if there is no primary and generating substance, how does the theory preserve the unity of the world and the unity of explanation? Heraclitus does not hesitate to proclaim the unity of the world in even stronger terms than his predecessors:

B108. Of all those whose *logoi* I have heard, none went so far as to recognize what the wise is, apart from all things.

B50. Listening not to me but to the *Logos*, it is wise to agree that all things are one.

B41. The wise is one thing: to know the thought which steers all things through all.

There is a unity behind the many appearances. But the many are not unreal. Rather, the unity is what holds the many together. The *Logos* can be seen as the law of transformations that dictates how one portion of one element turns into another, while an equivalent portion of the second turns into the first. Local instability preserves global stability.

One element is interchangeable with another. Wholes and parts are also interrelated:

B10. Collections:[65] wholes, not wholes; brought together, pulled apart; concordant, discordant; from all things one and from one all things.

There is an alternation of one and many. But everything is held together by the balance of change which we have already noted. Pervading the changes are rules of exchange which determine global patterns of order. Because of this we can say there is a divine order in things:

---

[64] That is, while it is true that all other basic substances arise out of fire, fire is not unique in giving rise to the others: they also give rise to fire. Hence fire does not have a special role in the system, except symbolically.

[65] Reading συλλάψιες, literally, "things grasped."

B67. The god is day night winter summer war peace plenty famine. It alters
just as <fire, which,> when it is mixed with spices is named according
to the aroma of each of them.

Moreover the unity arises precisely out of conflict:

B80. One must know that war is common and strife is justice and all things
happen according to strife and necessity.

Strife is embodied in the exchange of opposites, the upward and downward paths. In conflict one party dies as another is born. Without the
constant flux of life and death there would be no continuity, no order,
no world. The one constant in the world is not fire (as if that could
be constant!) but the law of exchange. That, I take it, comes closest to
what Heraclitus means to express by the Logos. It is a kind of Law that
nourishes all laws (B114), a message embodied in all messages, a truth
behind all truths. Its most concrete embodiment is the Law of Exchange
expressed in B31b: the balanced and proportionate exchange of basic
substances.[66]

Challenging GST, with its naive assumption of a generating substance,
Heraclitus criticizes its weaknesses and passes beyond them to a new conception. The world is orderly because of its order of changes. We need no
primary generating substance, only a flux of equal substances according
to law. The Problem of Primacy is telling: it shows that GST is built on
sandy foundations. When we remove them we find something more secure: a law of nature instead of a substance. We abandon a crude substance theory and create a sophisticated process philosophy. What is ultimately real is not a substance, which, according to GST, can never remain
the same anyway, but process itself, and the law that governs the process.
Substances are not enduring entities, but only phases of matter, temporary
states in the flow. The Law of Change itself is fundamentally real, and it
allows long-lasting or everlasting patterns to supervene on the flux. Thus
both below and above the material flux there is order, and beneath the
contraries there is a fundamental unity.

Conclusion

In the tradition of Plato and Aristotle, Barnes holds that Heraclitus is a
sturdy empiricist who, for lack of correct understanding, falls into the

---

[66] Cf. O'Brien 1990, 150: "Les transformations successives et mutuelles des ces trois élé-
ments laissent pourtant supposer qu'il doit exister une unité sous-jacente, laquelle unité . . .
ne doit s'identifier à aucun des trois termes 'opposés.' "

slough of incoherence. He accepts MM, for empirical reasons endorses Universal Flux, which entails the Identity of Opposites, which entails a rejection of LNC. On this view Heraclitus is a well-meaning failure. So, in general (although for sometimes slightly different reasons), say all the advocates of Heraclitus-F. I have argued, to the contrary, that Heraclitus understands Ionian philosophy as expressing GST. For dialectical reasons he feigns a commitment to GST, only to reveal its deficiencies. For GST entails Lawlike Material Flux, which in turn entails Transformational Equivalence. The two latter theses are coherent and defensible proxies for Universal Flux and the Identity of Opposites, respectively, and they are perfectly compatible with LNC. I have argued that the theses I have advanced for Heraclitus are supported by textual evidence and motivated by historical considerations. The present interpretation is, I claim, philosophically more sound, historically more relevant, and overall makes better sense of Heraclitus than Heraclitus-F.

As for Heraclitus-C, the present theory may be seen as an instance in some respect. However, some advocates of Heraclitus-C seem to downplay the flux in order to emphasize the constancy. That is a mistake: Heraclitus is quite committed to the flux, at the elemental or material level, and indeed it is an essential feature of his theory. Heraclitus is not an intellectual conservative, but a theoretical revolutionary. He goes beyond a theory of stuffs to a theory of process. Nevertheless, elemental flux does not entail disorder in the world or ignorance of the world. For there is a deeper unity, the law of transformation itself, which I take to be the essence of Heraclitus's Logos. This unity is itself cognizable, and its effects produce stable patterns of nature at a level above that of matter. Thus the world is knowable, both in its fundamental law and in its daily manifestations. Most mortals are ignorant, walking through life as sleepwalkers; but they need not be.

Heraclitus is the most articulate and philosophically sophisticated interpreter of Ionian philosophy. Using metaphors of birth and death implicit in talk of one substance coming to be from another, he reveals its hidden commitment to radical change. Using the theory of serial transformations of basic substances, he reveals its commitment to flux. Exploring the implications of material flux, he shows how it leads to the unity of opposites, whose essence is Transformational Equivalence. In short, he spells out what other theorists leave fuzzy and unexplained and problematic. He shows, in fact, that the correct interpretation of Ionian philosophy is GST. As an heir of Ionian philosophy he understands its implications like no one else. But he also appreciates its shortcomings, and takes pains to show how the only way to rescue Ionian philosophy is to abandon GST in favor of a theory that makes no claims of primacy for one basic substance over the others, and makes no assumption of a primeval chaos in which only the original substance existed.

If this account is true, the philosopher who has been both admired and reviled—but mostly reviled—since ancient times as the Mad Hatter of Ionian philosophy turns out to be the only sober man of the bunch. He understands Ionian philosophy intimately, as GST—strong GST—and furthermore he sees clearly what his predecessors miss: the theoretical tensions inherent in GST. His own solution to the problem is indeed a radical one: to refuse to recognize a generating substance, replacing it with a general law that governs the transformations of each substance equally. His alleged radical flux turns out to be only the flux of GST in the spotlight of philosophy, his unity of opposites only an appreciation of how the basic substances are interconnected through flux. And far from denying the Law of Non-contradiction, he employs it to overthrow GST.

In the last chapter we noted three problems that arise from GST: the Problem of Primacy, the Problem of Origination, and the Problem of Being. Heraclitus addresses two of these problems, if indirectly. To the Problem of Primacy, why one stuff should be prior to the others, he gives an ironic answer. Fire is the most fundamental substance because it is the least substantial: it is always changing, never the same. This provides the clue necessary for his radical answer to the Problem of Being: what is real is not any one stuff, nor any set of stuffs, but the system of changes itself, the process from one stage in the system to another to another and back again, in which the same order and ratio of transformation take place. Reality is lawlike process, the only changeless reality, for all stuffs are but temporary stages in the scheme. Since the world itself manifests this process, there is no need to seek an origin of the world: cosmogony is otiose, and the Problem of Origination needs no answer. Thus Heraclitus gives an ironic answer to the first question, rejects the second, and gives a radical answer to the third. He seems fully aware of the problems raised by GST and fully capable of rethinking the whole Ionian project in light of them. So radical is he that he consciously rejects substance theory in favor of process theory. He goes beyond physics to metaphysics, and a historically radical metaphysics at that.

In the end Heraclitus advocates a world order in which the material parts are always changing while the overall structure stays the same. As he is the first philosopher we know to use the term *kosmos* to refer to the world order, so he may be the inventor or at least the refiner of the concept:[67] what makes a world is not its matter but the ordering function that

---

[67] See Kerschensteiner 1962; there are indications that the term *kosmos* for some aspect of the world may antedate Heraclitus.

organizes matter. Of all the Presocratic philosophers before and after him who allow change at all, he is the only philosopher whose cosmos always was, is, and will be.[68] Ironically, the philosopher of flux builds the most stable world of all. The river of nature remains as its waters flow on:

B84a. Changing it rests.

[68] I exclude the Eleatics, who may be said to deny the existence of the cosmos. On Xenophanes see n. 55.

6

PARMENIDES' CRITICISM OF IONIAN PHILOSOPHY

6.1 Parmenides' Response to Heraclitus

WHAT CONNECTION, if any, there is between Heraclitus and Parmenides has long been disputed[1]. Of the four a priori possibilities: (a) that Parmenides influenced Heraclitus, (b) that Heraclitus influenced Parmenides, (c) that the two did not know or acknowledge each other, and (d) that they are influenced by a common source, only (b) and (c) seem likely. For, contra (a), Heraclitus likes to abuse his predecessors[2], and, contra (d), he tends to radically rework the material he inherits[3]. There have been, and continue to be, proponents of both (b) and (c).[4] While it seems attractive in some ways to dodge the question and thus deal only with textual certainties rather than historical contingencies, I believe that textual evidence is adequate to decide the question in favor of (b), and, moreover, to help determine the philosophical relationship between the two most philosophical Presocratics—and the two most ideologically opposed.

[1] The argument in this section is drawn from a longer study (Graham 2002a). The results are disputed by Nehamas 2002.

[2] Heraclitus B40, B42, B57, B81a, B106, B129. "Dieses bleiben die Ecksteine der Geschichte der Vorsokratiker: Heraklit zitiert und bekampft Pythagoras, Xenophanes und Hekataios, nicht Parmenides; dieser zitiert und bekämpft Heraklit" (Kranz 1916, 1174).

[3] E.g., he is at pains to deny the possibility of cosmogony at B30, the one doctrine common to all his philosophical forebears.

[4] Arguments for (a) start with Hegel 1971, 319ff., followed by Zeller, and revived by Reinhardt 1916; this view has mostly been abandoned, but see Holscher 1968, 161–65. The argument for (b) was first made by Bernays 1885, 1: 2.62, n. 1, and defended vigorously by Patin 1899; this view was accepted by Baeumker 1890, 54; Windelband 1894, 39, n. 2; Diels 1897, 68ff.; Ueberweg 1920, 1st Part: 95, 97, 99; Kranz 1916, 1934; Burnet 1930, 179–80, 183–84; Calogero 1977, 44–45; Cherniss 1935, 382–83; Vlastos 1955a, 341, n. 11, KR (tentatively) 183, 264, 272, Guthrie 1962–1981, 2.23–24; Tarán 1965; Coxon 1986; Giannantoni 1988, 218–20, and others. Diels 1897, 68, says of Bernays: "[S]eine Ansicht ist fast allgemein durchgedrungen," noting that only Zeller has resisted the interpretation; but in his revised edition of Zeller, 1919–1920, 684, n. 1, and 687, n. 1, Nestle abandons Zeller's view as obsolete. For (c) are Gigon 1935, 31–34; Verdenius 1942; Wilamowitz-Moellendorff 1959, 2.208–9; Mansfeld 1964, ch. 11; Marcovich 1965, col. 249; Stokes 1971, 111–27.

The fragments of the two philosophers have a number of verbal similarities. The question has been whether they are there by design or by accident. The case can be settled by noticing similarities of structure as well as of word choice. We have already noted Heraclitus's use of linguistic resources to charge his sentences with multiple, interacting meanings.[5] Here I would like to call attention to a few devices he uses to that end. Let us return to B12:

ποταμοῖσι τοῖσιν αὐτοῖσιν ἐμβαίνουσιν ἕτερα καὶ ἕτερα ὕδατα ἐπιρρεῖ.
potamoisi toisin autoisin embainousin hetera kai hetera hudata epirrei.

On those stepping into rivers staying the same, other and other waters flow.

To recall the points of our earlier analysis:[6] (1) The phrase *toisin autoisin* could agree with either *potamoisi*, "rivers," or *embainousin*, "those entering"; hence it is syntactically ambiguous. (2) The phrase *toisin autoisin* contrasts with *hetera kai hetera*, "other and other," creating a kind of antithesis. (3) The first four words have the same ending, a figure called *homoioteleuton*, and, moreover, imitate the sound of a river in a kind of word painting. (4) There is alliteration in the phrase *hetera kai hetera hudata*. There is, moreover, in the syntactical ambiguity of the first four words an interesting association. Dividing them into structural units, with the second and third words going together as a syntactical unity, we can construe the phase as either AB, C or A, BC: either "on the same rivers, those entering" or "on the same men entering rivers." Because the second unit can be construed with either the first or the third, it acts as a connective that binds the whole together into an ambiguous but indissoluble unity. It is as if Heraclitus has created a molecule out of an energy that alternates like electrons between one atom and another, belonging to both and neither. The dynamic unity he creates is quite unique. There is no other Greek author who employs—indeed exploits—the several elements in quite this way.

A rather common figure that has some elements of this Heraclitean trope is *chiasmus*. In chiasmus elements are crossed over (as in an X, the Greek letter chi) in the pattern ABBA—or expanded into ABCCBA, etc. Heraclitus himself uses chiasmus with great effect:

B25. *Moroi gar mezones mezonas moiras lanchanousi.*
For deaths that are greater greater portions gain.[7]

---

[5] Sec. 5.4.
[6] Sec. 5.4.
[7] Heraclitus's artfulness can be seen by contrast with a verbal imitation in Democritus B219: *mezones . . . orexeis mezonas endeias poieusin*, "greater desires make greater needs."

B22.   *Chruson gar hoi dizêmenoi gên pollên orousousi kai heuriskousin oligon.*
For gold hunters much earth do dig and find little.

In B25 we find alliteration (as in the latter half of B12), and in B22 we have syntactic ambiguity—for the sentence can be read: "For those who seek earth dig much and find little gold." On this reading people may not even be prospecting for gold—their ambitions are too limited. A further device in B22 is worth noting. Heraclitus has put "much earth" in the middle of his sentence flanked at the ends by "little gold." In other words, not only are the meanings and the sounds important, but also the very placement of the terms themselves can carry meaning. Indeed, we might suspect that Heraclitus's sentence has become a little cosmos itself with earth at the center orbited by golden satellites. We have, in any case, a complex structure in which we find a chiastic pattern with a very important element in the middle: ABCBA. Strictly speaking the C is not part of at least the Greek analysis of a chiasmus. But it plays an important role in Heraclitus, expanding the possibilities of the chiasmus to include a focal point, a kind of center of gravity.

More generally, we may see the pattern of B12 as instantiating the same structure: a central term with connections to both what precedes and what follows, binding them into a unity. The pattern may be found in as simple a structure as three words:

B119.   *êthos anthrôpôi daimôn*
Character of-man destiny.[8]

The second word can go equally with the first or the third; it becomes the syntactic glue that holds two meanings in a unity, until we realize character = destiny. Once we see A = C, we realize that the first and third terms are the same, or A = A, and we have something at least structurally equivalent to a chiasmus with only three elements. This is the grammatical equivalent of the "back-turning connection as of a bow or a lyre" (B51), a dynamic symmetry among contrasting elements. What is important here is that no other Greek wordsmith makes the same connections as Heraclitus, tying chiasmus, syntactical ambiguity, antithesis, and often many other figures, into a verbal harmony.

We turn now to one of many verbal parallels between Heraclitus and Parmenides:[9]

Heraclitus B34: Having heard without understanding they are like the deaf; this saying bears witness to them: present they are absent (*pareontas apeinai*).

---

[8] Literally, the guardian spirit.
[9] For a more or less complete list, see Graham 2002a.

Parmenides B4.1: *leusse d' homôs apeonta noôi pareonta bebaiôs*

Behold things though absent to-mind present securely.

At the level of word choice and thought, the connection is interesting but hardly compelling. Both authors create an antithesis between present and absent. But Heraclitus is talking about people, Parmenides about things, and in general we cannot, on the basis of the thought, establish more than loose parallelism. It remains open to the skeptic to explain the similarity as coincidence.

But when we look at the structure of Parmenides' line, we find something striking. Adverb–participle–noun–cognate participle–adverb: ABCBA, a chiastic structure. Moreover, the two participles are precisely the antithetical terms used by Heraclitus. And the middle term is a noun in the dative form similar to some used by Heraclitus, as in B119. In fact, the noun in the middle can be construed either with the first participle or with the second: absent from the mind, present to the mind. Hence mind has a focal position and governs the absence or presence of things. The goddess, whose words these are, is saying either to observe that things we are unaware of are present all the while, or to observe that things absent are present to the mind. The sentence is syntactically ambiguous. Because this sentence re-creates the unique formal and semantic elements of a unique Heraclitean figure, it is reasonable to assume that Parmenides is consciously imitating his structure. The pattern can fairly be said to constitute a kind of Heraclitean signature, a *sphragis* or seal of Heraclitean thought.[10] To ape it is to call attention to the background of the doctrines discussed, to summon Heraclitus through literary allusion rather than through direct apostrophe or mention.

But why go to all this trouble? Parmenides is consciously working within the conventions of epic poetry, drawing on its religious and mythological background, and invoking its authority. He avoids reference to any individual thinker, having the goddess who is his mouthpiece speak only of mortals. On the other hand, he makes use of the rich literary devices of epic poetry, including allusion.[11] Yet he is obviously interested in contemporary debates, else why the criticism of cosmologies and why does he need to construct an alternative cosmology? But precisely what is Parmenides' orientation to contemporary problems? Who does he see as his opponents, and why?

We might begin by looking at possible opponents for Parmenides to confront:

[10] See Kranz 1961.

[11] Bohme 1986, 35ff., supplies no less than fifty pages of poetic parallels for Parmenides' lines. Cf. also Bowra 1937; Mansfeld 1964, ch. 1.

(1) Mythographers and the common people who believe in their tales
(2) The Pythagoreans
(3) The natural philosophers, i.e., the Ionians in general
(4) Heraclitus in particular

Now the epic conventions Parmenides uses, as well as certain motifs, particularly of day and night, prefiguring his choice of Light and Night, as he calls them, as elements of his cosmology, might suggest that Hesiod is foremost in his mind. Hesiod begins his *Theogony* with a proemium or proem like the one we find in Parmenides B1.[12] He discusses the House of Night, which is rightly seen as a prototype of the place where the narrator of Parmenides' proem, the *kouros* or youth, meets the goddess.[13] It may be, then, that Parmenides is not merely adopting the trappings of epic, but targeting mythological assumptions.

As to (2), the hypothesis of a Pythagorean theory that Parmenides was objecting to has been influential. Proposed by Paul Tannery, followed by Burnet, Cornford, and Raven, among others,[14] and buttressed by certain ancient testimonies that made Parmenides a Pythagorean,[15] the interpretation held that Parmenides was debating specifically with the Pythagoreans, and further that Zeno would continue the debate after Parmenides. Unfortunately for this school of thought, students of Pythagoras have come to see that we cannot with any assurance ascribe to the early Pythagoreans any theory except that of the transmigration of souls.[16] Cosmological speculations associated with the Pythagorean tradition seem to have originated after Parmenides with Philolaus and to have been projected back to Parmenides without warrant.[17] In light of these developments, (2) no longer seems to provide a viable candidate for Parmenides' criticisms.

[12] "[B]ei einem Menschen von solcher Gebundenheit des Denkens als Darstellungsform für die Schilderung seines geistigen Erlebnisses allein die Strenge der epischen Kunst in Frage kam, was gleichbedeutend is damit, dass sein Vorbild nur Hesiod mit seiner Theogonie sein konnte" (Deichgräber 1958, 42; cf. Kranz 1916, 1159; Dolin 1962; Schwabl 1963a, 1963b; Mourelatos 1970, 33). On the structure of Hesiod's proem, see Friedländer 1914.

[13] *Theogony* 746–54, cf. the description of the underworld, 719ff., and the bronze gate, 732–33, KRS 244.

[14] Tannery 1930, 232ff.; Burnet 1930, 170ff.; Cornford 1939, chs. 1–3; Raven 1948, reflected in KR; other references in Tarán 1965, 68–69. As Burnet and KR were the dominant English-language textbooks on the Presocratics of their respective eras, the view was very influential.

[15] Diogenes Laertius 9.21 makes Parmenides a student of the Pythagorean Ameinias.

[16] Burkert 1972, followed by Huffman 1993. Earlier see Frank 1923; Cherniss 1935, 384–97; Heidel 1940; Vlastos 1953b; Furley 1967, 63–78.

[17] Burkert 1972, chs. 3–4; Huffman 1993, 202ff.

As to (3), there were a few early attempts to connect the Ionians with the Eleatics as their targets, notably by Gilbert and Cherniss.[18] There have also been a number of more recent attempts to identify Ionians such as Anaximander as potential targets, notably by Mourelatos and Curd.[19] While we cannot say that this hypothesis has been vindicated, neither can we say that it has been refuted. Since the demise of the Pythagorean hypothesis, there has been no major movement to replace it with anything. In recent times most scholars have been simply indifferent to the question.

In fact, much of the discussion of Parmenides' theory has taken place in a historical vacuum in which questions of motivation have simply been ignored. This vacuum was created by the assumptions of analytic philosophy that Parmenides was addressing timeless philosophical issues in a timeless way. The article that is perhaps most responsible for this reading, Owen 1960, encourages us to put aside historical for purely philosophical concerns.[20] So prevalent is this reading that "it is nowadays commonly supposed that Parmenides was a creative genius not much in debt to anybody."[21] The problem with abandoning a historical perspective is that philosophy seems to be a deeply historical enterprise. In the heyday of analytic philosophy in the early twentieth century, British philosophers were participating in the Revolt against Idealism, reacting to developments in mathematical logic and problems of existence raised by Meinong and others. Without a historical milieu, there would have been no movement of philosophical analysis. Even so cerebral a philosopher as Immanuel Kant tells us that he was awakened from his dogmatic slumbers—his adherence to a Leibnizian program—by reading the devastating critique of empiricism offered by Hume (at a time when Hume was hardly noticed by most philosophers). While we can understand virtually all of Kant's arguments without an appeal to the historical setting, we seem to miss a good deal of the significance of his critical philosophy if we ignore his primary motivation. And so the question of historical connections should remain an important one for Parmenides scholarship.

[18] Gilbert 1909; Cherniss 1935, 382–84. Gilbert's view links the two schools through Xenophanes, but his account of that philosopher is now obsolete. Cherniss focuses on a connection through Heraclitus; see n. 22 below.

[19] Mourelatos 1970, 56ff.; Curd 1998, 39ff.

[20] "Parmenides did not write as a cosmologist. He wrote as a philosophical pioneer of the first water, and any attempt to put him back into the tradition that he aimed to demolish is a surrender to the *diadochê*-writers, a failure to take him at his word and 'judge by reason the much-contested proof' " (Owen 1960, 68). Kahn 1969, 701–2, raises a mild objection to this assumption. While in one sense Owen is clearly right to emphasize the new questions Parmenides raised, the danger is that we may take his rejection of a historically inaccurate portrayal of Parmenides as a rejection of any historical evidence.

[21] Schofield 2003b, 44, who regrets this state of affairs, as do I.

As to (4), we have evidence that Parmenides was targeting Heraclitus. Does this invalidate the claims of (1) and (3)? I think not. I would like to propose the thesis that Heraclitus becomes a focal point for Parmenides of all that is wrong with earlier thought—primarily with Ionian natural philosophy, perhaps secondarily with mythological assumptions as well.[22] Thus far I have argued that Heraclitus recognizes and exploits the failure of GST to make good on its claim that there is some primary substance from which all others arise. Attacking the Problem of Primacy, Heraclitus shows not just that the Ionians have failed to support the primacy of any one substance, but also that the project is inherently flawed. If X turns into Y in such a way that Y turns into X, then neither stuff is structurally primary; furthermore, since they are all prone to change, no substance is permanent and everlasting. In place of the problematic stuffs of the Milesians, Heraclitus proposes that the ultimate reality is not any stuff, but the endless process itself, or, to be more precise, the law of transformation, the Logos. He proposes a radical solution to the problem, couched indeed in paradoxical language. He presents his sophisticated analysis, however, not as a straightforward thesis but as the solution to a conundrum. Heraclitus shows (rather than tells) what is wrong with GST. And he indicates the solution with oracular pronouncements.

Parmenides sees immediately the problems Heraclitus has uncovered. Heraclitus is to Parmenides as Hume is to Kant: an iconoclast, an implacable critic with a devastating analysis. What he does not see is that Heraclitus has a viable solution to propose. Taking Heraclitus's paradoxical pronouncements as contradictions, much as Plato and Aristotle would in the next century, Parmenides concludes that a radical solution is in order. Like Plato and Aristotle, Parmenides sees Heraclitus as caught in a philosophical trap. Forced to accept universal flux and the identity of opposites, he cannot but contradict himself. He becomes the reductio ad absurdum of Ionian philosophy: if one accepts the postulates of the Ionians, one will end up in contradiction. Heraclitus is, therefore, an object lesson in how not to philosophize. We must, accordingly, rethink the foundational principles of philosophy. From this point of view, Heraclitus is not to be taken as a lone figure, but as a representative of all that is wrong with Ionian philosophy—and perhaps mythological thinking as well, with its radical transformations. By alluding to Heraclitus, Parmenides can expose the errors of all previous thought.

Heraclitus had focused on the Problem of Primacy to show that, given the fact that stuffs are transformed into one another, what is everlasting in the universe is not substance but change. His answer, or nonanswer, to that problem leads to an answer to the Problem of Being: no stuff is

[22] Cf. Cherniss 1935, 382–84; Cherniss 1951, 336–39.

permanent, and hence the ultimate reality is not a stuff but the law of change itself. Parmenides, for his part, focuses on the Problem of Being: what is there to undergo the alleged changes of transformation? Heracli- tus was in some way an insider, a player in the Ionian game; he saw much that was of value in the project of Ionian science and sought to preserve as much as possible. Parmenides, on the other hand, will call into question the fundamental move of transformation and change. He will be an out- sider to the game who will seek to destroy the whole project. The problem is not that there is no primary substance, but that the whole notion of a subject that changes is incoherent. Rather than allow change to dissolve substance, we should banish change from philosophical discourse.

6.2 Parmenides' Criticism

Previously we have seen that Heraclitus makes far-reaching criticisms of Ionian philosophy. We have also seen that it is plausible to think that Parmenides is reacting to Heraclitus. Parmenides seems to think that Her- aclitus exhibits in an extreme form the shortcomings of Ionian philoso- phy. Hence, according to Parmenides, whatever is wrong with early Io- nian philosophy is also wrong with Heraclitean philosophy, and any criticisms he makes of the former will apply to the latter. Parmenides' poem consists of a proemium creating a mythological tableau in which a goddess addresses the narrator with a revelation about how the world is; an extended philosophical argument developing a strict ontology of being; and a "deceptive" cosmology for which the goddess claims no va- lidity. This odd combination of religious imagery, austere argument, and constructive cosmology presents major problems for interpretation. Mod- ern scholarship has decided that the proem is mere window dressing, usu- ally interpreted allegorically;[23] that the philosophical argument (known traditionally as the *Aletheia* or "Truth") presents the burden of Parmen- ides' message, which we have almost *in toto*;[24] and that the cosmology (the *Doxa* or "Opinion") is a straw man set up to be torn down as an example of what is wrong with cosmology.[25]

The fact that Parmenides finishes his poem with a cosmology shows what his immediate concerns are: a confrontation with the philosophical tradition that produces cosmologies, i.e., the Ionian tradition. The echoes

[23] Beginning with a crude version in Sextus *Against the Professors* 7.112–14; Cornford 1939, 30; Fränkel (1973, 351, n. 7): "The imagery is as impressive as it is readily intelligible to anyone acquainted with this kind of language"; cf. Bormann 1971, 64.

[24] E.g., Owen 1960.

[25] E.g., Owen 1960; Long 1963.

of Hesiod found in the proem indicate that the most philosophical of
mythological poets is also on his mind. It may be that Parmenides sees
the Ionian tradition as a continuation of mythological thinking: Hesiod's
theogony is the model for Ionian cosmology.[26] Indeed one can see many
features of philosophical cosmology as continuations of Hesiodic conceptions (some of these in turn expressing Greek cultural inheritances).[27] But
our concern is with the philosophical objections and their implications for
philosophy and scientific inquiry. Why is the cosmology of Parmenides'
predecessors unacceptable? And what are the implications of this criticism
for philosophy? In this chapter we shall look at Parmenides' criticisms of
the Ionian tradition both to see what is ruled out and to see what might
be thought of as positive contributions to scientific inquiry.

Parmenides' first explicit criticism of "mortal" thought comes in B6.
But before we can look at that fragment, we must consider the dichotomy
he sets up in B2, which will form the background for his criticism. The
goddess speaks:

> Come now and I shall speak, and do you attend hearing,
> which are the only two ways of inquiry [*hodoi . . . dızêsios*] for thinking:
> the one: that it is and that it is not possible not to be
> is the path of Persuasion (for she attends on Truth);
> the other, that it is not and that it is right it should not be,
> this I declare to you is an utterly inscrutable track,
> for neither would you know what is not (for it cannot be accomplished),
> nor could you declare it. (B2)

Using imagery of roads, the goddess distinguishes between two routes of
inquiry, one of which is the right one to follow, the other of which is
"inscrutable," providing no answers. What precisely Parmenides means
by his distinction is not immediately clear. But he seems to take up the
theme again in B6 and to complicate the picture:

> It is right to say and to think that what-is is, for being is,
> and nothing is not.[28] These things I bid you consider.
> From this first way of inquiry I withhold[29] you,

---

[26] This view is argued plausibly by Stokes 1962–1963; Plato seems to have Hesiod in
mind as a philosophical precursor at *Sophist* 242c–d; Aristotle sometimes ranks Hesiod
with the cosmologists, e.g., *On the Heavens* 298b25–9, but generally even in those cases as
reflecting popular ideas, *Metaphysics* 989a10–12. Homer too is credited by some as a precursor of cosmogony, ibid. 983b27–33.

[27] Stokes 1962–1963.

[28] This sentence is so full of ambiguities that a large number of alternative readings are
possible; see Tarán 1965, ad loc.; Wiesner 1996, 8–23, 74–76.

[29] Reading <εἴργω>. An interesting variation was recently suggested by Mourelatos 1999,
125–26: <εἶργον>, referring to the rejection of the route in B2. Curd 1998, 57–58, accepts

but then from this one, which mortals knowing nothing
wander, two-headed. For helplessness in their
breasts guides a wandering mind; and they are borne
both deaf and blind, dazed, undiscerning tribes,
by whom to be and not to be are thought to be the same
and not the same, and the path of all[30] is backward-turning
[*palintropos*]. (B6)

A great deal of effort has gone into trying to interpret the routes mentioned here. To most commentators there seem to be three routes, which
I shall abbreviate to the route of Is, the route of Is-not, and the route of
Is-and-is-not.[31] The most vexed question has been, what sense of the verb
"to be" is invoked here: is it the existential sense: "x exists"? Is it a predicative sense: "x is F"? Is it a veridical sense (possible in Greek): "it is the
case that p"? Is it a "fused" sense that combines several meanings?[32] The
existential sense provides a ready interpretation and has remained a leading contender. On this account the goddess rules out the route of Is-not
because what is nonexistent cannot be thought or expressed. On the other
hand, it is possible to make contact with the philosophical tradition
through the predicative sense, for the philosopher wishes to say x is really
F, relative to which to say x is not F is a completely uninformative utter-
ance.[33] In fact it is possible to translate the several senses in such a way
that they are equivalent: x is F = this F exists = it is the case that x is F.[34]

<αρξω> from Nehamas 1981—cf. Cordero 1979; besides the problem of two beginnings
pointed out by Mourelatos, there is also the problem that αρχω tends to mean "begin" only
in the middle voice: Smyth 1956, sec. 1734.5. There are a number of recent attempts to get
rid of the third way by identifying the condemnation in B6.4 with the way of Is-not; e.g.,
Tarán 1965; Cordero 1979. But see detailed criticism of this view in Wiesner 1996, 74–122,
who cites among others Simplicius as a witness against the reading.

[30] Ambiguous between "all men" and "all things," perhaps intentionally. See sec. 6.1
above. For competing interpretations see Tarán 1965, 66–68, who supports the latter reading; Coxon 1986, 186–87, supports the former.

[31] The conception of three routes, dominant since Reinhardt 1916, 36ff., introduced it,
has been questioned by Tarán 1965, 61, and rejected by Cordero 1979; Nahamas 1981;
Giannantoni 1988. Meijer 1997, 159–62, adds a fourth way.

[32] For the existential sense, e.g., Owen 1960; Tarán 1965, 175–201; Gallop 1979; Barnes
1982, 160–61; Wiesner 1996, 205–36; predicative sense (with an element of the "is" of
identity): Mourelatos 1970; Austin 1986; Curd 1998; fused meanings, with predicative and
existential senses basic: Calogero 1977, 6–10, who pioneered this approach in 1932; with
the veridical sense basic: Kahn 1969, supported by Kahn 1966, 1973; fused meanings: Furth
1968, Schofield at KRS 246 (predicative and existential senses conflated unjustifiably, Raven
at KR 270), Coxon 1986. For a helpful analysis of different approaches to the problem see
Mourelatos 1979.

[33] Mourelatos 1970, 74–80.

[34] Furth 1968.

It does not follow, however, that there is no difference to the argument which sense is taken to be fundamental.

Besides the existential view, the predicative view has been developed most completely. Although I find the predicative view attractive and I think it has much to teach us, it falls short as an interpretation at some key points in the argument. In the first place, in B2.6–8 offers a brief argument to the effect that the way of Is-not is untenable. This argument seems both to fit with the existentialist view and to provide the most intuitive reading of the argument. The argument can be given sense on the predicative view, but it seems to me that it assumes a great deal of prior understanding on the part of the reader. Furthermore, the way Parmenides moves from the formal mode to the material mode of expression, from "is not" (*ouk estin*) to "what-is-not" (*to mê eon*, lines 5, 7) seems to make better sense on the existential view than the predicative, simply because the predicate complement seems to be more glaringly absent when the participle is expressed. More important, only the existential view makes good sense of the argument of B8.6ff. that what-is cannot come to be because it cannot come from what-is-not, because what-is-not is not sayable or thinkable (B8.7–9). Here Parmenides invokes the argument from B2, and only on an existential reading does it make sense.[35] Furthermore, Parmenides goes on to equate "what-is-not" with "nothing" (*mêden*, B8.10), which seems an obvious move only on the existentialist reading.[36]

From a historical perspective, the existentialist reading makes good sense. Parmenides seems to have in mind Heraclitus's theory that the elemental bodies turn into each other in his criticism of the backward-turning way of mortals (B6.8–9). Advocates of the predicative view can cite statements in which Ionian philosophers give statements of the *phusis* or nature of a thing as the background for the predicative use of *einai*.[37] But at least such statements do not appear in the argument for the two routes. Furthermore, if the analysis I have given in the preceding chapters is correct, there are no fixed natures in early Ionian theory such that the appeal to them presupposes a timeless sense of being. There is very little to confirm the predicative reading in the text itself; much of the preparatory argument has to be imported from background material. To my mind, it is a strength of the present historical interpretation that it makes Heraclitus so prominent in the fragments, via allusions of course: simply by alluding to the cryptic philosopher of Ephesus, Parmenides can invoke a whole

---

[35] Mourelatos 1970, 98–100, gives a good general account of the argument on the basis of the predicative view, but his specific account of these lines is not compelling. For criticisms, see Gallop 1979. For further clarifications and articulations, see Mourelatos 1973; Mourelatos 1976.

[36] Cf. Furley 1973, reprinted at Furley 1989, 36–37.

[37] Mourelatos 1970, 60–63; Curd 1998, 43–47.

network of claims and assumptions that provide the target for his criticism. Without this, one is thrown back on mere conjecture to reconstruct the background of his argument, and one may be tempted simply to finesse the historical reconstruction, as many modern commentators do.

Parmenides does present his argument in an a priori, almost Cartesian way. He starts from a general distinction of two ways or routes, evidently two methods or sets of assumptions, the way of Is and the way of Is-not. He does not directly attribute the way of Is-not to anyone. It is difficult to imagine a philosopher building a system on the examination of nonexistent objects. (In later times one might imagine the view as a criticism of negative theology, or of a prodigal theory of intentional objects. But no plausible candidates present themselves in the historical context.) The way of Is-not is not likely to be an explicit philosophical method in its own right, but perhaps a method implicitly required by some philosophical position. Parmenides refers to adherents of mistaken views only in B6, when he accuses mortals of taking the way of Is-and-is-not. Thus it appears that the actual mistake made by mortals is the adoption of the third way. What then is wrong with the third way?

We have already seen that it is plausible to take Parmenides as reacting to Heraclitus. B6 is famously the most polemical confrontation with Heraclitus. Nothing in his argument need be taken as limiting his criticisms to Heraclitus only. Rather, it seems reasonable to think that for Parmenides Heraclitus makes explicit the problems that remain implicit in other thinkers, and perhaps in commonsense views as well, in just the way Hume could be thought to make explicit for Kant the problems that were implicit in other empiricists. B6 lines 4–7 vilify mortals, but do not explain why they are two-headed and helpless. Lines 8–9 provide the answer: mortals think that to be and not to be are the same and not the same. This answer provides one immediate and indisputable ground for criticism: mortals are committed to contradictory positions. But why are they?

It is here that our preliminary work on the influence of Heraclitus can pay off. Of course self-contradiction is grounds for criticism. But one must not merely accuse one's opponents of contraction; one must show how they contradict themselves. This Parmenides does not do. But the allusions which we are now entitled to recognize in Parmenides fill the gap: they remind us of remarks made by Heraclitus in which he seems to express himself in contradictory ways. Heraclitus stands condemned on the basis of his own utterances. Some of his sayings that Parmenides may have in mind are as follows:

> B88. As the same thing [in us] are living and dead, waking and sleeping, young and old; for these things having changed around are those, and those again having changed around are these.

B60. The way up and down is one and the same.

B51. They do not understand how being at variance with itself it agrees: back-turning[38] connection as of a bow or a lyre.

B10. Collections:[39] wholes, not wholes, brought together, pulled apart, concordant, discordant, from all things one and from one all things.

In B88 Heraclitus makes contraries to be the same, and he gives the reason: because they change into one another. The reason ties in with B60: there are contrary pathways of change (Heraclitus uses *hodos*, the same word as Parmenides for "way"), but they follow the same road, only in different directions. Parmenides B6.9 verbally echoes Heraclitus B51 (*palintropos*, "back-turning"), where Heraclitus marks the dependence of the stringed instrument on opposing tensions of its contrary members. But Parmenides also alludes to B60 by making the *path* (*keleuthos*) backward-turning: for change, cosmic or otherwise, is a path from one extreme to another, in which the subject must double back eventually to return to its original state.[40] Meanwhile, we see in B10 that contraries and contradictories can be freely mixed in a comprehensive opposition that constitutes the unity of all things.

Parmenides criticizes Heraclitus for conflating being and not-being, treating them as the same and not the same. Heraclitus's own words reveal the fact that he does regard unities as arising from opposites, including contradictories. And he regards opposites as being the same by virtue of their interconnectedness: opposites turn into opposites, which they could not do if they were wholly different. They are, then, the same and not the same: change presupposes difference, but it also presupposes sameness. Parmenides' appeal to the backward-turning path seems to confirm the fact that the model of sameness he is criticizing is precisely the model produced by Heraclitus, a model of change within a range of opposed values.[41]

---

[38] Reading *palintropos*. The case given for this against the alternate reading, *palintonos*, by Vlastos 1955, 348–51, seems convincing. It is his first argument that clinches the case: Hippolytus surely has a book in front of him, whereas Plutarch is quoting from memory. In his reply to Vlastos, Kirk (KR 193, n. 1) completely overlooks the point.

[39] Reading *sullapsies* rather than *sunapsies*, with Marcovich and Kahn.

[40] It is often objected (e.g., Stokes 1971, 116–17) that *pantôn* in B6.9 must be understood as masculine: the path of all *men*; but as we have seen, Parmenides is not adverse to allowing ambiguity to work in his poem, e.g., in B4.1, and I see no reason not to allow *pantôn* to remain ambiguous as between "all men" and "all things," in a continuing parody of Heraclitus which also allows for multiple dimensions of allusion.

[41] Commenting on B6.8–9, Schwabl 1953, 67, observes: "Dies aber heisst nicht nur, dass die Menschen Sein und Nichtsein nebeneinander annehmen, sondern dass sie, hin und her, eins aus dem anderen entstehen und eins in andere vergehen lassen; dass sie eine paradoxe

Heraclitus, as we have seen, exploits features of Ionian cosmology in a subversive way, to show how they point toward a different conception of nature, one in which process is prior to substance. In doing this, he stresses the interconnections of contraries, their mutual dependence. They prove to be the same and not the same: not the same because without difference there would be no change; the same because they are part of a larger system of variations. It is not Heraclitus's aim to subvert logic. Even prior to the invention of formal logic and the articulation of logical principles, we can recognize an intuitive grasp of concepts of inference, compatibility and incompatibility, truth, falsity, and contradiction. We may suppose Heraclitus is sensitive to those concepts precisely insofar as he exploits them to create paradoxes and puzzles. But no philosophical view of his need be taken as transgressing the Law of Non-contradiction. Rather he uses an intuitive grasp of the law to undermine the principles of GST. But Heraclitus does, as we have seen, see the birth of one element in the death of another. He is thus committed to a model in which one element comes into existence only when another perishes—just as were Anaximander and Anaximenes earlier. But whereas his predecessors focused on the unity of the original substance, Heraclitus focuses on the paradoxes of birth and death, existence and nonexistence, that make Ionian theory seem incoherent to an unsympathetic critic.

Parmenides reacts to Heraclitus's theory. He sees the impossibility of allowing what-is and what-is-not to be the same and not the same, though he does not necessarily understand Heraclitus's strategy. But he does see that something has gone dreadfully wrong in Ionian philosophy. What is problematic is brought out precisely by the paradoxes of Heraclitus; but how are we to understand those paradoxes? What diagnosis shall we make to avoid the errors that produce contradiction? I have suggested that Heraclitus's answer is to look for a unity of process at a level deeper than that of contrary manifestations of change. A possible Aristotelian answer is to look for a subject of change at a deeper level which can exemplify, at different times, opposite features.[42] But Parmenides does not recognize Heraclitus's solution—apparently seeing him as only presenting the problem, nor is he ready for Aristotle's solution.

To understand Parmenides' argument we must invert the argument. He starts by distinguishing the two ways, rejects the second, then notices a third way that arises from conflating the first and second, which he attrib-

---

Einheit und Nichteinheit von Sein und Nichtsein annehmen, indem sie die zwei auf der einen Seite scheiden, auf der anderen Seite aber gegenseitig voneinander ableiten." Coxon 1986, 187, sees "the same and not the same" of B6.8–9 as being explained by B4, and Anaximenes' theory of change as adopted by Heraclitus.

[42] *Physics* I, esp. chs. 6–8.

utes to mortals. It is that last point that motivates and accounts for the earlier ones. Parmenides makes his distinctions and criticisms precisely to block the problems generated by the third way. By way of diagnosis, he distinguishes the way of Is from the way of Is-not. Contradiction will arise only if we allow both ways. By eliminating one way, we eliminate the possibility of contradiction. Of course there are other ways of eliminating potential contradictions. The simplest one, and the one we use most often today, is simply to examine each several case and to adjudicate it a posteriori. In such a case, we assume, *p* will be true, not-*p* false. Or perhaps not-*p* will be true, *p* false. But we cannot decide a priori which of two (contingent) contradictory statements is true. Parmenides, by contrast, wants to decide the case a priori. One way of inquiry is true, one way false. Only the way of Is is acceptable. But what is the way of Is, and how is it acceptable?

6.3 Properties of What-Is

6.3.1 The Four Signs

As is now generally recognized,[43] Parmenides begins his B8 by rehearsing the properties of what-is he will prove in the following lines. He says that,

> only one tale is left of the way:
> that it is; and on this are posted
> very many signs, that [i] what-is is ungenerated and imperishable,
> [ii] a whole of one kind, [iii] unshaken and [iv] complete.[44] (B8.1–4)

Parmenides lavishes the most attention on proving (i); it seems to be the key to much else that he has to say. The proof, which occupies lines 5–21, follows easily from his rejection of the way of Is-not:

> what birth would you seek of it?
> Where, whence did it grow? Not from what-is-not will I allow
> you to say or to think; for it is not sayable or thinkable
> that it is not. (B8.6–9)

Coming to be presupposes a previous state of not-being for the subject. But since not-being was ruled out in B2, it follows that coming to be is ruled out now. Similarly, perishing presupposes a future state of not-

---

[43] Owen 1960, 76–77; Mansfeld 1964, 93ff.; Guthrie 1962–1981, 2.26–43; Tarán 1965, 191ff.; Mourelatos 1970, chs. 4–5.

[44] *ateleston* of Simplicius, kept by Diels, Kranz, and others, is contradicted by the argument. Some emendation is required: *oud' ateleston* (Brandis), *ēde teleion* (Owen 1960, 76–77), or possibly *teleēn* (Mourelatos 1970, 281, n. 5) or *teleston* (Tarán).

being, which cannot be countenanced. Furthermore, no need could have stirred it to come from nothing, either earlier or later (9–10). This is apparently an early invocation of the principle of sufficient reason: what reason could cause it to come to be at any given time?[45] But the emphasis is on the impossibility of not-being, as the reprise at lines 15–18 shows. Parmenides concludes by observing:

> Thus coming to be is quenched and perishing unknown. (B8.21)

He seems to turn to the fire imagery of Heraclitus, in which process and coming to be are identified with fire, to attack his theory. Up to this point there has been no fire imagery in Parmenides' poem, so from an internal standpoint the image of quenching is inappropriate or at least unprepared for. But if he has Heraclitus in mind, Parmenides utters a very clever riposte to his rival: the fire of generation is put out on this theory.

At this point, let us recall that according to the Standard Interpretation, Parmenides is responding to MM. But according to MM, there is no coming to be or perishing, only alteration of a continuing substance. If that is so, then this whole section of Parmenides' argument is directed at a straw man—or at most against Hesiod's theogony, which had already been discredited by Xenophanes and Heraclitus. If, however, the early Ionians were advocates of GST, Parmenides' argument is directed against a vibrant philosophical tradition.

Parmenides next argues point (ii):

> Nor is it divisible, since it is all alike [*homoion*],[46]
> nor is there any more here, which would keep it from holding together
>   [*sunechesthai*],
> nor any less, but it is all full of what-is.
> Thus it is all continuous [*xuneches*], for what-is cleaves to what-is.
>   (B8.22–25)

The argument is less than perspicuous. But the point seems to be the same one made more explicitly later:

>   for neither is it right
> for there to be anything more or anything less here or there;
> nor is there what-is-not, which might stop it from reaching
> its like, nor is there what-is in such a way that there would be of what-is
> here more and here less, since it is all inviolate. (B8.44–48)

---

[45] Barnes 1982, 187–88.

[46] Or: "all is alike." *Pan* can function as either a subject or an adverb, and Owen 1960, 58, argues that *homoion* functions as an adverb rather than an adjective. These different readings do not affect my argument significantly.

What-is could only be divided if we could differentiate it. But to do so, there would have to be recognizable differences, grounded in differences of being. But there is no instance of what-is-not to interrupt the complete consistency of being; nor can there be quantitative differences of being. Thus there is a quantitative uniformity of being, which precludes any qualitative differentiation. The most obvious target of this argument is Anaximenes, who explicates differences of being by reference to different concentrations (although many commentators overlook him).[47] Of course insofar as Anaximenes is seen as the paradigmatic Ionian philosopher, his position would seem to be fundamental to the Ionian project, and Parmenides' criticism of it correspondingly devastating.

Parmenides next makes his point (iii):

Further, motionless [akinêton] in the limits of great bonds
it is without starting and stopping, since coming to be and perishing
wandered very far away, and true faith banished them.
The same in the same by itself it remains
and thus it remains steadfast there; for mighty Necessity
holds it in the bonds of a limit, which confines it round about. (B8.26–31)

This argument depends much too much on opaque imagery of the bondage of necessity. But a thread of argument is discernible in the second line: since coming to be and perishing have been ruled out in (i), so have starting and stopping, which are preconditions of motion. The actual range of application of the present argument, however, remains obscure. Are we talking about change in general, or only motion in place? The imagery supports the latter, but the fourth line suggests Parmenides might have in mind several dimensions of sameness and hence of change. Although chains physically impede only motion in place, the chains of Necessity are metaphorical chains which could have application to other kinds of change.[48] Yet motion in place is the focus of the discussion, and Parmenides explicitly singles out such motion later in the argument as a type of change mortals refer to "changing place," topon allassein, B8.41. In view of the fact that Parmenides rules out coming to be and perishing by a

[47] Tarán 1965, 109, for one, denies that any particular theory is in question. Coxon 1986, 18 et ad loc., sees Anaximenes as a target in B4, B6, and B7, but says nothing about him in the lines of B8. Guthrie 1962–1981, 2.153–55, sees Anaximenes as one of several possible targets. Burnet 1930, 179, sees the connection.

[48] Note that beside the later ambiguity of kinêsis as between specifically "motion" and generically "change" (disambiguated by Aristotle, by fiat, Physics V.1), there are the implications of the epic use of kinêsis, of stirring up, causing commotion (Cunliffe 1924, s.v.). The classical philosophical use is vague enough, but Parmenides' use of epic style only adds to the uncertainty. The term akinêton appears again in line 38, but context there does not resolve the ambiguity.

separate argument, it appears that he wishes here to rule out change of place explicitly.[49]

Arguing for (iv) completeness, Parmenides says:

> Wherefore it is not right for what-is to be incomplete;
> for it is not needy; if it were[50] it would lack everything. (B8.32–33)

To be incomplete is to lack or need something. But what-is could only lack being, and hence it must not lack anything. Later Parmenides supports the property of completeness with a further consideration:

> But since there is a final limit, it is complete,
> everywhere like to the mass of a well-rounded ball,
> equally balanced from the center in all directions. (B8.42–44)

What-is is complete and perfect in the way a ball is perfectly balanced within its outer limits (cf. line 49). Here Parmenides gives us a vivid picture of the perfection of what-is, but his simile raises more questions than it answers. Is the point that what-is is actually spherical, having the shape of a physical ball? Or is it that it is as complete in its way as the geometrical shape of the sphere is spatially complete? Scholars are undecided on this question.[51]

6.3.2 Two Interpretations

Parmenides' four "signs" are clearly meant to represent properties exhibited by what-is. I shall call them Eleatic properties. In the preceding discussion I have not attempted to provide more than a superficial characterization of these properties. But a good deal hangs on just how they are to be interpreted and just how far the argument for them can be extended. In this section I wish to distinguish two sorts of interpretation—both of which have actual exemplars—of these properties. The first will be a minimal interpretation, the second a maximal interpretation. The question is, just how much can we infer from what Parmenides says?

According to a minimal interpretation, we are only identifying those properties that any thing would have to satisfy to count as a being. To be a candidate for being an entity, something would have to be (i) without coming to be or perishing, i.e., not be a temporarily existing thing; (ii) all alike, uniform through and through; (iii) motionless; and (iv) complete,

---

[49] See Mourelatos 1970, 115–20, for a helpful discussion.

[50] Or: what-is would lack everything. ἐόν can be read either as a circumstantial participle or as a substantive.

[51] E.g., against the spatial interpretation are Owen, Guthrie, Tarán, Mourelatos, Coxon; for it are Barnes 1979 and Schofield in KRS.

not lacking anything to achieve its full being. Any putative entity which did not satisfy those conditions would be ruled out as contravening the stricture against not-being. Only entities with a stable nature would qualify as candidates. On this view, what Parmenides would be doing is providing a meta-critique of theories of nature of the sort found in the Ionian tradition. He demands that the entity or entities in terms of which the explanation is to be developed should itself or themselves have a fixed nature. Only against such a backdrop would explanation be intelligible. In modern terms, Parmenides is demanding that we develop our theories in light of a rigorous *ontology*, i.e., a background theory based on determinate properties of fixed beings. On this view B8 elaborates strictly formal conditions to be satisfied by candidates for a role in an ontology. Parmenides is not telling what the world, the cosmos is like, but telling us what an explanation of the world would have to look like to count as a rational explanation.[52]

On the maximal view, not only does Parmenides set the conditions of explanation, but he sees those conditions as sufficient to determine the content of the only viable explanation of the world. The rejection of coming to be and perishing in (i) entails the elimination of past and future predications altogether (cf. B8.19–20). The elimination of differentiations in (ii), whether qualitative or quantitative, means that the only logically possible differentiations are between being and not-being; but since the latter is to be rejected, there can only be being, a completely homogeneous reality. The motionlessness argued in (iii) precludes any physical translation of what-is. And the limits referred to in (iv), whether they are actual spatial limits or only conceptual limits, ensure that the world consists in nothing else but pure being.

The minimal theory leaves open what precisely are the entities of the true ontology. There could be one, or two, or many. Thus the theory is compatible with monism, dualism, or pluralism. The maximal view, by contrast, allows only one entity: What-is, now understood not as whatever there may be, but as Being itself. The maximal view, then, presents a monistic system. Here it might be well to make a few distinctions for future reference. While classifications of systems into monistic, dualistic, and pluralistic are commonplace, too often philosophers are casual in the respects in which they classify. The same philosophical system can be monistic, dualistic, and pluralistic. For instance, ancient atomism posits only one kind of being in the atoms; it posits two basic kinds of reality: atoms and the void; and it posits an infinite number of individual atoms. Hence it is monistic in one sense, dualistic in another, and pluralistic in another. Similarly Descartes' dualism of mind and matter is compatible

---

[52] Examples of this view are found in Mourelatos 1970; Curd 1998; Cherubin 2001.

with a plurality of minds and bodies.[53] Curd has distinguished material, numerical, and predicational monism in a helpful attempt to clarify Parmenides' theory.[54] According to her, Parmenides' monism is a monism of predicates, not of matter or of individuals (numerical monism).[55] The maximal view is perhaps not well-served by the term Material Monism, which we have met in connection with the Ionians. While there seems to be nothing a priori objectionable to applying the term to the Ionians, it already involves an ambiguity: whose sense of matter are we referring to? Traditionally, the term has been associated with Aristotle's notion of the material cause. We have already seen reasons to question the applicability of that interpretation to the Ionians. Here, the term becomes problematic also insofar as it is not clear that Parmenides' being is material in either Aristotle's sense (what form contrasts with the matter?)[56] or in the modern sense (Parmenides' being may be associated with mental properties).[57] In its place I suggest a recognition of "essential monism," i.e., the unity of a single essence or nature in the world. The term "numerical monism" seems clear enough. I see no need to modify "predicational monism" either, though I will not have much use for it in my analysis. While it is potentially difficult to find some principle of individuation to apply to the Eleatic world, Parmenides' comparison of what-is to a well-rounded ball does invite the conception of the world as a single object contained within limits. Hence we can say that on the maximalist interpretation, Parmenides advocates both essential monism and (perhaps the strongest version of monism) numerical monism: there is only one individual in the world, the totality of what-is, and it is all alike.

Recent scholarship has become much more cautious about attributing monism to Parmenides than the ancient tradition.[58] But it remains possible that even if Parmenides does not argue explicitly for monism, the burden of his argument entails monism[59] in a strong form. For if there is nothing

[53] For one attempt to provide a more rigorous classification, see Broad 1925, 27–41.

[54] Curd 1998, 242–43.

[55] 253ff.; see also 65–75.

[56] The talk of limits may suggest a form-matter distinction. But we are hampered by the lack of a clear understanding of what Parmenides means by his limits.

[57] Based on B3, translated literally. For problems see Vlastos 1946, 67–68.

[58] Mourelatos 1970, 130–33; Stokes 1971; Barnes 1979; Curd 1991; Cherubin 2001. For a recent defense of monism, see Finkelberg 1988.

[59] The only place Parmenides seems to advocate monism explicitly is at the beginning of B8.6. But in its context with the end of the preceding line, ἐπεὶ νῦν ἐστιν ὁμοῦ πᾶν, it appears that ἕν may simply amplify ὁμοῦ πᾶν, meaning "uniform" or the like. Those who favor a stronger reading put a comma after πᾶν, but punctuation is a modern invention that of itself decides nothing. "First, the unity of Being was not, for Parmenides . . . the *initial* premise for argument or discussion. It represents the conclusion of a major argument, even if it served in turn as the premiss from which the further properties of Being could be deduced.

but what-is and what-is does not admit of further differentiations, then every kind of change and difference recognized by mortals is a mere name, as Parmenides seems to say.[60] There is only one reality, what-is, undifferentiated, unchanging.

Let us, then, articulate the principles of the two interpretations of Parmenides. The minimalist interpretation can be seen as asserting the following claims:

Weak Eleatic Theory

1. There is a set of substances {E_i} which are the basic substances.
2. The E_i are permanent existences.
   a. The E_i are (i) ungenerated and unperishing, (ii) homogeneous, (iii) unchanging, and (iv) complete. (Eleatic Substantialism)

Leaving the content of E, and the number of its instantiations, unspecified, we rule out any connection with not-being, and specify the Eleatic properties it must exemplify. These same principles can be taken as contained within the maximal theory, but more must be added:

Strong Eleatic Theory

1. (As above)
2. (As above)
   a. (As above)
   b. For any allegedly different substance S, $S = E$.
   c. Hence, there is only one substance. (Numerical Monism)[61]

In addition to (1) and (2), we construe the only possible basis of differentiation to be what-is-not, which is ruled out. Any two allegedly different substances must have only Eleatic properties. But since these are the same for all things, by what we would today call Leibniz's law, allegedly different substances are indistinguishable and thus identical. There is, then, only one kind of substance, and essential monism applies; further, there is only one particular substance of the kind, and numerical monism applies.

Second, the fact that Parmenides deduces several distinct things about Being from the proposition that it alone exists, that it is single, does not show that Parmenides was in any position to distinguish clearly between several kinds of unity." (Stokes 1971, 251).

[60] B8.38: the phrase τῶι πάντ' ὄνομ(α) ἔσται, the traditional reading, supports this interpretation. The phrase can also, with manuscript support, be read, τῶι πάντ' ὀνόμασται, with Woodbury 1958; Mourelatos 1970, 181–82 (reference to others' accepting the reading, 181, n. 37), with the translation "with respect to this thing have all names been spoken" (Mourelatos). On this view the text is much less deflationary to common sense. The latter reading is congenial to the minimalist interpretation. (Criticism of this view by Heitsch 1974, 156–57, applies only to an inadequate translation of the text.)

[61] It would be more rigorous to give Weak Eleatic Theory as simply Eleatic Theory, which is modified by (1b) and (1c) into Strong Eleatic Theory. But I believe the present nomenclature will be less confusing in the long run.

6.4 Deceptive Cosmology

At the end of B8 Parmenides' goddess introduces the deceptive cosmology which will stand as a touchstone against which current cosmologies are to be measured. The passage raises difficult questions.

6.4.1 An Eleatic Cosmology

Here I cease from faithful account and thought
about truth; from this point on learn mortal opinions,
hearing the deceptive order of my words.
For they made up their minds to name two forms,
of which it is not right to name one—this is where they have gone astray—
and they distinguished contraries in body and set signs
apart from each other: to this form the ethereal fire of flame,
being gentle, very light, everywhere the same as itself,
not the same as the other; but also that one by itself
contrarily unintelligent night, a dense body and heavy.
I declare to you this ordering [diakosmos] to be completely likely,
so that no judgment of mortals will ever surpass you. (B8.50–61)

Parmenides marks a clear transition from the *Aletheia* to the *Doxa*. In the third line he may be punning on *kosmos*, which signifies the order of words, but also the world order he will develop, a deceptive cosmology of mere words; the term is further echoed in line 60 in *diakosmos*, a term which can be used to refer to a cosmology.[62]

The crucial move for the cosmology is the recognition, to be precise the invention, of two forms (*morphai*), Light and Night, which manifest opposite properties. We clearly have a dualistic theory, advancing two entities which are defined, or at least distinguished, by their thoroughgoing opposition. Whatever properties light has, night seems to have the contrary. Each entity is identical with itself and contrary to the other entity in every particular. We shall focus on an analysis of the nature of the two entities later. At present a general characterization will suffice.

Although the goddess calls her account deceptive and identifies it with mortal opinion, there is clearly more to her story than simply a summary of current cosmological views.[63] Virtually all commentators note the fact

---

[62] The term appears in the titles ΜΕΓΑΣ ΔΙΑΚΟΣΜΟΣ and ΜΙΚΡΟΣ ΔΙΑΚΟΣΜΟΣ, both often attributed to Democritus, though the first is sometimes attributed to Leucippus, the latter to Democritus: Leucippus B1, Democritus B4b, c.

[63] This view, "canvassed by Zeller and modified by Burnet to a 'sketch of contemporary Pythagorean cosmology,' finds few adherents among modern scholars" (Long 1963, 90). Long's assessment is still true forty years later.

that the present account is meant to be better than the run-of-the-mill theory, for when the hearer has comprehended this account, no mortal will be able to surpass him. And, as we shall see presently, this account is not like Ionian accounts in significant ways, and hence is not a mere rehash of other views.

If we compare the characteristics of Parmenides' entities to those of GST, we see major differences. Contraries play a major role in some generating substance theories, e.g., contrary mechanisms determine changes to rare and dense in Anaximenes' theory. Contrary substances or powers combat each other and ultimately change into one another in Anaximander's theory. Heraclitus recognizes contrary features in different basic substances. But in none of these theories is contrariety a defining feature of basic substances. Similarly, though there are dualistic features, especially in connection with contraries, in Ionian theories, in none of them is dualism a basic fact. The one case that suggests a counterexample is that of Xenophanes, where water and earth are basic substances with at least some contrary features. But we have seen reason to suspect that water and earth can be transformed into one another, and so the apparent dualism turns out to be much more like a Heraclitean contrast than a Parmenidean dualism.[64] That is, while water and earth exhibit opposite properties and may occupy different poles of a transformation relationship, they likely constitute different stages in a unified system. But in Parmenides' cosmology, the contrariety is embodied in two distinct and independent principles. It is a fundamental and irreducible fact of the ontology rather than a temporary state of the basic entities.

With all the emphasis on Parmenides as the enemy of scientific philosophy, we must not fail to notice how innovative is his cosmology, which explains coming-to-be without allowing generation to its basic substances. No less than Heraclitus, who exploits the problems of GST to elaborate an ontology of process, Parmenides revises the basic principles of cosmology. Indeed, his revision is more radical than that of Heraclitus just because he revises all of the basic principles and not just some. For Heraclitus the structure of the Generating Substance Theory points toward the lack of a permanent generating substance. For Parmenides, problems with the concept of generation point the way to a theory of cosmic generation based on ungenerated substances with permanent properties.

We glimpse a system of explanation based on the Eleatic properties developed in the *Aletheia*. Light and Night are (i) ungenerated and imper-

---

[64] See above, sec. 3.4.3. For Xenophanes as a dualist who prefigured Parmenides' theory of elements, see Finkelberg 1997.

ishable as substances, being eternally present in the world as far as we can see. They are (ii) wholes of a single kind: each is identical with itself and nonidentical with the other. They are (iii) motionless, at least in some sense.[65] And they are (iv) complete, lacking in nothing. Were it not for a few crucial lines of Parmenides' exposition, we might think that his cosmology was a perfect realization of the principles of ontology developed in the *Aletheia*.

In all of these features Light and Night differ significantly from the basic substances of GST. For instance, Anaximenes' air is generated and destroyed within the cosmos as it is transformed into the several basic substances by the mechanisms of rarefaction and condensation. Portions of air at least potentially differ from each other in density and rarity, which forms the basis of substantial changes when the differences become great enough. Anaximenes' air and notoriously Heraclitus's fire are changeable substances that seem to be chosen for their role as generating substances precisely because of their adaptability and alterability. Thales' water can readily evaporate or freeze into solid ice; Xenophanes' earth can dissolve into mud and turn into water, which can form clouds and winds. Finally, Ionian generating substances seem to have some urge to change into something else, to fulfill themselves within an ordered cosmos.

Parmenides' cosmology, as unusual as it may be, seems to follow from certain principles of his metaphysics. The world is a mixture of stable beings of two different kinds. Since apparently any quality falls under either one or the other of his forms, any characteristic that a thing can have can be explained as deriving from the presence of one or other of the forms in the thing. Different qualities can be accounted for by different mixtures of the two forms. Indeed, the principles of his cosmology are consistent with the Weak Eleatic Theory according to which existent objects must conform to minimal principles of changelessness. Seen from the point of view of Weak Eleatic Theory, the cosmology is a constructive attempt to apply the insights of the *Aletheia* to cosmology.

Clearly Parmenides' cosmology is not a mere rehash of Ionian cosmologies. It is radically different, innovative in its principles and forms of explanation. His cosmology advances a new way of explaining the world—indeed a new paradigm of explanation. From the outset, according to the goddess, it is flawed. But she provides the best paradigm imaginable for scientific explanation. We shall return in the next chapter

---

[65] At B9.3 we learn that "all is full at once of light and dark night." It is not clear from the fragments exactly how the two forms mix and separate; but they are jointly present everywhere.

to Parmenides' model of explanation and its influence on later philosophy. For now, it suffices to note that his cosmology preserves and applies important features of his theory of reality. But what is the status of Parmenides' cosmology?

6.4.2 Cosmology or Anti-cosmology?

A number of views have been presented about Parmenides' cosmology, ranging from taking it seriously to rejecting it completely. The latter type of view has been prevalent in recent times. Perhaps the best and most influential statement is that of Long 1963. He presents four types of theories about the cosmology of the *Doxa*:

1. It is a "systematized account of contemporary beliefs."
2. It is "an extension of the way of truth."
3. It is "a second-best explanation of the world."
4. It is a false and untenable cosmology. (90)

Rightly pointing out that (1) has been abandoned, Long criticizes (2) and (3) to establish (4). He effectively points out the statements of the goddess that seem to show the cosmology she expounds is not intended to be taken at face value. The most obvious interpretation of her remarks is that (i) her cosmology is superior to all other cosmologies; (ii) her cosmology is based on a theoretical error; (iii) hence, a fortiori, all other cosmologies are erroneous.

This argument could be effective if we could show that there were some obvious sense in which the goddess's cosmology were superior to those already published, and that the kind of mistake made by the goddess's model cosmology were the kind of mistake any cosmology would inevitably make. As to the first condition, I shall argue in the next section that Parmenides' cosmology is indeed superior, surprisingly, on empirical grounds. As to the second, there is a difficulty:

> For [mortals] made up their minds to name two forms,
> of which it is not right to name one—this ıs where they have
> gone astray— (B8.53–54)

What precisely is the error mortals make? Unfortunately, the beginning of line 3 can be interpreted in at least three different ways:

I1. One of the two forms should not be named.
I2. So much as one of the two forms should not be named.
I3. A unity of the two forms should not be named.

On (I1), one of the forms (usually fire) is identified with being, the other with not-being, and we get a mixed cosmology like that of the third way, which is to be rejected. On (I2) both alternatives fail to qualify. (I3) is taken to mean that in the opinion of mortals, a unity should not be recognized. On (I3) mortals fail to recognize the unity that characterizes being.[66] All of these interpretations are compatible with monism, and could have as a consequence that any attempt to multiply principles is ruled out a priori. Without a plurality of principles it will be impossible to explain a plurality of phenomena. Hence any pluralistic theory is ruled out.[67] And the monistic theory that is true does not allow for phenomena to be explained. Cosmology is ruled out, and only an austere philosophy of absolute being remains.

At this point we can either admire Parmenides for his single-minded self-consistency, or wring our hands at his failure to address the questions that cause mortals, whether two-headed or not, to try to pursue the path of inquiry. To abandon explaining experience for some sort of theoretical unity is an indefensible move. If this account is true, then Parmenides is a bad philosopher. Aristotle concurs:

> Now to claim that (1) all things are at rest, and to defend this thesis disregarding sense perception is a case of intellectual failure—indeed it calls into question the whole of experience rather than some part of it, and not only in relation to the natural scientist, but in relation to virtually all the sciences and all judgments, since they all make use of motion. (*Physics* 253a32–b2)

If we reject the whole of experience, Aristotle implies, where shall we have a place to start or end our theorizing? Such a course stultifies all rational inquiry, whether philosophical or common, everyday inquiry. In fact, Aristotle thinks Parmenides was compelled by common sense to admit a plurality of beings.[68] Of course, it may be a historical fact that Parmenides is a kind of super-logician who prefers neatness and elegance to relevance and usefulness. But it would be good if we could show that we did not have to make such an invidious distinction.

[66] For discussion, see Tarán 1965, 217–26; Mourelatos 1970, 80–85.

[67] Coxon 1986, 220, argues, however, that B8.54 rejects not dualism or pluralism, but physical monism.

[68] "Parmenides seems to speak more circumspectly at times [than Xenophanes and Melissus]. For considering that beside what-is nothing that is not exists, he concludes that by necessity there is one thing that is, namely what-is, and nothing else. . . . [B]ut being compelled to follow appearances and supposing that what is one according to reason is many according to sensation, he goes on to posit two things as the causes and sources, calling them hot and cold, in other words fire and earth, of which he classifies the hot as what-is, the latter as what-is-not" (*Metaphysics* 986b27–987a2 = A24).

Long's argument has one serious gap. He holds that Parmenides' "cosmogony gives a totally false picture of reality" (91). Yet Parmenides never goes that far. It is true that he assigns his ontology of being to the *Truth* section of his poem. Yet the contrasting section is not called *Falsehood*, but *Opinion*. Parmenides certainly has the resources to talk about opinion as falsehood, but he does not.[69] Indeed, it is possible on his view that there is no falsehood, only a failure to say anything meaningful.[70] However that may be, a contrast between truth and opinion is quite different from a contrast between truth and falsehood—or between truth and nonsense. The most obvious parallel is Plato's contrast between knowledge, *epistêmê*, and opinion, *doxa*.[71] Yet that contrast is precisely the kind of contrast presented in Long's option (3), making opinion a second-best explanation with relative value. Now one thing is missing for turning Parmenides' cosmology into a Platonic-style theory: a clear-cut hierarchical ontology in which some types of being are superior to others.[72] Yet the theoretical consistency of Light and Night at least suggests some sort of ontological status. As in Plato's theory a lesser grade of cognition (opinion) would be correlated with a lesser level of being, while a higher level of cognition (knowledge) would be correlated with a higher level of being.

One historical precedent for physical knowledge as *doxa* provides at least a possible starting point for Parmenides' understanding of natural philosophy. Let us turn briefly to that precedent.

6.4.3 Xenophanes on the Limits of Knowledge

Parmenides clearly knew Xenophanes' writings and in part reflected them in his own composition.[73] I have already pointed out that I do not take Xenophanes to be the founder of the Eleatic school, nor to be his main inspiration.[74] But there is one important theme in Xenophanes that perhaps made a profound impression on Parmenides: his view of the limits of human knowledge. There is a wide range of possible interpretations of Xenophanes' epistemological views, from skepticism to a robust empiri-

---

[69] On the early meaning of *aletheia* and related terms, see Krischer 1965; Snell 1978. Yet these studies are perhaps a bit too subtle for the present case. There was already a well-developed opposition between true and false, which Parmenides could have exploited, but did not. See, e.g., Hesiod *Theogony* 27–28.

[70] Furth 1968, 118, 124, 128.

[71] *Meno* 97a–98a; *Republic* 476d–480a, 509d–511e.

[72] Best developed in *Republic* 509d–511e.

[73] Cf. esp. Parmenides B8.29 with Xenophanes B26.1; for four other lesser formal similarities, see Böhme 1986, 35ff.

[74] Sec. 1.3; secs. 3.4.2–3.

cism.[75] Here I cannot enter in detail into this controversy. But a quick review of his major statements is in order:

B34. Now the plain truth no man has seen[76] nor will any
know concerning the gods and what I have said concerning all things.
For even if he should completely succeed in describing things as they
come to pass,

nonetheless he himself does not know: opinion is wrought over [or:
comes to] all.

B35. Let these things be believed as being like true things.

B36. All things that they [the gods][77] have showed for mortals to look upon.

B18. The gods from the beginning have not revealed all things to mortals,
but in time by seeking they find what is better.

B38. If God had not created green honey, they would say
that figs are much sweeter.

Speaking from a religious perspective, Xenophanes points out the short-comings of human knowledge vis-à-vis divine knowledge. Human know-ers are limited by the comparisons they make between things they are familiar with (B38). They have not been granted a perfect knowledge of the world (B18) even if they are presented with appearances (B36). But human sensation does not confer knowledge, especially of the ultimate nature of the world, including natural and divine realities (B34). At best humans are left with appearances that approximate the truth (B34, B35). Yet clearly Xenophanes is not hesitant to inquire into the ultimate truths of the world, using, in part, information of the senses. For instance, he uses fossils of sea creatures found on land to infer that at one time the sea covered much of what is now dry land.[78]

Clearly for Xenophanes recognizing the limits of human knowledge does not preclude studying natural philosophy. Yet the inferences made on the basis of analyzing experience, and the theories built on them, do not seem to amount to certain knowledge. In some ways the modern reader must be impressed by Xenophanes' circumspection: for the first time in early Greek philosophy, clever analyses and inferences are not thought sufficient to provide any final proof. But Xenophanes does not despair: even though we mortals do not have all things revealed to us, in time we find out what is better. He does not tell us precisely how—

[75] Surveyed in Lesher 1992, 161–66.
[76] Reading ἴδεν with Snell.
[77] See Lesher 1992, 176–77.
[78] Hippolytus *Refutation* 1.14.5–6 = A33.

whether there is a secure method or a sure path to reach knowledge. But somehow he trusts that we can make progress in our study of the world. Another thinker of the West, Alcmaeon of Croton, who was writing perhaps during the old age of Xenophanes, expresses a similar point of view:

> Concerning invisible things, concerning mortal things, the gods have certainty, but humans must learn by inference. (Alcmaeon B1)

Like Xenophanes, Alcmaeon distinguishes between the knowledge of the gods and human knowledge; and like Xenophanes, he pursues human knowledge in the only way it can be pursued, through evidence and inference.

While it is difficult to say precisely what Xenophanes' positive theory of knowledge is, or indeed how well he could articulate such a theory, we can recognize that a belief in the limits of human knowledge is not incompatible with a belief in the possibility of progress in knowledge. Indeed, the empiricists of the modern period explicitly accepted both beliefs.[79] What is important for our story is that Xenophanes presented something like the *status quaestionis* on knowledge of the natural world. His conclusion was that there was no secure knowledge of the world.

For Parmenides, then, to say that there is no knowledge of the perceptible world, but only *doxa*, was not, from a historical perspective, to make a new attack on cosmology. Parmenides' innovation was to say that there *was* secure demonstrative knowledge of *something*—namely of what-is—independent of the senses and the failures of knowledge they bring with them. Relative to Xenophanes' theory, Parmenides was not bringing down cosmology, but rather building up knowledge in a new realm, what we now call ontology. It is not clear that Parmenides' views of cosmology constructed according to the best possible methods are more damaging than Xenophanes' views. What Parmenides' *Aletheia* offers, by contrast, is a standard of knowledge that is utterly unattainable by *Doxa*. If, like Xenophanes' figs, the truths of cosmology seem so much the worse in Parmenides, it may be simply by an invidious comparison to the green honey of ontology.[80]

Yet can we accommodate both ontology and cosmology in Parmenides? What is needed is a model according to which the statements of both Parmenides' *Aletheia* and his *Doxa* are true. I offer what follows as a very tentative approach to reconciliation.

---

[79] See, e.g., Locke *Essay* IV.iii, "Of the Extent of Human Knowledge," with ch. xii, "Of the Improvement of Our Knowledge"; David Hume, *An Enquiry Concerning Human Understanding,* secs. IV–V, VII, XII.

[80] "[L]ittle notice has been given to the close correspondence between the 'sceptical' conclusion of [Xenophanes] fragment 34 and the attack on mortal *doxa* from which Parmenides

6.4.4 A Structuralist-Pragmatist Reading of the Doxa

In *A General Course in Linguistics*, Ferdinand de Saussure opposed the traditional conception of language as a series of names representing discrete ideas or experiences of the world—a view we meet in Aristotle's philosophy of language.[81] Saussure stressed the fact that we do not encounter the world as a discrete set of experiences; we cannot, prior to or independent of language, identify items of the world. On the contrary, the world, or rather our experience of the world, is a kind of continuum which must be divided up by language. But how can language do that? By offering a system of distinctions that applies to experience. Yet the system itself is arbitrary, and different languages divide up experience in different ways. Each language sets up a series of contrasts by means of which we can articulate experience into discrete units and express our responses to other members of our language community.[82]

The crucial points of this theory are the holistic approach to the world, or to experience, and the arbitrary character of language. In saying that the distinctions made by language are not given prior to experience, Saussure is not saying that they are false or impossible. Rather he is stressing the fact that they are not the only possible way of articulating experience. Given a system of semantic distinctions, there are right and wrong ways of describing a given experience, and there are right and wrong ways of communicating that experience to others.

It is striking that Parmenides stresses the arbitrary character of human language and cosmology. Parmenides talks about the names "mortals have established" (B8.39). "They made up their minds to name two forms" (B8.53). They name things by association with light and dark (B9.1–2). Parmenides even seems to conclude his tale with another reference to the act of naming:

> Thus you see according to opinion these things arose and now are,
> and hereafter when they have been nurtured, will they pass away.
> And men imposed on them a distinguishing name for each thing. (B19)

The conventions of naming seem to be inseparable from *Doxa*. While the goddess speaks of the first way of inquiry as a way of truth, she never calls the *Doxa* a way of falsehood. She promises to relate the opinions of mortals, but she does not call them lies or falsehoods, nor does she claim

---

begins his own account" (Lesher 1992, 183); see ibid., 182–84, for a general comparison; cf. also Coxon 1986, 224–25.

[81] *On Interpretation* 1. See also criticisms of Augustine's theory of language in Wittgenstein 1953, secs. 1–5.

[82] See Saussure 1966, esp. 101–22.

that they are mere illusions. Indeed, she seems to say that she will declare these opinions in the best light possible:

> But nevertheless you shall also learn this, how beliefs
> should have been acceptable, all things just being completely.[83] (B1.31–32)

The verb of the second line, *chrên*, seems to mark a contrary-to-fact statement.[84] But would the account be acceptable if mortals started with things that are completely? Or is it the case that things of the sort they deal with cannot be completely? In any case, at some level the account she will give must be acceptable.

If the choice we mortals make of principles is arbitrary, it does not follow that all are equally bad. Those that deal with beings that are completely, or are as completely as is possible, will be more acceptable than those, presumably, in which things-that-are are confounded with things-that-are-not. And there would be room to distinguish between rival schemes on the basis of how successful they are in solving problems.

Here we must go beyond the structuralism of Saussure, according to which there is no way to evaluate linguistic schemes against each other. A further step can be taken in the theory of pragmatism. While pragmatism rejects the possibility of a direct confrontation of experience with reality, it allows for an approach to reality. Some theories are more useful than others for achieving certain ends. By pursuing those theories, we approach more nearly to an ideal of truth. On a pragmatic reading, a theory of the world that allowed us to control it better would be a better theory than another which had no beneficial consequences. Like structuralism, pragmatism allows us to articulate and navigate through a holistic reality. Pragmatism goes beyond structuralism in offering a criterion against which we can evaluate the relative success of theories.

While Parmenides stresses the arbitrary character of cosmology and of everyday distinctions in general, he does not claim that the theory the goddess proposes is false. And he does at least indicate that her cosmology is the best theory available. It remains possible that Parmenides is offering a twofold theory (as he offers a twofold tale) of reality. There is the theory of what-is, which can be known a priori by deduction from first principles. And the theory of things in the world, which can be known only by a second-best cognition drawing on the general principles of what-is. Cosmology, then, is an inferior realm of inquiry that does not admit of absolute certainty. But it is not impossible, if it is pursued in the right way.

The remarks in this section are offered only as a possible sketch of a model for understanding the *Doxa* as a positive contribution. Let me

---

[83] See the translation recently proposed in Mourelatos 1999, 125, "how it were right for the things-that-seem to be reliably, if only all of them were in every way." Reading περ ὄντα in 32.

[84] Mourelatos 1970, 205–7, 216.

stress that my own account of the pluralists, which I will present in the next chapter, does not depend on the success of the model. It could be that Parmenides completely rejects the validity of his own cosmology. In that case the account I give of the pluralists will show them to have missed the point of Parmenides' cosmology. But it well may be a historical fact that they did miss the point. I suggest that they did not miss it; but even if the suggestion is wrong, the larger picture I present may be accurate.

The case for the value of Parmenides' cosmology would be strengthened if it could be shown that Parmenides took it seriously enough to make contributions to empirical cosmology. It can be shown that he did make such contributions.

6.5 Parmenides' Scientific Discovery

One curious footnote to Parmenides' cosmology has not been properly attended to.[85] It is the fact that, despite his emphasis on the inadequacy of what we would call scientific knowledge, he makes a contribution that is perhaps greater than that of any of his predecessors to science. He recognizes a simple, and apparently obvious, fact about the heavenly bodies:

B14. [moon is a body] shining by night,[86] wandering around earth with borrowed light,

B15. ever peeking toward the rays of the sun.[87]

From these brief words we gather that (1) the moon does not produce its own light; (2) it always looks toward the sun; and, we may infer, (3) the moon gets its light from the sun. At first glance we may suppose this is a commonplace observation not worth comment except for a wordplay.[88]

To us it scarcely seems possible that an observer of the heavens could be unaware that the moon gets its light from the sun. But the greatest observers of the heavens in antiquity, the Babylonians, who had been recording heavenly portents carefully since the seventeenth century BC, did not, so far as we can tell, discover the fact,[89] nor did the Egyptians.[90] Then who did?

---

[85] This section is based on Graham 2002b.

[86] Or: shining like night, according to Coxon 1986, 245–46.

[87] Although there is no evidence the lines were connected, they make perfectly good sense together, and for convenience I join them. The message of the fragments is supported by Aëtius 2.28.5 = DK 28A42, on which see below.

[88] The phrase *allotrion phôs*, "borrowed light," is a pun on Homer's *allotrios phôs*, "foreign man" (Homer *Odyssey* 18.219). On possible connections in thought, see Coxon 1986, 245.

[89] See n. 102 below.

[90] It has recently been claimed that the Egyptians made this discovery (Beatty 1997–1998), but the evidence seems to me to be weak and questionable.

We are told by a late source that Parmenides did not originate the view:

Thales first said [the moon] is illuminated by the sun. Pythagoras and Parmenides likewise. (Aëtius 2.28.5 = Parmenides A42)

But there are problems with this report. We know virtually nothing about Thales' views of the heavens. But if he discovered this important relationship, we would expect it to appear in the works of his successors and Parmenides' predecessors. Yet we find no hint of it. Indeed, what we know of his cosmology—the earth is like a raft floating on an infinite sea—suggests that he could not have held the view. For, as we shall see, it entails that the heavenly bodies pass under the earth. Furthermore, Thales' immediate successors, Anaximander and Anaximenes, show no awareness of Thales' alleged discovery.[91] As to Pythagoras, his alleged contributions to cosmology and astronomy have been invalidated: our early sources can assure us only that he founded a religious brotherhood with a view of transmigration.[92] This leaves us with no predecessors. In fact, all the early Ionian thinkers had models of the heavens in which the moon had its own light.

In the generation after Parmenides both Anaxagoras and Empedocles have theories in which the moon gets its light from the sun.[93] Moreover, both of them give explanations of eclipses that are in principle correct, for the first time, with Anaxagoras being given credit for the discovery.[94] Clearly the correct explanation is predicated on the assumption that the moon gets its light from the sun. The historical evidence, then, converges to support the claim that Parmenides first made the simple but profound observation that the moon is illuminated by the sun.

Now Parmenides' discovery is a pivotal one for astronomy precisely because it has far-reaching implications for the understanding of celestial phenomena. They are as follows: (1) The heavenly bodies are spherical. If the phases of the moon reflect the sun's light, the moon must be spherical. It does not follow deductively that any other heavenly body is spherical. But it provides inductive evidence of the sort the Presocratics were always ready to seize upon in generalizing. (2) The orbit of the sun is

[91] One passage seems to support the view that Thales understood the source of the moon's light: "The earth lies in the middle occupying the place of the center being spherical; the moon is a pseudo-luminous body illuminated by the sun; the sun is not less than the earth in size and of completely pure fire" (Diogenes Laertius 2.1 = Anaximander A1). But this sentence seems to refer to the views of Anaxagoras, which Diogenes has confused, probably because he misinterpreted his own abbreviation: Mejer 1978, 22, 25–28.

[92] Burkert 1972 shows that most information on cosmology and astronomy goes back only to Philolaus, who lived after Parmenides. See also Huffman 1993.

[93] Anaxagoras B18, A42; Empedocles B42–43, 45–47.

[94] Plutarch Nicias 23 = Anaxagoras A18; Plato Cratylus 409a–b = A76. On the priority of Anaxagoras, see O'Brien 1968. On the evidence of Plutarch, see Görgemanns 1970, 35–38, and n. 67.

above that of the moon. This is seen from the darkening of the moon at the time of the new moon. And it is confirmed by the solar eclipse. While there can be other reasons for assigning relative distances from earth to sun and moon, Parmenides' discovery at least provides a secure confirmation.[95] (3) The heavenly bodies continue in existence. This now seems superfluous, but in the theories of Xenophanes and Heraclitus, the sun was new every day. Thus Parmenides' observation rendered certain astronomical theories obsolete. During the middle of the lunar month the moon is visible all night; hence the sun is beneath the earth illuminating it all night, even though it is invisible itself. (4) The paths of the heavenly bodies pass under the earth. This is demonstrated by, for instance, the relative positions of the sun when it disappears from view while the moon is still illuminated. Anaximenes' theory dictated that the sun went around high mountains to the north, not under the earth, while Xenophanes viewed sun and moon as traveling parallel to the infinite plane of the earth.[96] These theories are rendered obsolete. (5) The heavenly bodies interact causally. This fact is verified for the first time. (6) Eclipses result from the blocking of the sun's light by the moon (solar eclipse) or the earth (lunar eclipse). The explanation of eclipses does not immediately follow from the discovery of the moon's illumination. Other assumptions must be made. So if secondary sources do not attest the correct explanation of eclipses for Parmenides—as they do not—this is not to be taken as evidence against his account of the source of the moon's light. On the other hand, once the riddle of the moon's light is solved, it is a relatively easy step to account for the relative rare phenomenon of eclipses.

All of these implications make possible a new era in astronomy.[97] To be sure, we do not yet have a mathematical astronomy. But Parmenides' discovery rules out many theories and opens up avenues of exploration for later astronomers. Ultimately, it shows that heavenly bodies are continuous solid bodies of spherical shape that travel in (at least roughly) circular orbits around the earth. This provides the foundation of ancient scientific astronomy and the dominant model for understanding the heavens until early modern times. One further implication may have been drawn by Parmenides himself. Pursuant to implication (1), Parmenides seems to have inferred that the earth is spherical. The inference is less obvious than others, simply because the earth was not thought of as a heavenly body, but as a sui generis foundation. And the evidence for Par-

[95] One reason for putting the moon below the sun is that it is supposed to be carried westward more slowly because it is less caught up in the vortex that carries the stars. This view is attested later for Democritus: Lucretius 5.621–36.

[96] Aëtius 2.24.9 = A41a.

[97] All six implications are present in Anaxagoras: see Hippolytus *Refutation* 1.8.6–10 = A42, and Anaxagoras's prominence assured their wide diffusion.

menides' view is ambiguous.[98] But on balance he seems to have made the connection—one which was not followed by his earliest successors.[99] Ultimately, fourth-century science would vindicate the view, transforming geography and geology.[100] Parmenides is also credited with identifying the morning star and the evening star as a planet, Venus.[101] This fact was well-known to Babylonian astrologers centuries earlier, but no evidence is found that it was known to the Greeks before Parmenides.[102]

Parmenides emerges as the premier figure in early Greek astronomy. What are we to make of this? Certainly Parmenides was not merely going through the motions in his deceptive cosmology. He was bringing striking new insights to bear and laying the principles for a more adequate account of the heavens. That it was finally adequate he does not assert, and rather implies the opposite. But that he could not only criticize speculative cosmologies from a theoretical point of view, but also improve them from an empirical perspective shows his intellectual breadth. And it may indicate a reason for which his successors took him not as an iconoclast but as a reformer. For this, I shall argue, contrary to standard opinion, is just what happened.

6.6 Parmenides' Response to GST

In chapter 4 I developed three problems that GST faced: (1) the Problem of Primacy, (2) the Problem of Origination, and (3) the Problem of Being.

---

[98] Theophrastus, from Diogenes Laertius 8.48, said Parmenides' earth was *strongulos*, "round"—ambiguous between round of a two-dimensional figure and spherical, of a three-dimensional. But Theophrastus also said Parmenides was the first to say it was "round"—and he knew that, e.g., Anaximander's earth was a round disk; hence the report will only make sense if he attributed to Parmenides a spherical earth. (See also Burkert 1972, 304–5; against this reading, Morrison 1955, 64).

[99] Especially convincing is Panchenko 1997, showing that Anaxagoras was already criticizing a theory of the sphericity of the earth; the only possible target for Anaxagoras's argument is Parmenides.

[100] Of course, Plato in the *Phaedo* and Aristotle would defend a spherical earth. Earlier, Hippocrates of Chios, ca. 430, would seem to support the view: see Burkert 1972, 305, and n. 28; 314, n. 77, on his dates.

[101] Aëtius 2.15.4 = A40a.

[102] Observations of (among other heavenly bodies) Venus, recognized in both morning and evening appearances, go back to Tablet 63 of the Enūma Anu Enlil, a collection of classical astronomical texts dating to the seventeenth century BC, published in Reiner and Pingree 1975. For a general treatment of Babylonian astronomy, see Neugebauer 1957, ch. 5, who denies that the Babylonians had the ability to predict eclipses in early times; see also Neugebauer 1975, 2.602–4. He points out that there is also no evidence they could explain eclipses or even the phases of the moon (ibid., 1.550). Scientific information from Babylon seems to have been available to Greek researchers in a systematic way only in the later fourth century BC: Lloyd 1979, 176–80.

We saw in chapter 5 that Heraclitus responds to the Problem of Primacy, but rejects cosmogony and hence the need to answer the Problem of Origination. He has a radical answer to the Problem of Being which leads him to abandon any effort to find the real generating substance, in favor of the Logos that controls natural processes.

Parmenides, by contrast, spends no time on the Problem of Primacy: the question of which basic substance is the generating substance does not arise for him. He recognizes the Problem of Origination when he asks why what-is should arise at one time rather than another (B8. 9–10). Yet the question has no intrinsic interest for him, but comes up only to show that if one does not recognize the stability of what-is, one will be forced to answer impossible questions. If the question is unanswerable, the presuppositions that entail it are to be rejected.

Parmenides directs his whole effort toward articulating an answer to the Problem of Being. In B2 he identifies the object of inquiry as what-is. In B8 he details the properties of what-is. The first one is its inability to come to be or perish. Taking pride of place, this property provides the exact antithesis to the property of the basic substances in GST. According to (strong) GST, the basic substances are transformed into one another in a series of orderly changes. Heraclitus, as we have seen, calls attention to this feature by stressing the fact that the birth of one substance is the death of another. Heraclitus excludes transformation from the outset. Parmenides' concern with this feature makes perfect sense if he has as his target GST as interpreted by Heraclitus. By blocking elementary transformations, he overthrows GST, the leading, and probably only, philosophical account supporting rational cosmology.

The second Eleatic property, consistency—being all alike—rules out allowing one portion of what-is to be different from another. There cannot be more here and less there, or any difference of kind. As we have seen, to rule out quantitative difference blocks Anaximenes' account of the differences among the basic substances. Rarefaction and condensation are precisely specifications of more or less of what-is in different places. To rule out qualitative differences blocks more general theories, such as, presumably, that of Heraclitus. If what-is is the same everywhere and always, there can be no different substances, nor can there be any basis for differentiation. If what-is is completely uniform, the kind of explanation that is most central to GST will not work.

The third Eleatic property, being unmoved, is, as we have seen, ambiguous. It may signify lack of motion in place, or possibly change in some broader sense or senses. The survey of changes in Parmenides B8.40–41 provides a set of changes that Parmenides considers problematic. But since change in general and locomotion in particular are not unique to GST, this feature does not have any special power against GST.

The fourth Eleatic property, being complete, is suggestive as directed against GST. The Problem of Origination arises from the question of why the original substance, if it is the preferred state of affairs, should differentiate itself into other things in the first place. (According to Theophrastus a residual motion causes differentiation—perhaps tying motion to incompleteness.)[103] But if, Parmenides can object, the original substance is complete or perfect, what could cause it to change? By hypothesis it constitutes the whole of reality in the primeval state. Why should it need or lack anything that would drive it to change itself (cf. B8.33)? Furthermore, if we take the presence of limits literally, the boundless original substance cannot qualify for completeness because of its limitless extension. Thus we see that completeness is motivated by perceived weaknesses in GST.

In summary, we can see Parmenides' analysis of what-is as motivated by the failures or weaknesses of GST. Parmenides' theory makes good sense as a criticism of an actual theory, specifically strong GST.[104] Parmenides' concern with cosmology, as seen in the *Doxa*, confirms his intent: he sees himself as competing with the Ionian cosmologists (perhaps not distinguished from the mythological cosmology of Hesiod, as seen in the proemium). He finds an egregious failure in the paradoxes of Heraclitus, which he takes to be contradictions endemic in the Ionian project; he diagnoses the failures as stemming from a misconception of the object (or objects) of inquiry; he remedies the putative problem by clarifying what the object of inquiry must be, and consequently what properties it must have; and for good measure he throws in an improved cosmology based on principles that avoid at least the worst errors of contemporary cosmological theory.[105]

Conclusion

Having set out to refute Ionian theory as developed by Heraclitus, Parmenides locates its failures in contradiction, its penchant for saying that to be and not to be are the same and not the same. The only way to escape

---

[103] Hippolytus *Refutation* 1.6.2 = Anaximander A11; Simplicius *Phys.* 24.31–25.1 = Anaximenes A5.

[104] A point grasped by Gigon 1968, 104, speaking of Anaximenes: "Der Eine Ursprung verwandelt sich in Alles und der unendliche Prozess der Verdünnung und Verdichtung steht dem Begriff des reinen Werdens schon ganz nahe. . . . Anaximenes gibt als Erster die konkrete Anschauung des endlosen Wandels der sichtbaren Dinge und wird darum in einem entscheidenden Punkt der Hauptgegner des Parmenides."

[105] "Diese Kritik passte ausgezeichnet auf Heraklit, doch geht die Stelle wohl viel eher überhaupt gegen alle Philosophen, die ein Werden durch das Ineinanderubergehen von Gegensätzen erklären, und trifft somit im Grunde die ganze ionische Tradition" (Schwabl 1953, 67).

from this contradiction is to reject the negative way. When we reject that way, we see that there are signs determining the features of what-is. A deceptive but superior cosmology can be derived by conferring on contrary substances the attributes (the signs) of Eleatic being. The inadequacy of the cosmology has something to do with the positing of two substances. But is the inadequacy a complete failure, or does it simply mark the difference between a kind of a priori certainty of first principles and a kind of a posteriori uncertainty of empirical knowledge? Depending on how we answer that question, and on how rigidly we interpret the *Aletheia*, we can see in the *Doxa* either the utter failure of any possible cosmology or the relative success of the right kind of cosmology. There is at least a prima facie case to be made for the *Doxa* as a success story within the realm of an inferior cognitive enterprise.

In any case, Parmenides' radical ontology makes good sense as a reply to GST, whereas it would make no sense at all as a reply to MM. The coming to be and perishing of the basic substances of GST are just the sort of processes that Parmenides seeks to exclude, while MM admits of no such processes. GST, especially as radicalized by Heraclitus, makes the serial birth and death of elements the driving force of the cosmos. Parmenides rules out such changes once and for all, and turns cosmological explanation in a different direction.

7

ANAXAGORAS AND EMPEDOCLES

ELEATIC PLURALISTS

A NEW STYLE OF scientific theory appears after Parmenides. Anaxagoras, Empedocles, and later the atomists account for natural
phenomena by appealing not to some single primeval substance
that generates all things, but to some set of permanent substances that
interact in such a way that all things arise from them. The set of substances always consists of a plurality; hence the new practitioners of cosmology are known as pluralists. They also continue to pursue the aims
and ends of earlier Ionian philosophy. Hence they have been called neo-
Ionians.[1] What accounts for the new style of explanation, how does it
build on its Ionian roots, and how does it go beyond them?[2]

7.1 The Standard Interpretation

Our understanding of the relationship between the neo-Ionians and their
predecessors has changed hardly at all since the early twentieth century.
There have been squabbles about details of the succession and fine points
of interpretation. But the general conception of what the neo-Ionians are
up to is virtually the same now as several generations ago. The story is
as follows.

The early Ionian philosophers had explained the world by positing a
single substance such as water or air, which was transformed into other
substances and ordered into a cosmos. They were monists in the sense of
proposing a single original substance from which all things come, but in
their theories the original substance comes to manifest contrary qualities
such as hot and cold. Parmenides sees serious shortcomings in the views

---

[1] The term is coined by Barnes 1982, ch. 15; see above, sec. 1.3.2.
[2] For an introductory treatment of this topic, see Graham 1999.

of the Ionians.[3] Drawing a distinction between being and not-being, he argues that not-being cannot be thought or known. He goes on to rule out the possibility of coming to be and passing away, then all change and differentiation, finally arriving at a strict essential monism (being is the only type of entity) and numerical monism. Change and difference are ruled out, along with any possibility for accepting the evidence of our senses concerning the phenomenal world, and along with any possibility that scientific study could account for it.

The pluralists, realizing that the possibility of scientific explanation, indeed of any explanation of human experience, is at risk, desperately attempt to halt the march of Eleatic logic. Seeing that monism renders change and difference impossible, they posit the existence of a plurality of beings. These beings conform, so far as is possible, to the strict requirements of Eleatic being: they are everlasting, changeless in their nature, complete and self-identical. But there is a plurality of them. And that plurality allows for the possibility of difference, change, and ultimately the existence of the cosmos. Cosmology and scientific study can accordingly be rescued from the Eleatic elenchus.

On this view the pluralists are at once indebted to Parmenides and opposed to him. They accept his general principles that what-is should be permanent and self-identical, that not-being should be ruled out.[4] But they oppose his overall conclusions and seek to make room in philosophy for change by attacking Parmenides' monism and, moreover, his claim that every other kind of change is as incoherent as coming to be and passing away. They maintain a plurality of substances which change in some minimal way, ultimately in location.[5] We may say that their positive project is to identify some model of change which does not entail coming to be and passing away, but which allows for at least the appearance of change in the world. On this model the constituents of reality will remain fixed

---

[3] In this detail the historiography of Presocratic philosophy has evolved. Tannery 1930, 232–47, seconded by Burnet 1930, 183ff., and followed by Cornford 1939, ch. 1, and Raven 1948 held that Parmenides and the Eleatics were responding to a kind of mathematical atomism formulated by the Pythagoreans. Subsequently, it was shown that there is no evidence that Pythagoreans ever held such a theory (see Vlastos 1953b, 1967, 376–77). It is generally held that his targets were Ionian philosophers such as Heraclitus (Cherniss 1935, 383; Guthrie 1962–1981, 2.23–25). Owen 1960, 61, 68 et passim, attempts to distance Parmenides from the Ionian tradition. But, though Parmenides is critical of the tradition, he adopts its usages in developing his own cosmology, indicating that he is consciously responding to the tradition at the same time he is attempting to go beyond it. Coxon 1986, 19, 219, still sees Pythagorean influence in Parmenides.

[4] See passages cited in next section.

[5] While the atomists would explicitly emphasize the importance of locomotion, the rearrangements of elements envisaged in the systems of Empedocles and Anaxagoras presuppose locomotion of those elements.

forever, but their relationships are free to change in specified ways so as to produce a cosmos or ordered existence.

Unfortunately for the pluralists, or at least for the first generation of pluralists, Anaxagoras and Empedocles, in the early to mid-fifth century,[6] their project is a philosophical failure. For they fail to address the real source of the problems of cosmology, the analysis of reality into being and not-being, with the subsequent acceptance of the former and rejection of the latter. They do not diagnose the problem correctly and can offer only partial measures in the solution of the Eleatic paradoxes. They posit the existence of a plurality of beings. But they do not argue for them, nor do they offer a detailed rejection of Parmenides' monism. They simply beg the question. Nor do they justify the possibility of some kind of change that does not presuppose a change from what-is to what-is-not or vice versa. They merely advance some kind of change as unproblematic and acceptable.

The atomists, perhaps responding to more detailed objections by Zeno and Melissus, confront the problem more directly than do the earlier neo-Ionians. They simply deny that not-being is to be equated with nonexistence and posit the existence of not-being, which they identify with the void. With the void they have the ground for differentiation between the plurality of beings, namely the intervening space, and for motion, namely empty place in which atoms can move. Their bold and innovative approach refutes the position of the Eleatics and lays the foundations of modern science.

7.2 Questions about the Standard Interpretation

The Standard Interpretation has much to recommend it. It takes into account the order of philosophical development as nearly as we can ascertain it. It provides a clear motivation for pluralist views. It allows us to appreciate the apparently ongoing debate between the Eleatic school, represented by Zeno and Melissus, and the pluralist school. It takes advantage of clear allusions to Parmenides' argument in both Empedocles' text (frequently) and in Anaxagoras's (occasionally). It puts philosophical debate in the context of a perspicuous dialectic between opposing views. The interpretation has recommended itself to the great majority of scholars.

---

[6] Empedocles and Anaxagoras are almost universally taken as pluralists. However, Paxson 1983 argues for a holistic interpretation of Anaxagoras, in which the components of the whole are not independent realities, and hence Anaxagoras is not a pluralist; but this seems an extreme interpretation.

Yet there have been signs of a nascent dissatisfaction with the Standard Interpretation.[7] And, as we shall see, there is good reason for dissatisfaction. Let us consider some of the chief passages which exhibit a connection between Parmenides and the early pluralists:

> there is no birth [*phusis*] of any of all
> mortal things, neither any end of destructive death,
> but only mixture and separation of mixed things
> exists, and birth is a term applied to them by men. (Empedocles B8)

> when things being mixed in the form of man arrive to aether
> or in the form of the race of wild beasts or of bushes
> or of birds, then <men call> it being born
> and when they are separated, ill-fated destruction;
> what is proper they do not call it, but by custom I speak so myself. (B9)

> Fools! For not far-reaching are their thoughts,
> who expect what was not before to come to be
> or that something dies and perishes completely. (B11)

> For from not being at all it is impossible for something to come to be
>   [or: be born]
> and for what is to be destroyed is impossible and unheard of.
> For always it will be there, wherever anyone ever sets it. (B12)

Coming to be and perishing the Greeks do not rightly understand; for no thing comes to be or perishes, but from existing things it is mixed and separated. And thus one would rightly call coming to be mixture and perishing separation. (Anaxagoras B17)

Parmenides' rejection of coming to be and perishing is strongly supported in these passages, sometimes in language reminiscent of Parmenides' own expressions, as in Empedocles B12. Many echoes of Parmenides' thought and language also appear in Empedocles' verses, and indeed one supposes that Empedocles chose the medium of dactylic hexameter verse to convey his thought precisely because Parmenides did. Thus there is ample evidence to attest to a strong connection between Parmenides and Empedocles, and sufficient evidence to show that Anaxagoras was influenced by Parmenides.

What is missing from the body of potential evidence of the relationship between Parmenides and the early pluralists is any hint of a *disagreement* with Parmenides. The passages above suggest enthusiastic support for at least certain Eleatic doctrines. They do not bespeak dissent. Nor do the

[7] Mourelatos 1987, 128–30.

pluralists in question ever seem to be defensive about their project or their relationship to Parmenides. Where is the dispute? Where is textual evidence that the pluralists were reacting negatively to Parmenides, desperately trying to save appearances from a perceived destructive attack? The one passage which scholars have pointed to for evidence is the opening lines of Anaxagoras's treatise:[8]

Together were all things, boundless both in multitude and in smallness. (B1)

The initial words *homou chrêmata panta ên*[9] are allegedly a parody of Parmenides' *nun estin homou pan, hen* (B8.5–6), as well they could be.[10] And if there were extensive textual evidence of similar criticisms in Empedocles and Anaxagoras, we could confidently follow the interpretation. But at present there is no confirming evidence. The reading that makes the opening lines into a tacit criticism is based on a set of assumptions such as that Anaxagoras takes Parmenides to be a monist in a strong sense, so that his own pluralism will contrast with Parmenides' position. But do we know that Anaxagoras is reading Parmenides as a monist? We do not. The present passage is the only piece of textual evidence they can cite, and it seems more like a straw grasped at than a rock to build a strong foundation on. In fact, there is another possible source of inspiration for Anaxagoras in Parmenides:

> *pan pleon estin homou phaeos kai nuktos aphantou*
> all is full at once of light and dark night (B9.3)

This passage presents two basic realities together, offering a model for Anaxagoras's primordial chaos. If this is the passage Anaxagoras has in mind when he writes his opening lines, he intends no criticism of Parmenides; rather he is carrying out the program implicit in the *Doxa*. The fact that Anaxagoras starts his account with a pluralistic mixture and no polemic seems to suggest that he is following the latter course; otherwise, he would leave himself open to the scorn of Parmenides and his successors. He would at least need to justify his departure from Eleatic principles with a belated methodological apology—of which there is no evidence in the transmission of his theory.

In short, while there is ample textual evidence that Empedocles and Anaxagoras endorse views of Parmenides,[11] there is no secure evidence that they consciously *reject* anything he says. Here is a very serious gap

[8] Raven in KRS 358; Boeder 1966–1967, 47, 53.

[9] For the word order, see Rosler 1971, followed by Sider 1981 and other recent editors.

[10] KRS 358: "[Anaxagoras B1] shows at the outset how extreme was the reaction of Anaxagoras against the Eleatic monism."

[11] In the case of Empedocles the number of verbal parallels is overwhelming. For Anaxagoras, see Furley 1976, 63–64.

in the evidence for the Standard Interpretation, one which no advocate of the interpretation ever seems to notice. With the Presocratics we can always suppose that the passages in which they said something were not preserved. But given the interest of the doxographical tradition in dialectical interactions among the schools, it seems unlikely that the tradition would not preserve some trace of a criticism by the early pluralists of Parmenides. Yet we have nothing. On the contrary, there are traces in the tradition of a strong positive connection between Empedocles and Parmenides:

> Theophrastus says [Empedocles] was an enthusiast of Parmenides and an imitator in his poetry. For Parmenides too composed a treatise *On Nature* in verse.[12]

> Empedocles of Acragas, who was born not long after Anaxagoras was an enthusiast and a follower of Parmenides and even more of the Pythagoreans.[13]

On the very solid authority of Theophrastus we find that Empedocles is a fan of Parmenides. And although according to Simplicius in the second passage, he is not less a devotee of the Pythagoreans, his commitment to Parmenides is unqualified by any disagreements.

Far from taking Parmenides as the destroyer of cosmology, the early tradition sometimes sees him as assuming an ambiguous stance toward the world. According to Aristotle:

> Parmenides seems to speak more circumspectly [than Xenophanes and Melissus]. For considering that beside what-is nothing that is not exists, he concludes that by necessity there is one thing that is, namely what-is, and nothing else. . . . [B]ut being compelled to follow appearances and supposing that what is one according to reason is many according to sensation, he goes on to posit two things as the causes and sources, calling them hot and cold, in other words fire and earth, of which he classifies the hot as what-is, the latter as what-is-not. (Aristotle *Metaphysics* 986b27–987a2)

Aristotle sees him as being "compelled to follow appearances," willy-nilly committed to a physical theory that allows for cosmology and physical explanation. If Aristotle can interpret Parmenides so, it is not outlandish to think of Anaxagoras and Empedocles reading him as sympathetic to, and even supportive of, cosmology.

In general, the case for making the pluralists critics of Parmenides is quite problematic. For given the terms of the debate envisaged by the Standard Interpretation, the pluralists must not simply disagree with Par-

---

[12] Diogenes Laertius 8.55 = DK 28A9.
[13] Simplicius *Physics* 25.19–21 = DK 28a10.

menides, but also reply to those principles or premises of his argument they disagree with. In other words, they should raise objections to his argument. Not only do we have no disagreements, we also have no sign of what principles or premises they disagreed with or on what grounds they disagreed. The focus of the Standard Interpretation, from ancient times, has been on the monism-pluralism debate. But modern scholarship has called attention to the fact that monism is not a premise of Parmenides' argument but only (if anything) a conclusion to be drawn from it.[14] That means that it is not sufficient merely to oppose monism: the opponent must locate among those premises that generate monism the one or ones which are false, or the flaw in reasoning that leads from the premises to the conclusion. What are those premises the pluralists opposed, and on what grounds? Do we have any hint? Indeed, do we have any hint they opposed any part of Parmenides' argument itself, as contrasted with its conclusions?

In fact we have extensive excerpts from Empedocles and pivotal, if not extensive, fragments from Anaxagoras, as well as a considerable number of doxographical reports. The evidence of a disagreement with Parmenides is missing. The evidence of objections to his arguments is missing.[15] Any trace of a general negative attitude toward the Eleatic is also missing.

Furthermore, on the Standard Interpretation we are forced to make negative assessments of Anaxagoras and Empedocles. They invoke pluralism against Parmenides' monism; they defend alteration against his denial of change. But they do not seem to have an argument for their position. Driven by the need to save the appearances, they simply deny key parts of Parmenides' theory. They do not argue. They do not even confront his position. They simply assume the negation of some key principles without ever acknowledging that they do so. They disagree with Parmenides, not as philosophers must disagree, with objections and counter-arguments, but rather as naive opponents disagree, assuming whatever they need to make their own intuitively held views fly. As Barnes (1982, 442) puts it:

> [Melissus might reply,] "they [Empedocles and Anaxagoras] do not explain how their analysis constitutes a defence; and they do not indicate where they think the Eleatic arguments against generation fail.[16] Their position may not be internally contradictory; but it amounts to no more than an unargued rejection of Eleatic metaphysics."

[14] Mourelatos 1970, 130–33; Barnes 1979; Curd 1991.

[15] Pepple 1996 speaks of "a lost fragment of Empedocles," but gives only very general inferential evidence for it.

[16] *Sic.* If they think any of Parmenides' arguments fail, it is not the arguments against generation that they disagree with, but rather arguments against other kinds of change.

I have sympathy with Melissus' hypothetical retort; and I believe that the neo-Ionians never apprehended the power of the Eleatic deduction. . . . The neo-Ionians threw off the intellectual paralysis with which Parmenides had threatened Greek thought: they manfully attempted to tread again the scientific road, and they took many progressive steps even if their feet remained shackled by Elea. . . . For all that, the neo-Ionian revival is fundamentally a flop: it does not answer Elea.

The early pluralists are regularly faulted for their lack of dialectical candor. Meanwhile, in a procedure unique in the history of philosophy, the atomists are praised for denying Parmenides' rejection of not-being, although they themselves offer no argument for their own position:

> Atomism is in many ways the crown of Greek philosophical achievement before Plato. It fulfilled the ultimate aim of Ionian material monism by cutting the Gordian knot of the Eleatic elenchus. (KRS 433, unchanged from KR 426)

The early pluralists ignore Parmenides' arguments and we blame them for begging the question; the atomists blatantly deny Parmenidean theses and we praise them for their dialectical chutzpah! Apparently, begging the question is virtuous so long as one is brazen enough, and so long as one is on the side of the best explanation of the phenomena, as determined by posthumous research.[17] But here the distinction can only be between bad philosophy and worse philosophy, and the story of the pluralists must provide a dark chapter in the history of philosophy. The early pluralists fail miserably, and the atomists fail spectacularly.

We see, then, the following problems for the Standard Interpretation. (1) There is strong textual evidence for a positive response by the early pluralists to Parmenides. (2) There is no firm textual or doxographical evidence for a negative response to Parmenides, and some evidence for continuity from Parmenides to the pluralists. (3) There is evidence in the doxographical tradition that Parmenides is not seen as purely critical of cosmology. (4) There is no sign of a diagnosis by the early pluralists of Parmenides' argument which might support objections or criticisms. (5) The interpretation makes the early pluralists out to be unphilosophical, incapable of carrying on a rudimentary debate according to what should be intuitive rules of informal logic.

Recently, challenges to the Standard Interpretation have appeared. Mourelatos sees the pluralists as pursuing an independent philosophical program that requires no direct reply to Eleatic criticisms.[18] This, however, seems excessively generous to the pluralists. After all, they acknowl-

---

[17] Barnes 1982, 343–44, is critical of naive enthusiasm for the atomists as forerunners of modern atomism.

[18] Mourelatos 1987, 127–34. See earlier West 1971, 219.

edge at least the insights of Parmenides in a number of cases, and are willing and even eager to criticize positions of thinkers with which they disagree.[19] Curd has argued that (i) Parmenides should be taken as advancing something like Weak Eleatic Theory, and hence (ii) the pluralists are justified in seeing themselves as his successors.[20] Although I find the Weak Eleatic Theory reading attractive, Parmenides' text seems to me to go beyond it to something like Strong Eleatic Theory. Yet Parmenides himself seems to suggest there is some advantage in his own cosmology, taken as an account of appearances; accordingly, he may endorse Strong Eleatic Theory as his account of reality, but leave room for Weak Eleatic Theory as the best way to deal with appearances. In any case he does not eschew cosmology altogether, as Melissus does.[21] Thus I accept a modest version of (i), and I agree with Curd that (ii) the early pluralists do see themselves as Parmenides' successors, and have some reasons to do that. The textual evidence, of which there is a generous amount, supports the new reading, and for the standard reading there is only the questionable claim that the pluralists' criticisms of Parmenides are not preserved. But since even Aristotle and Theophrastus seem unaware of the alleged criticisms, we are justified in discounting them. And if even so astute a reader of early Greek philosophy as Aristotle could conclude that Parmenides made concessions to everyday experience, we need not be loath to ascribe such a reading to Anaxagoras and Empedocles.

One general methodological point needs to be made. Typically the attention of interpreters of Parmenides and his tradition focuses on reconstructing what his theory was, and largely from that reconstruction inferring what the response of his successors to him was. But as John Palmer has recently pointed out in his study of Plato's reception of Parmenides, the immediate question for an interpreter interested in the influence of Parmenides is not what we might expect. Paradoxically, it does not matter what Parmenides' theory actually was:

> [I]t is simply a mistake—one might term it the "essentialist fallacy"—to privilege Parmenides' intended meaning as the determining factor in his subsequent influence. . . . Instead, I assume that what Parmenides himself may have intended is largely irrelevant to an account of his influence on Plato. (Palmer 1999, 9)

---

[19] Anaxagoras B8 and B17, Empedocles B8–B9, B11–B15, seem all to contain criticisms of philosophers as well as of unreflective people.

[20] Curd 1998. See now Sisko 2003, who attempts to give Anaxagoras a strategy for dealing with Parmenides' monism; his account seems to me to make a number of unjustified assumptions.

[21] A point stressed recently by Palmer 2004.

What matters instead is how Plato himself understood Parmenides' theory. If his understanding departs from ours, we might fault him for his historiography, but we had better pay attention to his interpretation, for it will form the real basis of his reception. Of course in the case of Plato we have rich resources from which to reconstruct his reception of Parmenides, whereas we have relatively meager evidence in the case of the pluralists. Nevertheless, the general point is well taken: what counts is not how modern interpreters now construe Parmenides, but how philosophers of an earlier time construed him. And the latter is a historical question that does not follow immediately from an adequate philosophical analysis of his theory.

7.3 The Elemental Substance Theory

Before we can decide between competing hypotheses about the pluralists' aims, we must examine the structure of their theories. Superficially there are major differences between the respective theories of Empedocles and Anaxagoras, and certainly those differences represent major disagreements on some points. But there are also major areas of agreement, and it is on these areas that we shall focus in the present discussion. For, as we shall see, their global strategy seems to be the same one even if their local tactics are different.

7.3.1 Empedocles' Theory

Empedocles posits the existence of four *rhizômata* "roots" that came to be known as the four elements.[22] The four materials of which everything is composed are earth, water, air, and fire. In addition two cosmic forces are present, personified as Love and Strife, that represent the attraction and repulsion of unlike bodies.[23] The four elements and the two forces

[22] See Curd 2002, 147–53, for a recent account, with which I am in general agreement.

[23] Although in introducing them Empedocles arrays them with the four elements, he clearly treats the two forces as playing a uniquely active role in structuring the cosmos. Critics from Aristotle (*Metaphysics* 985a21ff., *On Generation and Corruption* 333b19ff.) on have found Empedocles to be inconsistent in the way he applies Love and Strife in different situations—for Love *separates* portions of elements from each other and Strife *combines* portions of elements. Aristotle, I believe, misses the significance of Empedocles' terms. In the political arena Strife is always between unlikes (hostile classes); and Love in the political context replaces Strife when it harmonizes unlikes into a viable polity. The cohesion of like members (rich with rich, poor with poor) requires no special force, only their mutual opposition to unlikes. Politically speaking the reign of Strife is faction and class warfare, the reign of Love peace and domestic tranquillity. See Graham 2005. On the sexes, see the following note.

are everlasting entities. The combinations of elements give rise to the multitude of other stuffs that exist in the world, according to the ratio of the respective elements in the mixture. Thus blood is composed of equal measures of the four elements (as it must be to allow for perception of the four elements, B98), while bone is composed of two parts earth, two parts water, four parts air (B96). The ratios in question are evidently arrived at in an unscientific way: blood must have equal measures to satisfy the role assigned to it a priori in Empedocles' theory; the ratio present in bone seems to be arbitrary. Nevertheless, Empedocles does clearly see the ratios as determining the character of the resulting compound. A small number of everlasting realities generates an indefinite number of phenomenal stuffs by their interactions. The conception is elegantly illustrated in an extended simile:

> As when painters decorate offerings,
> men well trained by wisdom in their craft,
> who when they grasp colorful chemicals with their hands,
> mixing them in combination, some more, some less,
> from them provide forms like to all things,
> creating trees and men and women,
> beasts and birds and water-nourished fish,
> and longlived gods mightiest in honors. (B23.1–8)

As a few pigments can be mixed to produce many colors and, by extension, depictions of many types of bodies, so a few stuffs can in principle produce all the multifarious objects of the world.

Empedocles seems to envisage the power of Love as necessary to unite different elements in a unity.[24] During certain periods of the cosmic cycle, compounds that exist as forces of attraction and repulsion operate together to allow for the existence of compounds of limited size. At one point in the cosmic cycle, Love predominates to such an extent that all the elements are brought together into a perfectly homogeneous mixture in which no differences are discernible. The cosmic mixture Empedocles denominates the Sphere because of its radial symmetry (B27, B28). The harmony of mixture is shattered as Strife invades the Sphere, and the elements separate out, eventually to form the cosmos. One stage that seems to be present in the cycle (although its existence is controversial) is a complete separation of the elements into concentric spheres of earth, water, air, and fire; so complete is the separation that no compounds exist while Strife dominates completely. Both before and after the complete separation, there seem to be compound substances, including living

---

[24] E.g., B22, B35. In the former the personification of Love as Aphrodite suggests the attraction of unlikes (male and female).

things.[25] The four "elements" recognized by Empedocles are clearly what we would call elements in a sense univocal with the elements of modern chemistry: they are substances with characteristics properties of their own that exist as permanent features of the world, and that are not further reducible to simpler bodies of the same type, but which by interacting produce complex substances with different properties.

Of course some disclaimers must be made about modern elements: we now understand them to be not everlasting but to have been created by nuclear fusion in the heart of stars. But they are not able to be created or destroyed by chemical means (only by nuclear processes). Furthermore, they are not absolutely simple, but are analyzable into subatomic particles (which are in turn analyzable into other subatomic particles, etc.). This analysis, however, takes place not at the level of chemical interaction but of atomic physics. Individual atoms are atoms of elements having their basic properties determined by the number of protons in the nucleus (and consequently the number of electrons in the shells), and all molecules are composed of these atoms. Thus, although modern chemical atoms are not strictly speaking everlasting or simple, they are in some important sense permanent fixtures of the universe and the simple building blocks of matter. They have fixed properties, and they combine to generate complex bodies with different but derivative properties. Empedocles' elements have these same characteristics and hence are elements in a sense consistent with our modern conception of a chemical element.

7.3.2 Anaxagoras's Theory

Anaxagoras envisages a primeval chaos in which "[t]ogether were all things, boundless both in multitude and in smallness" (B1). There is a limitless number of substances mixed together in small quantities which make up the chaos. The cosmos arises as a cosmic Mind begins a revolution that separates different substances into different places like a cosmic centrifuge. Where Empedocles postulates a small number of elements

[25] I am following the standard account of Empedocles' cosmic cycle. See O'Brien 1969; Bignone 1916, 545–603; Minar 1963; Wright 1981; Graham 1988; O'Brien 1995; Inwood 2001; also Barnes 1982 (criticized by Brown 1984). According to a dissenting view, there is only one creation of living things, or zoogony, rather than two: Tannery 1887; Arnim 1902; Hölscher 1965; Solmsen 1965; Bollack 1965–1969; Long 1974; Schofield in KRS 288, n. 1; Osborne 1987. Martin and Primavesi 1999 claim that new material from the Strasbourg Papyrus confirms the standard account, but that remains controversial; see Laks 1999. Yet I think Trépanier 2003 uses the new material to good effect in defense of the standard account. See now Trépanier 2004, 184–92. Though in fact I support the standard account, it is not crucial to my thesis which view is correct. (The single zoogony view would be preferable if a coherent one could be constructed; thus far I have not seen one that has seemed to me coherent.)

which combine to form compounds, accounting for the diversity of natural bodies and phenomena, Anaxagoras postulates a large number.[26] Anaxagoras seems ready to postulate a basic substance for every phenomenally distinct substance, e.g., earth, air, aether (which he mentions by name), flesh, blood, bone, wood, iron, etc.[27] His substances do not combine to form compounds; rather, one substance becomes manifest as it separates out from another. We may understand him as picturing every stretch of matter as a mixture (like the original chaos, though not so thoroughly mixed, B1, B6) of all substances, but in varying concentrations. Change takes place when the components of a mixture are redistributed so that new concentrations emerge in different parts of the mixture. In each case the emergent stuff was already present in the mixture, but was not manifest because it had not attained to sufficient concentration to dominate in the mixture. When it dominates, it changes from being latent to being manifest (B12 ad fin.).

According to Anaxagoras, everything is in everything.[28] This seems to mean that every phenomenal substance contains portions of every real substance. This allows him to account for a change in which one type of thing apparently turns into something very different. For instance, if water is heated in a pot, thereby turning into steam or air, what must have happened is that the airy parts of water came to predominate. Perhaps they were transferred from the fire which heated the water. In any case, no new entities were created, but preexisting portions of air came to characterize the material that was in the pot. Similarly, if a man eats bread and gains weight as a result, what happened was that in his digestive tract portions of flesh were extracted from the bread; they were there all along, but were not perceptible because other features predominated.

In order, apparently, to preserve the principle that everything is in everything, Anaxagoras also stipulates that matter is infinitely divisible.[29] Thus, however small a portion of matter we take, it can contain trace amounts of all kinds of matter. And however much of one kind of matter we remove from a sample—for example, however much flesh we remove from bread—there can still be more there. Extraction never removes all of one

---

[26] It is often assumed that his "boundless" (*apeiros*) multitude (B1) is infinite; but the term is less exact in the fifth century than it will be in the fourth century, e.g., in Aristotle's mathematical-physical discussions of infinity in *Physics* III. It is simply not clear that Anaxagoras is committed to an infinite number of substances, though he possibly might be.

[27] B1, B2, cf. B15 mention explicitly only air, aether, and earth. Melissus B8 gives Empedocles' four elements plus iron and gold, possible concessions to Anaxagoras.

[28] This axiom is reiterated in B1, B4a (weakened version), B4b, B6 (four times, with different nuances), B11, and B12 (three times, including once in weak form). It is far and away the most important principle recognized by Anaxagoras.

[29] B3 with B6.

kind of matter, nor does what is extracted ever consist of one pure kind of stuff. Everything remains in everything, allowing for any kind of change to take place.

Anaxagoras's theory is unlike modern chemical theory in that it does not provide for the creation of compounds from elements. Nevertheless, it too has similarities to that part of chemical theory that deals with mixtures—with colloids, suspensions, and solutions. Molecules from one or more substances can mix with those of another, either in trace amounts so as to register no phenomenal properties, or in sufficient quantities to characterize the solution, e.g., the salty taste in salt water, or to dominate in the mixture, e.g., watery wine versus wine-flavored water. The boundless number of basic substances are elements in the sense of being the basic building blocks of matter. They are everlasting and have fixed properties, which they manifest whenever they occur in sufficient quantity in a mixture.

7.3.3 The Concept of Elements

The theories of both Empedocles and Anaxagoras seem to instantiate a broad pattern of explanation in which entities we could call elements, without doing much violence to our modern scientific use of the term, explain the phenomena of the world. A fixed (but perhaps indeterminate) number of elements exist in reality. They are everlasting and exhibit fixed properties. Their interaction, whether by compounding or by a looser kind of mixture, accounts for the changing phenomena of the world. Moreover, there are forces which drive the interactions—Love and Strife in the case of Empedocles, Mind in the case of Anaxagoras. We see the ingredients of a new paradigm of explanation, different from anything used by the early Ionians, for natural phenomena.

The key to the new paradigm is the concept of *elements*—changeless substances that by their combinations can combine or mix to produce different results. The word for element—*stoicheion* in Greek—does not appear in the context of physical explanation until the fourth century BC.[30] The Greek term originally denotes a letter of the alphabet, as does the Latin translation, *elementum*.[31] As letters combine to make up syllables and syllables to make up words, so the simple stuffs combine to make up compounds or mixtures.[32] The resulting theory is very different from

[30] The first use of the term *stoicheion* to mean "element" is found in Plato *Theaetetus* 201e1, according to Eudemus from Simplicius *Physics* 7.13. For its linguistic use as "letter" see Plato *Cratylus* 424e4.
[31] See the brilliant study by Diels 1899.
[32] See Plato *Theaetetus* 201eff.

a theory in which one stuff is transformed into other things. In the theory of transformation—GST—there is a historic starting point for change but no constant identity of any stuff. What was now fire, characterized, e.g., by hot and dry qualities, later becomes water, characterized by the cold and wet. The original substance does not preserve its character: it has no continuing essence, or nature. By contrast, elements never lose their own character, essence, or nature. Water is water, fire is fire, air is air. They may combine in such a way that the conglomerate acquires new characteristics, or so that the characteristics of only one constituent are manifest. But the elements themselves are changeless forever more.

If the elements are changeless, what accounts for phenomenal change? It can only be the changing configurations of the elements. By changing their relative quantities or their connections, the elements produce different appearances to the senses. But what in turn produces the changing configurations? Ultimately it must be some sort of change of place of the elements, such that some piece or portion comes to be where it was not before. Thus there must be some kind of change of position, typically unobserved, which results in the phenomenal changes we observe.

But what drives the changes that take place? For Empedocles it is a tension between Love and Strife, the attractive and repulsive forces of the universe. Love tends to bring different elements together, Strife tends to keep them apart. At times when both forces are present in the world, certain things join together. For Anaxagoras, Mind starts the cosmic vortex. But according to the testimonies of Plato and Aristotle,[33] Mind does not play a major role in explanation. In general the motion produced by the vortex seems to account for the initial changes, and secondary changes result from the primary differentiations. Some sort of mechanism accounts for any given change, with the exception of the original push supplied by Mind, which produces a physical effect on the primeval chaos.

Thus we see a new system of explanation in which elements are the basis of the world. I shall call this theory the Elemental Substance Theory (EST). According to EST: (1) there is a set of basic substances which (2) are permanent existence. (3) These substances, which are the elements, combine to produce derivative substances. (4) There is some mechanism that controls the production of derivative substances and (5) a set of forces that governs the mechanism. And (6) the world comes to be as a result of the forces acting on the elements.[34]

---

[33] Plato *Phaedo* 97b–98c; Aristotle *Metaphysics* 984b8–22, 985a10–21.
[34] A formal characterization will be given in the next chapter.

7.4 Parmenides and Origins of the Elemental Substance Theory

7.4.1 *Eleatic Criticisms of Cosmology*

Scholars universally recognize that in some sense the pluralists are responding to the Eleatic challenge. As we have seen, the emphasis on their engagement in a desperate rearguard action to save natural philosophy from Eleatic criticisms seems misplaced. But what *is* going on in the pluralists' theory? In the following chapter we shall look in more detail at the formal aspects of EST and more formal connections to Eleatic theory. For now, however, we may note some striking similarities between EST and the *Doxa*.[35]

In the latter half of his poem, Parmenides develops a cosmogony and cosmology designed at least in part to exceed anything in existence. The goddess promises the youth that armed with her theory, "no judgment of mortals will ever surpass you" (B8.61). Unlike the Ionian theories preceding Parmenides, in which one stuff was transformed into the other stuffs and things of the world, Parmenides posits the existence of two entities with fixed natures, which, by mixing together, produce the phenomena. The Light and Night of the *Doxa* are changeless beings which produce different effects by different ratios of mixture. Parmenides' entities seem to prefigure the elements of the pluralists precisely in constituting a set of changeless beings which interact to produce multiple effects. Certainly they are a novel departure from the kinds of explanatory entities found in earlier theories.

There are some key differences between pluralist theories and the account of Parmenides' *Doxa*. In the first place, there are only two beings in Parmenides' theory, producing a dualism rather than a strict pluralism. Parmenides could argue that, since his two entities contain between them all the contraries needed to account for the diverse properties of things, only two items are needed. By a principle of simplicity, such as Ockham's razor, only two entities are needed, and any recourse to more than two is ruled out.

In the second place, it is not clear that Parmenides' two entities are to be thought of as stuffs or material realities. He calls them, respectively, Light and Night. But we may note that his terms suggest he has in mind something like light and darkness, two very unsubstantial beings. Indeed, according to Karl Popper's insightful analysis, he may have in mind precisely the interplay of light and darkness, which, while it can make great differences appear to the eye, does not itself amount to any substantial

---

[35] Cf. Curd 2002.

reality.[36] Light and darkness, moreover, echo Parmenides' allusion to the House of Night, an image from the proemium of Hesiod's *Theogony*, found in Parmenides' own proemium:

> the maiden daughters of the Sun
> hastened to escort me, leaving the House of Night
> for the light. . . .
> There stand the gates of the paths of Day and Night. (B1.8–11)

Day and Night are personified in Hesiod,[37] as also in Parmenides' mythological proemium. But of course Day and Night are precisely the sorts of beings that the natural philosophers had wanted to reject in their positivistic accounts of the world. Day, for instance, would be just the time when the sun is shining, night the time when the sun is not shining; there are no independent realities Day and Night that we need to acknowledge. Indeed, Heraclitus had made this a point of his polemic against the mythographers:

> The teacher of the multitude is Hesiod; they believe he has the greatest knowledge—who did not understand day and night: for they are one. (B57)[38]

Day and night are not real beings—not *eonta*—but the product of the actions of other things. Parmenides picks precisely these items—or something very similar—to reify as his ultimate principles. Are they then antisubstances or antielements rather than substances and elements? And is Parmenides saying that Heraclitus and his colleagues have not succeeded in going beyond Hesiod and the mythographers in their allegedly modern style of explanation?

This may be so, and Light and Night may be part of the deception embodied in Parmenides' deceptive cosmology. But on the other hand, they may be seen as offering a totally new style of explanation. For they are constant and immutable in their respective essences, and they explain phenomena not by generating them but by mixing together. We see for the first time a kind of ontological chemistry opposed to the phase states of GST. Parmenides' *Doxa* offers a radical departure from conventional explanation.

Whatever Parmenides' entities are, they are unchangeable in their natures. Being unchangeable, they will not be dynamic substances. They will be static in their own, Eleatic, essences. We need, then, some external

---

[36] Popper 1998, 72–73, argues that Light is a "no-thing" for Parmenides, whereas Night is a thing. I would suppose that both are non-things.

[37] Hesiod *Theogony* 744–54.

[38] Cf. B67: "God is day night."

force or forces to account for their continuous interactions. Parmenides had suggested just such a force:

> And in the midst of these (heavenly rings) is the deity who steers all things. For she rules over frightful childbirth and copulation of all things, sending the female to mingle with the male and again contrariwise the male to mingle with the female. (B12.3–6)[39]

The noun for copulation, *mixis,* and its cognate verb are destined to become standard terms for combination of elements. The sexual image comes to stand for the chemical connection: two elements join to produce an offspring. The goddess of Love presides, as does Eros in Hesiod.[40] But now the mythical figure stands for a cosmological force of attraction, reflected in Empedocles' Love. Before Parmenides, the Ionians saw the changes of the world either as resulting organically from the original substance (as Anaximander's boundless) or as resulting from a mechanism that was vaguely connected with the governing power of that substance (Anaximenes' air, with GST-4–5). But now the notion of power or force is distinguished from the substances which make up the world. The separation of force is surely a third major innovation of Parmenides' cosmology.[41]

In recent times we have been conditioned to read the *Doxa* in light of the *Aletheia,* Opinion in light of Truth, thus denigrating the insights of the *Doxa* as mere illusions. But what if we read the influence the other way around, the *Aletheia* in light of the *Doxa?* Then the principles of being enunciated in B8 look forward to an application in the cosmology. The principles of cosmology would be Eleatic principles adapted to the needs of cosmology. Hence we might see in the interactions of Light and Night the key to a philosophically adequate cosmology: we must work with principles that have unchangeable essences and interact so as to produce phenomena.

Because the boundaries between what is new and improved in the *Doxa* and what is wrong and deceptive are left vague, the pluralists have a great deal of room to maneuver. Parmenides tells us something is wrong, but he also shows us what would be right. Suppose then that what is wrong is the choice of elements. Light and Night cannot really qualify as entities because—what? Perhaps they are not self-sufficient, but mere congeries

---

[39] The actual structure of Parmenides' cosmos, with its *stephanai* or rings, remains highly controversial. For different interpretations, see Tannery 1930, 238–42; Patin 1899, 598–626; Morrison 1955; Tarán 1965; Pellikann-Engel 1974, 87–99; Finkelberg 1986; Bollack 1990. My argument does not depend on any particular interpretation of cosmic structure.

[40] Hesiod *Theogony* 120: Eros is the fourth being to be born, after Chaos, Earth, and Tartarus. Also mentioned in Parmenides B13.

[41] The notion of *dunamis* as power or force has not been adequately studied in its earliest philosophical occurrences. But see Souilhé 1919, 1–70; Carteron 1923.

of qualities. What is needed is something substantial, concrete—some stuff. But there must be more than one to allow the mixture of natures. Perhaps there need to be more than two, because no substance is a mere set of contraries. The early Ionians were on the right track in identifying basic substances, but wrong to see them as genetically related to one another, as being transformed into each other. Whatever basic substances there are will be unchanged in their natures forever.

Hidden in the *Doxa*—or perhaps manifest, if one is eager to find it—is a formula for creating an adequate cosmology. A plurality of self-sufficient substances must combine under the influence of a force of attraction to produce the phenomena. The integrity of the elements is preserved, yet the reliability of the phenomena is not impugned. Appearances do change, driven by the changing relationships of fundamentally changeless elements. Parmenides has provided the key to understanding the world about us. The *Doxa*, on this reading, reveals the inadequacies of human explanations. At the same time it embodies a program for correct explanations. Parmenides is a critic of cosmology to be sure. But he is not a destructive critic: he offers a constructive criticism for the sake of a viable program. He is, after all, the great reformer of cosmology. And Anaxagoras and Empedocles are his disciples.

7.4.2 Principles of Eleatic Cosmology

In fact, one may see in Parmenides' cosmology the principles for a new kind of explanation. We have already noted that the forms of the *Doxa* seem to instantiate the properties of what-is in B8 (sec. 6.4.1). In order to adequately explain the phenomena of the world, we must further posit entities that meet the following conditions:

Principles of Eleatic Cosmology
I. Explanatory entities must have permanent and unalterable natures.
II. They must be (ontologically) distinct from and independent of one another.
III. Hence, there must be a plurality of them.
IV. They must be equal.

More problematic is a final implication of Parmenides' model:

V. They must be opposite in character to each other.

All of these features are emphasized in Parmenides' *Doxa*. The element of Light is "everywhere the same as itself, not the same as the other" (B8.57–8), fulfilling the first two conditions. Mortals have "made up their minds to name two forms," fulfilling (III), and "they distinguished contraries in body and set signs apart from each other" (B8.53, 55–6), fulfilling (V).

Furthermore, "all is full at once of light and dark night, both equal, since neither has no share in it" (B9.3–4),[42] illustrating (IV) equality.

The preceding principles follow naturally from applying the Eleatic properties to the problem of cosmology. If what-is is (i) without coming to be and perishing, (ii) all alike, (iii) unmoved, and (iv) complete—and we take Parmenides to be promoting Weak Eleatic Theory—then we will have to account for the apparent changes of the world in terms of unchanging realities. These unchanging realities will not come to be or perish, they will be all alike at least in their own character, they will be unchanging in character, and they will be self-sufficient in their own right. What this will come down to is that instead of one stuff, say air, turning into many other things at will, the explanatory entities of an Eleatic theory will have to stay the same. But we can explain change only by supposing that there is a plurality of static entities, each exemplifying properties that distinguish it from the others. The changes of the world will be a product of changing relationships of unchanging beings. These beings will be equal because there is no hierarchy among them, no way to give preference to one over another. They are all equally things that are, and consequently they have an equal title to primacy.

It is possible, indeed, that the feature of equality was especially attractive to the first generation of pluralists. Anaxagoras was a guest of a leader of the democracy in Athens, and perhaps a refugee from the autocratic rule of the Great King.[43] Empedocles was himself a leader of the democratic party in his native Acragas.[44] Early in the fifth century BC the medical writer Alcmaeon of Croton had identified monarchy with sickness and equality of rule with health.[45] The connection of physical theories with political ideologies must remain elusive and speculative, but the connection between (weak) Eleatic pluralistic theories and the ideals of equality could not have hurt the prospects of those theories during the half century after the Persian War when democracy was triumphant over autocracy.

We shall see that all the principles of Eleatic cosmology reappear in the cosmologies of Anaxagoras and Empedocles in a fairly straightforward way, with the exception of (V) the opposition of the entities. Even opposition plays a role in the elements of the pluralists, however, if not the transparent dualistic role found in Parmenides' cosmology. In any case it is a small step from the contrary entities of the *Doxa* to the elements of the

---

[42] The final phrase is difficult to interpret; see options in Tarán 1965, 163–64. In any case, the important issue for my argument is the equality expressed in the former phrase.

[43] Plutarch *Pericles* 4.6, 32.1–2, 5, puts him in Pericles' circle and suggests he became a target of political attack because of his association with Pericles.

[44] Diogenes Laertius 8.64, 66, Plutarch *Against Colotes* 1126b.

[45] Alcmaeon B4; on the political and conceptual importance of *isonomia* see Vlastos 1947, 1953a.

pluralists, to which we shall look in the next section. But now let us turn to a brief consideration of one relatively positive response to Parmenides' principles.

7.4.3 Empedocles' Appropriation of Eleatic Principles

In the case of Anaxagoras, we do not have enough material to say how specifically he saw his relationship to Eleatic principles. But in the case of Empedocles, we have enough to at least get a glimpse of his response. We have already seen (sec. 7.2) that Empedocles several times endorses the first Eleatic property (from Parmenides B8) that there is no coming to be or perishing. But what of the other three Eleatic properties?

The third property is that what-is is immovable, *akinêton*. Empedocles seems to provide a complex response to this property. Describing the cycle in which the many combine into the one and then divide apart again, he says:

> Thus, inasmuch as they are wont to become one from many,
> and in turn with the one growing apart they produce many,
> they are born and they do not enjoy a steadfast life;
> but inasmuch as they never cease continually alternating,
> they are ever immobile [*akinêtoi*] in the cycle. (B17.9–13 = B26.8–12)

The elements alternate between joining together and being submerged in the Sphere, and separating to exist in their own right in the cosmos. Thus in one sense they are born and die and do not enjoy a permanent ("steadfast") existence; yet in another sense they are part of the recurring pattern of things and in this sense they are *akinêtoi*. They are, for all their changeable states, permanent fixtures of the cosmic cycle, and indeed permanent and ultimate ingredients of the world:

> And besides these nothing comes to be or ceases to be,
> for if they perished thoroughly, they would no longer be.
> What could increase this totality, and whence would it come?
> Into what would it perish, since nothing is void of these things?
> But these are the very things that are, which running through each other
> come to be now this, now that, yet always continually alike
> [*homoia*]. (B17.30–35)

In a deeper sense, the elements do not come to be or perish; they are the building blocks of the world. "These are the very things that are," which constitute the world. They enter into different combinations, but they are continually alike, *homoia*.

The last term brings us to the second Eleatic property, likeness or homogeneity. Empedocles seems to mean that each of the four elements is through and through like itself—it is all of one character or nature. Each element is like one of Parmenides' quasi-elements from the *Doxa*, "everywhere the same as itself" (B8.57). Empedocles' elements "run through each other" so that they may disappear from view or lose their phenomenal properties, but they seem to keep their inner nature, their essence, as though strands in a rope or separate roots in a root system. In this sense, too, they are always present and could be called *akinêtoi*.

Overall, we should not overlook the fact that Empedocles is willing to find one sense in which his elements are born and die and become different, and another sense in which they are permanent, immobile, and all alike. If, as is often noted, we do not find a sophisticated semantic program in Empedocles to deal with the problems of Eleatic being—as we do in Plato's *Sophist* and Aristotle's *Physics* I—we do nevertheless find a semantic strategy. Empedocles is not afraid to disambiguate notions such as immobility and likeness, and to find a sense in which his dynamic cosmology can instantiate the Eleatic properties. He seems ready to take a cue from Parmenides' own cosmology as to how to apply Eleatic principles. And again, so far as we see, Empedocles does not seem to think of the Eleatic properties as obstacles to cosmology, but only constraints to be satisfied.

The one Eleatic property that Empedocles does not seem to address in the extant fragments is that of completeness. Completeness is at least depicted in the image of the well-rounded ball of Parmenides B8.43–44. Empedocles produces a state of the cosmic cycle in which all the elements are perfectly blended in a homogeneous sphere:

> but equal <to itself> in every direction and completely boundless,
> a rounded Sphere rejoicing in circular solitude. (B28)

Thus in a certain sense Empedocles builds the perfection of Parmenides' what-is into a stage of the cycle itself. Yet the strict Eleatic would complain that the Sphere is merely a temporary state of the universe, not the whole of it. We do not know how Empedocles would respond. But he could well point out that Parmenides' what-is is *tetelesmenon* (B8.42), literally "completed" or complete as the result of an action of completion. Empedocles' own Sphere certainly can qualify as the completion of a process. Moreover, there is a kind of completeness in the closed cycle Empedocles envisages, which is invariable in its order and draws on fixed elements.

One dimension of Parmenides' argument that Empedocles does not seem to address even obliquely is the problem of time. Parmenides bases his rejection of coming to be in part on the impossibility of temporal variation (B8.5, 19–20), which for him presupposes what-is-not. What precisely Parmenides means by his rejection of time differences is a complex issue.[46] Yet it is at least possible that Empedocles sees repeated occurrences of events "as time rolls round" (B17.29) as fulfilling timeless relationships and expressing timeless truths. His willingness to recognize multiple senses of words, with the suggestion that there are multiple contexts in which they may or may not apply, should give him the means to divert Parmenides' seemingly unanswerable attack on temporal duration.

In general, Empedocles seems to be adept at thinking about Eleatic properties and constraints in creative ways that allow him to operate within an Eleatic framework without having to face the full weight of the Eleatic challenge. Again, there is no sign that Empedocles sees himself as having to reinvent cosmology to rescue it from attack; rather he seems to see himself as applying correct principles to give an adequate account of cosmology.

7.5 Two Theories of Elements

No pluralist followed precisely in Parmenides' footsteps in developing an ontology. That is, no one posited just two entities manifesting contrary qualities. This may result from the pluralists' understanding of what was wrong with the admittedly inadequate cosmology of the *Doxa*. But Parmenides' own diagnosis of the failure is opaque, and we cannot recover the precise reasons for which the pluralists insisted on a larger number of explanatory entities.[47] We may speculate that they saw Parmenides as offering the principles of a viable theory, but condemning the particular version. The very fact that the two entities Light and Night were so thoroughly opposed to each other might be the reason for their inadequacy: they are not independent realities, but mere dependencies, unspecifiable in their own terms.[48] Certainly they lacked the concrete character of stuffs advanced by earlier Ionian theories. In any case, the first generation of pluralists, Anaxagoras and Empedocles, would opt for a larger number of entities with more complex natures. Their elements would be the entities of the *Doxa* infused with a robust sense of realism.

One further preliminary is the question of historical priority. The ancient sources are agreed that Anaxagoras was slightly older than Empedocles. But about their time of composition Aristotle is ambiguous, and some scholars have argued that Anaxagoras wrote first, while others have

---

[46] See Owen 1966; Mourelatos 1970, 103–10; it is a central theme of Collobert 1993.

[47] Parmenides B8.53–54; above, ch. 6, sec. 4.2.

[48] Mourelatos 1973, esp. 40ff.; Curd 1998, 104–10, at 109: "Light and Night, with their intertwined natures, are not independent entities; rather . . . the nature of each is to not be the other."

argued for the contrary.[49] Even the importance of the question is controversial: philosophers tend to want to leave contingent questions aside, while classicists and historians insist that the contingent questions are the decisive ones. The question cannot be taken as settled at this point, so any decision made here will have to remain provisional. I make the following assumptions. First, the conditions of open intellectual debate seem to have allowed for the two thinkers to be aware of each other's work even before final publication, whatever that might have been. This would not have assured their consensus about important issues, even if they admired each other, for instance as Galileo and Kepler did. Anaxagoras was clearly active in Athens, for a period of thirty years, we are told.[50] The question of priority revolves around whether those years were closer to 480–450 BC, or 460–430. While there are arguments to be made both ways,[51] it seems to me that the meteorite that fell at Aegospotami in 467 shows that Anaxagoras' theories were well-known at that date;[52] the fact that his theory of the Nile floods was known to Aeschylus shows that his work was well-known in Athens in the 360s;[53] and the fact that Socrates is not associated with him as a student, colleague, or critic[54] shows that he was long gone from Athens well before 430. Hence I lean toward giving historical priority to Anaxagoras, without insisting on it. I turn first to Anaxagoras as a theorist.

7.5.1 Anaxagoras: Elements without Emergence

Above I gave a preliminary account of Anaxagoras's theory of matter. Much remains controversial about the theory. Most fundamentally, there is a disagreement over what the basic realities are and how they interact. A now venerable account, first advanced by Paul Tannery[55] and still pop-

---

[49] Aristotle *Metaphysics* 984a11–13 with Ross 1924, 1.132.

[50] Diogenes Laertius 2.6.

[51] For the earlier activity in Athens, see Taylor 1917; O'Brien 1968; Woodbury 1981; for the later (traditional) view, Mansfeld 1979–1980. See also Davison 1953, 39–45; Cappelletti 1979, 12–13, who advocate two stays in Athens. I support his early residence in Athens.

[52] *Marmor Parium* and Pliny *Natural History* 2.149 = A11; Plutarch *Lysander* 12 = A12. Guthrie 1962–1981, 2.302, assumes Anaxagoras developed his theory in response to the event; but then it is difficult to see how he could be thought of as predicting it.

[53] Aeschylus died in 456/5 and produced his last plays in 458. Seneca *Natural Questions* 4a.2.17 = DK 59A91; Aeschylus fr. 300 Nauck, *Suppliants* 497, 561. Furthermore, Herodotus (2.20–22) knows his theory of the Nile floods along with those of Thales and Hecataeus, but no later theories, e.g., of Diogenes of Apollonia or Democritus (see Graham 2003c, 294, n. 12).

[54] Taylor 1917, 83–85; Woodbury 1981, 296–98; KRS 354.

[55] Tannery 1887, 280ff., followed by Burnet 1892, 288–90, who later clarified the qualities as opposites: 1930, 263–64.

ular,[56] has it that there are two categorically distinct sets of realities, roughly qualities and stuffs, or contraries and "homoeomeries," to use Aristotle's term for the latter.[57] Qualities such as hot and cold, wet and dry, are the constituents of stuffs such as earth and flesh. Thus by combining qualities in different ratios, one can generate all the stuffs. On this theory Anaxagoras achieves a significant theoretical simplification, producing an infinite number of stuffs from a limited number of qualities. Furthermore, he avoids the redundancies of having qualities and stuffs play equivalent roles.

The textual evidence for this, as well as competing interpretations, is unfortunately very thin. The most important statement is found in B4b:[58]

> Before these things were separated, when they were all together, not so much as a color was manifest. For the mixture of all objects prevented it, [i] of the wet and the dry, of the hot and the cold, and of the bright and dark, [ii] and much earth was in it and [iii] seeds countless in number which were not at all like one another. For no one of them appeared like to any other. Since these things were so, one must grant that all objects are present in the totality.

One can argue that Anaxagoras does distinguish here between (i) qualities, (ii) stuffs, and (iii) seeds. But instead of saying that, e.g., earth is a product of the qualities, he seems to treat earth as also an ingredient of the mixture, not as something of a logically or ontologically different type. Certainly in B1 and B2 air and aether seem to be treated as stuffs that are primary constituents of the primordial mixture.

What then of the seeds? Here there is a wide variety of interpretations of these entities that are mentioned only here and in B4a. Are they particles of stuffs, or portions, or structural features, or actual biological seeds?[59] The other passage that mentions them tells a few facts about them:

---

[56] Advocates include Giussani 1896; Cornford 1930; Vlastos 1950; Mugler 1956; and recently Schofield 1980, 107–21; Inwood 1986, 25–30 and n. 29; Spanu 1987–1988.

[57] Bailey 1928, 551–56, argued that the term must have been found in Anaxagoras. But see, to the contrary, Mathewson 1958, 78; Guthrie 1962–1981, 2.325–26. Aristotle *On Generation and Corruption* 314a18–20 explains that homoeomeries are things whose parts are synonymous with the whole (e.g., flesh is composed of flesh, bone of bone), and attributes this view to Anaxagoras. But Aristotle's homoeomeries are also nonelemental, and it is not clear whether Aristotle is saying that Anaxagoras's stuffs are what *Anaxagoras* maintains are homoeomerous, or whether he is saying that Anaxagoras's stuffs are what *Aristotle* would call homoeomeries. I believe only the latter interpretation is justified.

[58] With Jöhrens 1939 and recent editors such as Sider 1981, I distinguish the two paragraphs of B4 as parts (a) and (b), since we cannot be sure the second followed immediately on the first (see Simplicius *Physics* 34.21, 156.4).

[59] There are a great many divergent interpretations of the seeds, which appear only in B4a and B4b. Are they particles of stuff (Peck 1926; Stokes 1965), substances in contrast

Since these things are so, one must believe that there are many things of all kinds in all the composites, and [or: namely] seeds of all things having all sorts of figures, colors, and tastes; and that men too are compounded, and every other kind of creature that has soul. (B4a)

Certainly we would like to know more. But from what little Anaxagoras says, it appears unlikely that he has biological seeds in mind. For biological seeds can at least look all alike rather than being all different from each other (contra B4b), and their shapes, colors, and tastes are quite irrelevant to their character as seeds and quite unhelpful in determining what kinds of seeds they are (contra B4a). Furthermore, the following remarks about men and animals seem peculiarly inept following a discussion of seeds, insomuch as Anaxagoras seems to stress their composite character rather than their natural development from their biological starting points. In addition, if Anaxagoras has portions of things in mind, why does he speak of "figures" (ideai)? Portions are determined quantitatively rather than by their shape and appearance, and they may be dispersed in a mixture. It seems then that Anaxagoras has something more concrete in mind—either particles or structures or both.

To return to Anaxagoras's overall discussion of qualities, stuffs, and seeds: while he speaks of them in different terms, it is not clear that he treats them as categorially different. Rather he seems to conflate them at a physical level. In B8 he says:

Things in the one world-order are not isolated from each other, nor are they cut off by an axe—neither the hot from the cold nor the cold from the hot.

In this remarkable echo of Parmenides B4.2, Anaxagoras declares that the hot and the cold cannot be separated. The reference to an axe, whether it seems playful or serious, suggests a deep category mistake on his part: separation is not a physical but a logical or conceptual impossibility. But if he thinks of the hot and the cold as physically inseparable, he must think of them as physical realities that mix and separate (in a limited way) like stuffs.[60] The mixture of B4b is not a mixture of qualities to produce stuffs, but a mixture of qualities, stuffs, and (perhaps) trace particles of stuffs.

What then is the point of Anaxagoras's physical theory? It is not to reduce the number of phenomena or apparent realities to a manageable

---

to opposites (Strang 1963), qualitatively differentiated portions (Schwabe 1975), structural elements (Teodorsson 1982; Mourelatos 1987, 155–58), or biological seeds (Furley 1976; Schofield 1980; Lewis 2000)?

[60] Even *nous*, which has properties to distinguish it from material stuffs proper, e.g., the fact that it does not mix with everything, has location and is found in some things: B11, B12, B14.

number; it is not a reductionistic theory.[61] Rather it is to preserve the point that Anaxagoras reiterates endlessly: everything is in everything. In the beginning everything was in everything, so completely that no differences were observable. Then Mind set the world in motion, so that things began to separate out. But they never separate out completely because things are endlessly divisible and nothing can be completely separated from anything else. Thus even now everything is in everything—even though this does not *appear* to be true. Anaxagoras's aim is not to simplify the phenomenal world, but to complicate the theoretical world. Everything we think we perceive in the world—earth, fire, flesh, wood, hot, cold, red, green—is really there. Even when we do not perceive the full range of appearances in a location, their seeds are there. If we define wood by a range of qualities it has, we will not be able to account for hot and cold wood, wet and dry wood; we must even allow hotter and cooler fires.[62]

Anaxagoras's way of saving the appearances is to say they are always present. If water turns into steam or something hot becomes cold, the steam was already in the water, the cold in the hot. There is nothing new anywhere, no emergent or supervenient property. There are only manifest realities and latent realities, determined by their relative concentrations in a local mixture. By blocking emergent entities Anaxagoras is blocking coming to be, and hence adhering to an austere Eleatic physics (if in a prodigal ontology). He must allow some sort of change in concentration (about which he seems to say little), presupposing change in location. But there is only one emergent property in his theory: manifestness. And that property seems to be paradigmatically phenomenal.

What of the charge that on this reading Anaxagoras has an extravagant ontology in which virtually everything we perceive is real? He may say simply that each one has an equal claim to being. Since ultimately they are all beings, they do not violate the Eleatic strictures. What we must avoid is allowing not-being into the ontology; to multiply beings is no crime.

But how can we define and understand things if we cannot specify their components? This is a problem only to a philosopher with a shopkeeper mentality who thinks that everything can be defined, packaged, and labeled. We can identify everything that is ostensively. This here (pointing) is white, that there is black; this here is wood, that there is metal, and so on. We can give distinguishing marks, as Locke does for gold, but we will be giving only nominal essences. We learn by an experience that is much more powerful than words and definitions what natural kinds things fall into. "Appearances are a vision of the invisible" (B21a). On Anaxagoras's

---

[61] See Potts 1984; Graham 2004a.
[62] Cf. variations of heat in Diogenes B5.

theory the microstructure of the world is simply a reiteration on a smaller scale of the plurality of things we encounter in the sensible world.[63] (I shall have something more to say in defense of Anaxagoras's approach in the next chapter.)

A final problem: how can Anaxagoras reiterate stuffs at every level? If, to take a simple model composed of elements A, B, and C, I perceive A, it will supposedly have B and C in it. But now B will have to have A and C in it, and A will have to have B and C in it; we have an infinite regress. Yes, he can say, we do. But it is not a vicious regress. Everything is infinitely divisible. What we have is a pattern identified in mathematics as fractal geometry: a pattern that repeats itself endlessly at every scale. It is bizarre and mind-boggling, but it is not contradictory or impossible, for it is a fact that there are fractal structures in mathematics, some of which are at least partially instantiated in nature.[64]

Even in differences of scale, nothing new emerges. Everything is in everything. The world is full of being.

When Anaxagoras applies his physics of matter to cosmogony and large-scale cosmology, he combines traditional features with innovations. His world begins with a cosmic rotation, as had Anaximander's. But Anaxagoras supplies a push from an external agent: *nous*, the cosmic mind (B12). Now that matter has been divested of intelligence, Anaxagoras makes *nous* a being in its own right, one that does not mix with everything as do stuffs:

> Everything else has a portion of everything, but mind is boundless, autonomous, and mixed with no object, but it is alone all by itself. If it were not by itself, but were mixed with something else, it would have a portion of all objects, if it were mixed with any; for there is a portion of everything in everything, as I said earlier. And the things mixed with it would hinder it from ruling any object in the way it does when it is alone by itself. For it is the finest of all objects and the purest, and it exercises complete oversight over everything and prevails above all. (B12, beginning)

Here, as is often noticed, *nous* acquires special properties that distinguish it from ordinary stuffs. But it is still treated as a physical entity with extension. Incidentally, we find at least one limitation on universal mixture: "everything" does not include *nous*.

---

[63] It can be objected that Democritus, who did not think the macroscopic world was a mere reiteration of the microscopic, approved B21a (Sextus Empiricus 7.140). But the sentence must mean something quite different for Democritus, for his atoms do not have phenomenal properties such as color and taste. He must take it to mean that we infer the properties of the unperceived world from the perceived.

[64] See detailed treatment in Graham 1994, 101ff.

The cosmic revolution creates a centrifuge which separates things out:

> And of the whole revolution did mind take control, so that it revolved in the beginning. And first it began to revolve from a small revolution, now it is revolving more, and it will revolve still more. . . . And the things that were to be—what things were but now are not, what things now are, and what things will be—all these did mind set in order, as well as this revolution, with which the stars, the sun, the moon, the air, and the aether which were being separated now revolve. For this revolution made things separate. And from the rare is separated the dense, from the cold the hot, from the dark the bright, and from the wet the dry. (B12, continued)

> And when mind began to cause motion, as a result of everything being in motion there was a separation, and as far as mind caused motion everything was segregated. And with everything being in motion and being segregated the revolution caused an even greater segregation. (B13)

Although mind is said to comprehend all things, it seems to act by means of the revolution, which itself "made things separate" and seems to have a cumulative effect, as B13 indicates. It is unique to Anaxagoras to posit an ever-increasing vortex.

The cosmos arises from the effects of the vortex:

> the dense, <the> wet, <the> cold, and the dark came together here, where earth[65] now is, while the rare, the hot, the dry, <and the bright> retreated to the further parts of the aether. (B15)

Here clusters of contrary qualities much like those found in Parmenides' cosmology separate into those at the center and those at the periphery. But Anaxagoras's contraries do not preclude the presence of stuffs—indeed they may be just an alternative description of them:

> From these things being separated the earth is compacted. For from clouds water is separated, from water earth, and from earth stones are compacted by the cold. These stones move out more than water. (B16)

As we noted in chapter 3, this account is essentially that of Anaximenes, down to a mechanism of compression. This mechanism cannot be strictly appropriate for Anaxagoras, for whom (as far as we can see) qualities (including quantities) and stuffs are independent. Strictly speaking, a portion of, for instance, the water that was in cloud separates out, a portion of the earth in the water, and so on, but water is not compressed cloud. The term "separated" in the first two sentences of B16 is correct, while the term "compacted" describes only an appearance. But Anaxagoras can

---

[65] Reading γῆ with manuscripts (Diels's <ἡ γῆ> is unnecessary: see Sider 1981, ad loc.).

claim that cold or pressure causes separation, *as if* the mixtures were actually compressed. Accordingly, he preserves the steps of a sixth-century cosmogony with an Eleatic physics. The underlying mechanisms are redescribed, but the apparent behavior of matter remains the same.

Thus Anaxagoras can preserve much of earlier cosmology and cosmogony while making a radical change in the physics or chemistry of matter. The appearances are preserved while the basic theory is altered to eliminate emergent properties and substances.

7.5.2 Empedocles: Elements with Emergence

Empedocles gives a summary of his principles early in the argument:[66]

> I shall speak a double tale: at one time they grew to be one alone
> from many, and at another time it grew apart to be many from one
> Fire, Water, Earth, and the lofty expanse of Air,
> destructive Strife apart from them, like in every direction,
> and Love among them, equal in height and width.   20
> See her with your mind; do sit with eyes dazzled by
> her who is thought to be innate in mortal limbs.
> By her they think kindly thoughts and accomplish harmonious acts,
> calling her by the names of Joy and Aphrodite,
> whom no mortal man has perceived twisting   25
> among them. But do you hearken to the undeceptive train of my speech.
> For these things are all equal and of the same age
> Each rules its own domain and has its own character,
> and they rule in succession as time rolls round.
> And besides these nothing comes to be or ceases to be,   30
> for if they perished thoroughly, they would no longer be.
> What could increase this totality, and whence would it come?
> Into what would it perish, since nothing is void of these things?
> But these are the very things that are, which running through each other
> come to be now this, now that, yet always continually alike.
> (B17.16–35)   35

The four elements plus the two powers are "the very things that are" (line 34). Everything else is perishable and dependent, but the elements and powers are eternal entities responsible for the existence and makeup of the world. The are "equal and of the same age" (line 27) and, like citizens in a participatory democracy, each one rules in its turn. Like a good Elea-

---

[66] The Strasbourg Papyrus, by marking line 300 of the manuscript, shows that the first line of B17 is line 233, presumably much of the early lines being taken up by a proemium or prologue; see Martin and Primavesi 1999, 160.

tic element, each "has its own character" distinct from the rest (28). These beings cannot perish or increase in number (30–33). According to the principles of Eleatic cosmology, the elements (1) have their own unalterable natures, (2) which are distinct from each other, (3) there is a plurality of them, and (4) they are equal.

By combining the elements under the influence of Love, the many compounds of the world arise:

> But come, gaze on this witness of previous words,
> if anything in them was lacking in form:
> sun, shining to sight and everywhere hot,
> and immortal things which are soaked in heat and blazing beam,
> and rain, dark and cold in everything,   5
> and from earth flow out thick and solid things.
> In rancor they are all distinct and separate,
> but they come together in love and are attractive to each other.
> From them all things that were, all that are, and all that will be later
> are sprung—trees, men, women,   10
> beasts, fowls, water-nourished fish,
> and long-lived gods foremost in honors.
> For these are the very things that are, which running through each other
> become different; for blending alters them. (B21)

Here Empedocles invokes the sun, perhaps the heavenly bodies,[67] rain, and the earth as concrete exemplars of fire, air, water, and earth. As long as Strife dominates ("in rancor") they remain apart; but when love influences them, the elements come together to make animals, plants, and even the "long-lived"—but not immortal—gods (for only the elements and powers are immortal). The moral of the story, in line 13, repeats almost verbatim the statement at B17.34 of the theme that only the elements and powers are everlasting and unchanging in their own natures.

But there is a step between the elements and the biological specimens that populate the world: the compound stuffs that make up the bodies of the specimens. Empedocles holds that these are directly composed of the four elements, or some subset of them:

> Pleasant earth in well-wrought funnels
> got two of its eight parts of glittering Nestis,
> and four of Hephaestus; and white bones were produced,
> joined by the marvelous glue of harmony. (B96)

> Earth met with these in most equal measure,
> with Hephaestus, rain, and blazing aether,

---

[67] See Wright 1981, ad loc.

> dropping anchor in the perfect harbors of Cypris,
> either a little greater or less among more parts,
> and from them came blood and other kinds of flesh. (B98)

Here we are given chemical formulas: bone is two parts earth, two parts water, and four parts fire, or $2\ E + 2\ W + 4\ F \rightarrow 1$ bone, or 1 bone = $E_2W_2F_4$. We see an example in which whole-number ratios of elements combine to make compounds with their own properties, just as in modern chemical theory. We also see that not all four elements need be present, contrary to what is the case in Anaxagoras. In the case of blood, the ratios are 1:1:1:1, with all elements present. As for the kinds of bonds that exist between the elements, that is not clear. In B96 the image of funnels suggests crucibles of a foundry, or perhaps a mixing bowl; yet Empedocles says the elements are joined by glue. In B98 we get the mixed metaphors of meeting and anchoring in a harbor. Different passages suggest glue, nails or rivets, mixing, mingling, blending, cooking, molding, and growing together.[68] Ancient secondary sources speak of microscopic parts which join together by juxtaposition.[69] But the evidence is derivative and may reflect an attempt to generalize from a few examples.[70] Hence it remains unclear whether Empedocles advocates a particulate microstructure, or whether he might allow a kind of fluid mixture among elements.[71]

Besides organisms and elements, Empedocles notoriously treats the parts of animals and plants as capable of at least temporary existence by themselves, during one stage of the cosmic cycle.[72] Thus we can distinguish a stage of composition between the compounds and complete organisms. Indeed, Empedocles seems to allow for all the levels of organization that Aristotle later recognizes: elements, homoeomerous or homogeneous parts, an-homoeomerous parts (organs and limbs), and organisms.[73]

For our purposes the important fact is that Empedocles recognizes at least two levels of beings, the primary realities: earth, water, air, and fire, influenced by the contrary powers of Love and Strife, and a large (perhaps infinite?) set of stuffs that are derived from the elements: the compounds.

---

[68] B8, B9, B17.34–5, B21.14, B22, B33, B34, B35, B56, B71, B73, B75, B86, B87, B90, B95, B96, B98.

[69] Aristotle, Aëtius, and Galen in A43.

[70] E.g., B33 with the rivet metaphor.

[71] Wright 1981, 34–40, and Mourelatos 1987, 166–78, see Empedocles as advocating a mechanical account of composition in which small parts are physically connected. Hence, he is often thought of as a precursor of atomic theorists: Kranz 1912, 1954. Gemelli Marciano 1991, in a sweeping survey of the evidence, argues against a mechanical interpretation of mixing. I tend to agree, and to be suspicious of a theory based on one set of images in Empedocles' bag of mixed metaphors.

[72] B57–61, Aëtius 5.19.5 = A72.

[73] Aristotle *Parts of Animals* II.1.

The elements are everlasting or "immortal," the compounds temporary or "mortal." Hence there is a two-level ontology cleverly mapped to the two-level social structure of mythology, contrasting immortal gods and mortal creatures. In B23 (cited above, sec. 7.3.1) Empedocles compares the creations of nature to the paintings of artists, who mix colors to produce the manifold forms of living things. Since on some Greek accounts there are four primary colors,[74] the simile is especially apt. If we take the simile seriously we may infer also that the images are not really real, but only temporary illusions. They are only colors on the canvas.

What precisely is the status of compounds? It is difficult to give a straightforward answer in contemporary terms, partly because of the poetic diction Empedocles uses, which is colorful but often figurative and hence imprecise; and partly because the whole question of complex ontological relationships was new and would in any case be difficult to explain. Although Empedocles often stresses the marvelous character of the changes wrought by the action of Love on the elements, he always gives an explanation for the phenomena in terms of the elements and powers.[75] The compounds are mortal, that is, temporally limited, and clearly derivative and hence dependent upon the immortal components and powers. We seem to be justified in saying that they are *nothing but* the elements in certain configurations, and that those configurations occur only by the intervention of Love. The compounds are, then, products of the elements and reducible to them. Their properties and behaviors are deducible from the properties of their component elements in the ratios and arrangements they enter into. Empedocles never goes so far as to say the compounds do not exist; but he does seem to hold that their existence is merely contingent on the proper conjunction of elements. In modern terms, we might think of Empedocles' chemical theory as a kind of identity theory: a compound is identical to a certain configuration of elements. It is a real, if temporary and contingent, existent, capable of entering into complex relationships (for instance in animal bodies). But it is not an equal and independent entity with the four elements and two powers, for it can be accounted for without remainder in terms of relations between the elements. Its mortality reflects its ontological dependency.

Empedocles allows for emergence because his compounds are different from his elements (in a way Anaxagoras's mixtures are not), and their properties are not the properties of the elements (again in contrast to Anaxagoras). In Anaxagoras, what you see is what you get; in Emped-

---

[74] Including Empedocles himself, who recognizes white, black, red, and yellow: Aètius 1.15.3 = A92.

[75] B23, B35.16–17, B96, with Mourelatos 1987, 138–43.

ocles it is not.[76] In Empedocles a piece of wood is not (mostly) wood, but some ratio of earth, water, air, and fire; its weight, color, texture, capacities, etc., are not those of any element, nor a simple sum of all of them. Wood is a unique substance with a unique set of properties determined by subtle interactions among the elements configured in a determinate way. Wood is a new phenomenon supervening on a certain conjunction of elements. It is identical to and inseparable from the conjunction of elements. In general, there is a set of phenomenal realities derived from the elementary realities; the elements are finite in number and everlasting, while the phenomenal entities are infinite in number (perhaps finite, but much more numerous than the elements, in type) and ephemeral.

In his cosmogony, Empedocles reacts to the problem of origination very differently from Anaxagoras. Whereas Anaxagoras had posited a cosmic mind to begin the separation process, Empedocles has two contrary forces, Love and Strife. According to him, at a point at which Strife has separated the four elements completely, apparently into concentric rings, Love, now confined to the center of the cosmos, begins to expand, forcing Strife to retreat, and causing the elements to mix and compound with one another:

> When Strife reached the innermost depth
> of the vortex, and Love comes to be in the middle of the circle,
> there all these things combine to be one thing alone,
> not suddenly, but joining together willingly each from its own place.
> When these things are mingled the myriad races of mortal things flow out,
> and many things stayed unmixed alternating with things that were
>   being blended,
> which Strife held still suspended; for not perfectly
> yet had he withdrawn to the uttermost bounds of the circle,
> but some of his limbs remained within while some had withdrawn.
> As far as he would run ahead, so far would advance
> the gentle immortal onset of blameless Love.
> And suddenly those things grew mortal which before were wont to
>   be immortal,
> and what was before unblended became mixed, exchanging paths.
> When these things are mingled the myriad races of mortal things flowed out,
>   fitted with all sorts of shapes, a marvel to behold. (B35)

More and more compounds "flow out" until presumably the full range of stuffs and creatures we are familiar with are in existence. But Love continues to combine things until the middle-sized objects are absorbed and finally everything coalesces into a single compound, the Sphere:

[76] Mourelatos 1987 treats the question of emergence in the pluralists insightfully. Anaxagoras still has a gap between appearence and reality, but a much smaller one than Empedocles.

then were not descried the swift limbs of the sun

. . .

thus it was set in place with the tight covering of harmony
a circular Sphere rejoicing in joyful solitude. (B27)

At this point Love occupies the sphere, with Strife apparently pushed out to the periphery of the universe. But the harmonious state does not last:

But when great Strife was nourished in his limbs,
he leapt up to lay hold of his office as time rolled round,
which had been fixed for them by a broad oath of mutual succession. (B30)

The Sphere is broken up:

All the limbs of the god shuddered in turn. (B31)

The elements begin to separate out into their own proper places. As they separate, they produce living things, which become increasingly articulated. Thus there are two zoogonies, or creations of living things, one while Love is increasing toward the Sphere, and one when Strife is increasing toward its complete rule. And as Strife expands toward the center of the cosmos, Love retreats to the very center, and the compounds formed when Strife was beginning its reign dissolve into their components in separate strata.[77]

The cycle is endless. Hence there is no beginning point, only a ceaseless alternation between Love and Strife, combination and dissolution, with mortal compounds, including living things, being produced in parts of the cycle. A cosmic Sphere reminiscent of Parmenides' sphere exists at one point of the cycle. Thus Empedocles has an alternation between a complete union and a cosmos.

7.6 Empirical Advances

Thus far we have looked almost exclusively at theoretical issues in constructing a pluralistic cosmology. But empirical evidence has a role to play too. Anaxagoras is associated with several important observations that after him became part of the Ionian tradition. These observations deserve more space than we can give them here, but at least we can acknowledge them (see sec. 6.5 above on Parmenides' contribution to empirical research).

---

[77] On the relative positions of Love and Strife at the time when Love begins to expand, see ch. 8, n. 15.

7.6.1 Eclipses

According to Hippolytus, "The moon is eclipsed when the earth blocks it, or variously one of the bodies below the moon; the sun is eclipsed when the moon blocks it at the time of the new moon" (*Refutation* 1.8.9 = A42). This rather modest notice marks the first correct account of eclipses—in the history of the world.[78] The insight that makes it possible is the recognition that the moon is illuminated by the sun: "The moon does not have its own light, but gets it from the sun" (ibid., 8), or in Anaxagoras's own words, "The sun imparts to the moon its brightness" (B18). But this insight comes from Parmenides and is echoed in Empedocles.[79] Now to recognize that the moon gets its light from the sun does not immediately entail an explanation of eclipses, nor does it ensure that the person who accepts the account of the moon's light will grasp the cause of eclipses. It is a contingent fact that the moon is the right size and distance to block the sun's light over a small portion of the earth, when its position lines up exactly with the earth and sun. And similarly it is a contingent fact that the earth is the right size and distance to block the sun's light to the moon when the three heavenly bodies line up. But once one sees that the moon gets its light from the sun and that the moon is a solid spherical body, one is in a position to propose the correct hypothesis about how eclipses occur. It appears that Anaxagoras was the first person to correctly account for eclipses, and he did it at least in part by applying Parmenides' insight about the moon's light. Anaxagoras's theory rightly came to be seen as a paradigm case of a scientific discovery.[80] It confirmed Parmenides' insight and rendered obsolete almost all previous theories of the heavens.

7.6.2 The Size of the Sun and Moon

Several reports attribute to Anaxagoras the curious view that the sun is larger than the Peloponnesus.[81] It has plausibly been suggested that this claim has an empirical basis: an eclipse visible in the Peloponnesus (there was one in 478 and another in 463) led him to make the inference. Assuming that the moon was blocking the sun's light throughout the Peloponnesus, the moon must be at least that large and the sun larger.[82] Here Anaxagoras is using his theory to infer other facts about the heavenly

[78] See Graham 2002b.
[79] Parmenides B14–15, Empedocles B45, 47, 42, 43.
[80] Aristotle *Posterior Analytics* 89b25–31, 90a7–14, 93a30–b6.
[81] Hippolytus *Refutation* 1.8.8; Aetius 2.21.3 = A72; Diogenes Laertius 1.8 = A1.
[82] Thus West 1971, 233, n. 1; Sider 1973.

bodies, and correctly so (although he is not in a position to consider the optics of shadows).

7.6.3 The Meteorite of Aegospotami

One of the most famous stories about Anaxagoras is that he predicted a stone that fell from the sky near Aegospotami, around 467; the stone was about the size of a wagon and became a tourist attraction for many years after.[83] Of course it is not possible to predict meteorites (although meteor showers are now predictable). Whatever the exact sequence of events, the meteorite seems to have been connected with Anaxagoras's name almost immediately. A few short years earlier the event, one supposes, would have been considered a divine omen. That it was not seems to indicate that an alternative explanation was handy, namely that of Anaxagoras (earlier theories had typically not posited heavy objects in the heavens).[84] Subsequent theories of the heavens make use of stones revolving about the earth.[85] This perhaps lucky coincidence provided data for scientific theories and put constraints on the theories themselves.

Clearly Anaxagoras was interested not only in abstract theory but also in empirical evidence which revealed the nature and size of the heavenly bodies. Scientific hypotheses were starting to yield results that could accumulate and become part of a progressive explanation of natural phenomena.

Conclusion

Anaxagoras and Empedocles developed cosmologies in the Ionian tradition. They began with basic principles, showed how the phenomena could in principle be derived from the principles, then elaborated a cosmogony in which one stage led to another. They explained the phenomena of the world, astronomical, meteorological, biological, and anthropological, and used empirical data when they could identify it. The superstructures of their theories were much like those of pre-Parmenidean Ionians. But the foundations were radically different: they worked with a set of ele-

---

[83] Pliny *Natural History* 2.149–50 = A11; Plutarch *Lysander* 12.1–2; see West 1960.

[84] The only earlier theory that allegedly has heavy bodies above is Anaximenes', and in that case the report may confuse his views with Anaxagoras's: Aetius 2.13.10, Hippolytus *Refutation* 1.7.5. Bicknell 1969, 68–69, thinks Anaximenes' heavenly bodies have solid centers; but this clashes with other accounts, including Hippolytus's, by which fiery bodies float on air. See criticisms at KRS 156.

[85] E.g., Diogenes' theory: Aetius 2.13.5, 9 = A12; Leucippus: Diogenes Laertius 9.32 = 67A1.

ments with fixed natures, that by combining and recombining can produce the manifold appearances of the world.

According to the Standard Interpretation, Anaxagoras and Empedocles are reacting negatively to Eleatic theory: they are at most anti-Eleatic pluralists. But if the present analysis is correct, we may more accurately understand the first generation of cosmologists after Parmenides as they saw themselves: as *Eleatic pluralists*.[86] They had no need to defend cosmology against Parmenides, for he had not, so far as they were concerned, overthrown the project. They had only to carry out the project he had adumbrated. And thus we see why there are no defensive moves against Parmenides in Anaxagoras and Empedocles. They were his disciples, not his opponents, building on his foundations and elaborating his program.

---

[86] See Wardy 1988, who, however, applies this term to the atomists as well as Anaxagoras and Empedocles; I see the atomists as less Eleatic, as will become clear later.

8

THE ELEMENTAL SUBSTANCE THEORY AS

AN EXPLANATORY HYPOTHESIS

WE HAVE SEEN that the neo-Ionians can be said to share a common conception of what it is to explain the natural world: they both accept the Elemental Substance Theory (EST). We must now see what formal claims are made by the theory and how it relates to other theories we have studied.

8.1 EST Formalized

We may see EST as comprising several principles, which I number for comparison with GST:

The Elemental Substance Theory (EST)

1. There is a set of substances $\{E_i\}$ which are the basic substances.
2. The E_i are permanent existences.
   a. The E_i are (i) without coming to be and perishing, (ii) all alike, (iii) unchanging, and (iv) complete. (Eleatic Substantialism)
   b. The set has a plurality of members. (Pluralism)
3. Derivative substances S_j are a product of relation R_k of E_i.
   a. (Definition) The E_i are *elements*.
4. There is a mechanism M that controls the production of R_k.
5. There is a set of forces $\{F_l\}$ that governs M.
6. (a) The world comes to be through the orderly application of the F_l to E_i, and (b) continues to exist through a balance of forces.

As in GST, points (1)-(2) define the ontology of EST, (3)-(5) the etiology or causal relations, and (6) the cosmogony and cosmology of EST. EST-1 defines a set of basic substances, which, according to EST-2a, have Eleatic properties, and by EST-2b are several in number. (In the next section I shall have more to say about the Eleatic content of EST.) The conditions in EST-2a distinguish the basic substances in EST from those in GST: those in GST are subject to transformation into other basic substances; those in EST never change in their own character.

While Ionian theories before Parmenides recognize a plurality of basic substances, they do not recognize a fixed nature for them such that they remain in their nature and that they must be alike bearers of constant properties through and through. For the early Ionians the derivative substances are just the basic substances which arise out of the single generating substance. By contrast, EST must distinguish between a set of phenomenal substances (the S_i of EST-3) and a set of elemental substances (the E_i) such that the phenomenal substances depend on the whole set of elemental substances. In the case of Empedocles this dependence is clear enough: there are just four elements, and perhaps an infinite number of compounds made from them. The two sets have different members. It is more difficult to see, but also true in the case of Anaxagoras. For, since according to him every substance is in every substance, e.g., phenomenal water must consist of elemental water, *plus* elemental earth, *plus* elemental salt, *plus* elemental air, etc. Thus for Empedocles,

$$E1 \qquad 4Blood = 1E + 1W + 1A + 1F,$$

and

$$E2 \qquad 8Bone = 2E + 2W + 0A + 4F,$$

assuming quantities can be summed.[1] For Anaxagoras (though he gives us no actual analyses) we might have

$$A1 \qquad 100W = 98W + 1Salt + 0.01A + \ldots,$$

where all elements will appear on the right side, to represent a sample of seawater. Now in fact the equation does not compute, because the W on the left and the W on the right do not have the same quantities; since elements cannot come into existence or go out of existence, we must recognize that the W on the left stands for an apparent amount of phenomenal water, not an actual amount of elemental water. Hence we should write something like W_p >for "phenomenal water," and recognize that there is a difference between what is on the left side and what is on the right side of the equation. Hence, in general:

---

[1] Empedocles B96 clearly allows for a quantitative formula: "Pleasant earth . . . got two of its eight parts of glittering Nestis, and four of Hephaestus."

$$q_p S_p = (q_1 E_1, \ q_2 E_2, \ \ldots, \ q_n E_n),$$

where q designates the appropriate quantity and the series represents the components of the phenomenal substance. But if the respective quantities of the elements bear a determinate relationship to the quantity and nature of the phenomenal substance, then

$$q_p S_p = R_p(E_1, \ E_2, \ \ldots, \ E_n),$$

where R_p designates the ratio of the respective quantities of the elements.[2] Simply put, the phenomenal substance is a function of the arrangement of elemental substances. (We should presume that the ratio in any given case includes a range of variations, since unlike Empedocles' model, Anaxagoras's model does not require precise whole-number ratios.)

Changes in phenomena result from changes in the arrangements of elements, which in turn are governed by outside forces—attraction and repulsion in the case of Empedocles, the force of the vortex motion (caused by Mind) in Anaxagoras. In addition to the elements themselves and their relationships, we must posit the existence of some mechanism of change. According to EST-4, a mechanism controls the interrelations of elements. When the elements are mixed in a certain way, a compound or a new phenomenal substance appears.

One of the distinctive features of EST is the separation of forces from matter. Whereas in GST in some sense the forces of the world were embedded in the generating substance or its successor substances, in EST the force is physically and ontologically distinct from the elements. Indeed, the founders of EST seem to take a step backward toward mythology in personifying the forces—Anaxagoras presents a cosmic Mind, and Empedocles Love and Strife—as quasi-mythical deities. Yet the forces seem to act regularly rather than capriciously,[3] and in general the trappings of mythology are just that—mere trappings, window dressing. While Anaxagoras seems to take his Mind seriously, according to the testimony of Plato and Aristotle he does not call on Mind to do more than set the cosmos in motion, leaving mechanical interactions to sort and organize the stuffs.[4] Thus despite appearances to the contrary, the worlds of Anax-

[2] Empedocles seems only to admit whole-number ratios, while Anaxagoras's views on infinite divisibility (B6) suggest infinitely various proportions (cf. B12 ad fin.).

[3] In Empedocles the elements come together in chance combinations (e.g., B59), but it does not follow that the two primal forces act in a chance way; rather, Love always combines and Strife always separates. For Anaxagoras, Mind seems to act with some kind of foresight (B12), and it affects everything it acts on (B13).

[4] See below, sec. 8.4.1.

agoras and Empedocles evolve in regular and virtually mechanical stages. Yet the forces are not merely effects of, for instance, contact between elements: they are independent causes of contact and interaction. And so we must recognize that EST-5 adds to the ontology of EST as well as specifying causal relations.

EST-6 emphasizes the fact that the theories of force and matter are applied to produce cosmogonies much like those of GST. A primitive disorganized state is gradually articulated into a world system in which the features of the present world emerge. Physical theory ultimately is at the service of cosmogony and cosmology. In fact, the most surprising thing about EST may be its conservatism in cosmology. Anaxagoras, for instance, takes over a good deal from Anaximenes' GST theory in his own EST cosmology.[5]

8.2 EST and Eleatic Theory

Now EST-1 and -2a are just the principles that constitute Weak Eleatic Theory. EST can be viewed as a logical extension of Weak Eleatic Theory. If the neo-Ionians accept Weak Eleatic Theory as the correct interpretation of Eleatic Theory, then they will see no need to argue against the Eleatics when they develop their own cosmology in accordance with EST. Indeed, we might see the cosmogony and cosmology of the *Doxa* as conforming to all the principles of EST, with one qualification. For, in accordance with EST-2, all phenomena are composed of a mixture of Light and Night, presumably in a determinate ratio that controls the properties of the result. And the goddess referred to in Parmenides B12 as controlling all things may represent a force of attraction governing mixture, and producing the results which constitute the world (EST-4).

The only respect in which the cosmology of the *Doxa* does not conform to EST is in -1b: there is not strictly speaking a plurality, but a dualism of elements. It is not clear, however, whether this represents a theoretical conflict between the neo-Ionians and Parmenides or only a difference of application. If we stipulate that a plurality consists of at least two elements, there will be no conflict. In fact the neo-Ionians seem to think two is not enough. But is that because they think two could not possibly be enough, or because two are not likely to be enough? An a priori argument against two elements might go as follows. In order to make allowances for all the appearances we experience, a theorist would need to have as inputs to the theory all contraries. For instance, if we had only cold elements, we could not account for hot. So if we posit one cold element, we must posit a hot element also. This seems in fact to be the procedure of Parmenides: Light is hot, dry, light, rare; Night is cold, wet, dark, dense (B8.56–9). But if we posit such elements we will find that they conflict with the principle of independence (noted above).[6] For one element will be defined in terms of another, and the two will not be independent, unconnected realities. We must, then, begin with a set of substances that is rich enough to include all qualities, in such a way that no one is a mere inverse of another.[7] We shall need more than two elements.

Another argument might be aimed at the particular elements Parmenides has chosen, but not at the number of elements per se. As Descartes appeals to the principle that there must be at least as much reality in the cause as in the effect, the theorist might suppose that there must be at least as much reality in the element as in the compound or mixture of elements.[8] But Light and Night are manifestly mere appearances, standing for light and dark. Such ephemeral and insubstantial characters could not ground a robust ontology. We need, then, something concrete as a basis. The question then arises as to what principles could meet the demand. Whatever the answer, and whatever the particular criteria to be used, we may note that there is a similarity between the answers of Anaxagoras and Empedocles on the one hand and Anaximenes on the other in this: the choices the neo-Ionians make are at least like the set of basic substances of GST. The pluralists seem to agree that the ultimate realities should be stuffs of the kind found everywhere in the world. In the case of Empedocles, the elements are just the subset of Anaximenes' stuffs that seem to comprise the major stretches of the world: earth (elemental earth), sea (water), sky (air), and heavenly bodies (fire).

To note the unsubstantial character of Light and Night may be to note Parmenides' resistance to an interpretation such as that extended to him by the proponents of EST. Even if, however, we reject the view that Parmenides intends the *Doxa* as a model for cosmology, we may nonetheless be struck by the continuity between Parmenides' cosmology and that of the pluralists. Whether or not the route from the *Doxa* to pluralism is one Parmenides plots, it is in a certain sense a more natural path than is usually recognized. For it accounts for the similarity between Parmenides' cosmology and those of Anaxagoras and Empedocles.

---

[6] See sec. 7.4.2, principle (II).

[7] On this see Mourelatos 1970, 132–33; Curd 1998, 105–10.

[8] What Barnes calls the Synonymy Principle, which can be found among the Presocratics: 1982, 119, 88.

8.3 EST with and without Emergence

In the last chapter[9] I observed that there is a major theoretical difference between the theory of Empedocles, which allows for emergence, or the appearance of properties and states which do not characterize the basic entities, and that of Anaxagoras, who does not allow for emergence. The problems of emergence, and the relative value of the two theories, can be explained in light of the formal model of EST developed above. As noted in the first section of this chapter, the two theories have much in common, so that both can be considered instances of EST.

There are several differences in the two theories, which can be seen in the concrete cases of formulas E1, E2, and A1 above. In Empedocles there are only four elements, whereas in Anaxagoras there is an indefinitely large, possibly infinite number of them. In Empedocles, there are some chemical formulas of compounds in which the amount of one element is zero, whereas in Anaxagoras this cannot happen by the principle of Universal Mixture (UM). In Empedocles, the elements seem to combine only in whole-number ratios (as in modern chemistry), whereas in Anaxagoras it is doubtful that such determinate ratios can be found. For there are trace amounts of everything in every mixture, and there is no way, apparently, of determining the relative values of the elements, nor of assuring that they exemplify whole-number ratios.[10] Finally, for Empedocles what appears on the right side of a formula does not appear on the left (unless perhaps of a trivial identity of the form $E = E$), whereas for Anaxagoras one of the items that appears on the right side of the formula will always appear on the left side.

The last difference is the most important for the present discussion. We can see the denial of emergence in Anaxagoras precisely in the fact that what appears on the left also appears somewhere on the right. There are no new substances and no new properties. (For all we know properties such as Hot are treated like substances, and will appear on the right side as well as the left.) Everything that appears in the phenomenology is already present in the ontology. This of course is both the advantage and the disadvantage of Anaxagoras's system. If there are no emergent characters to deal with, in one sense there is no coming to be, only a recycling or reshuffling of existing characters. The system is maximally Eleatic, we might say. On the other hand, Anaxagoras presents a most prodigal ontol-

---

[9] Sec. 7.5.

[10] "So the quantity of things being separated is not discernible, either in word or in deed" (B7).

ogy, and it is not yet clear whether his system is coherent, or indeed whether it explains anything at all.

8.3.1 Pluralism with Emergence

We need to examine Anaxagoras's theory in more detail precisely because of the extraordinary features it presents. But first let us look at Empedocles' theory, which has the advantage of familiarity. Thanks to its similarity to modern chemical theory, it can serve as a point of comparison with and a foil to Anaxagoras's more exotic theory. For Empedocles, items that appear on the left side of the formulas are compounds, those on the right are elements. Compounds are made up of whole-number ratios of elements and are always able to be not only analyzed but also decomposed into their component elements. The elements are everlasting, the compounds ephemeral; the elements are basic, the compounds derivative; the elements are real, the compounds somehow unreal.[11]

Because there is a finite number of elements, each of which is familiar from experience, Empedocles can characterize them:

> Sun, shining to sight and everywhere hot,
> and immortal things which are soaked in heat and blazing beam,
> and rain, dark and cold in everything,
> and from earth flow out thick and solid things. (B21.3–6)

Here, in his roundabout way, Empedocles characterizes fire as bright and hot, air as (likewise) bright and hot, water as dark and cold, and earth as perhaps heavy and hard. With each element we can associate a characteristic set of attributes. Since the elements and their proper attributes are everlasting, these attributes almost constitute *essences* for the elements. In the heyday of essences, Aristotle will distinguish between the essence of a thing (*to ti esti, to ti ên einai*) from its proper or invariable attributes (*idia*); the essence tells what a thing is (e.g., man is a rational animal), while the proper attribute always attaches to the thing without defining it (e.g., man is "risible": able to laugh). This distinction perhaps is not important for our purposes. What is important is that, for the first time, Empedocles can assign permanent attributes to each basic substance. By contrast, Anaximenes or some other proponent of GST could assign at most temporary characteristics to transformable stuffs.

But Empedocles is forced to say that the essential features of his elements sometimes disappear from view. For "these [elements] are the very things, which running through each other / become different; for blending alters them" (ibid. 13–14). The new characters that appear in a compound

---

[11] B17.30–35, B21.9–15, B26.

emerge and come to qualify the conjunction of elements differently from the way they were qualified individually. Blood, bone, and other substances emerge from a combination of the elements. New substances occur on the left side of the formula, increasing the number of entities in the system. But are these new substances real? No; at least they are not fundamental realities, but at most only derivative ones. In Empedocles' mythological idiom:

> And suddenly those things grew mortal which before were wont to
>   be immortal,
> and what was before unblended became mixed, exchanging paths.
>   (B35.14–15)

The immortal elements become mortal by entering into ephemeral compounds. The compounds do not have permanent essences, but are themselves merely temporary states of the elements, which are eternal and real entities with their own essences. There is at least a clear hierarchy of beings in Empedocles, with the elements being fundamental and independent, the compounds derivative and dependent.

Because, according to Empedocles, derivative substances can be fully accounted for in terms of four elements, we may say that Empedocles is a *reductionist*. He can, for ordinary purposes, operate with a countless number of substances, as many as Anaxagoras countenances. But when it comes to determining what things really are, he recognizes only four substances; all the others are products of the four elements in combination. This move gives Empedocles a powerful system of explanation. But it makes him vulnerable to the charge that he allows coming to be in the new arrangements of things. The alternative is a theory without emergence of new properties or substances, and without reduction of some properties and substances to others.

8.3.2 Pluralism without Emergence

In Anaxagoras's theory we lack the powerful tools of Empedocles' theory: we cannot reduce a large, perhaps infinite set of substances to a minimal set of elements.[12] And we cannot, on the other hand, start from a small number of elements and generate the vast diversity of the world by making it emerge from the combinations of the elements. Formally, we can see the lack of emergence by the fact that anything that can appear on the left side of a formula giving an analysis of a substance (as in formula **A1**) will appear also on the right. This analysis raises a number of questions and problems for interpretation.

---

[12] See Graham 2004a.

a. If an analysis of a given substance reiterates the substance in the analysans, what have we learned about it? Is not the analysis circular and pointless?

    1. If the analysis is supposed to be, or include, a definition, the definition fails because it is circular.

    2. If the analysis is supposed to provide an understanding of what the given substance is, it fails because it merely lists components, each of which needs explication and each of which is inexplicable.

b. If the number of elements is infinite, or even merely indefinitely large, how can any analysis actually be carried out?

    1. No exhaustive list of elements can be given.

    2. Consequently no summary of the quantities present can actually be given.

If, for instance, we read the label on a package of food to find out the ingredients, and found all chemical elements listed, without any quantities, we would not know anything useful about the product. What, then, is the point of Anaxagoras's pluralism?

We have seen already that the principle of Universal Mixture, UM, is the main thesis of Anaxagoras's theory. His odd form of analysis is a direct expression of the theory. But how can there be knowledge of the world if everything is in everything? I have already given a preliminary reply to this kind of objection above.[13] By virtue of the fact that what predominates in the mixture characterizes it, we can point to this table and say it is wood, to that drink and say it is water, to that file cabinet and say it is steel, and so on. The theory allows for ostensive definition: anything like that is . . . wood, or water, or steel. But, the objector can reply, you have not told me anything about the object; you have not explained it.[14] Anaxagoras must reply that he has explained it in the only way he can: by showing it. What more can the objector ask for than a confrontation with reality? The wood or water or steel is ultimately real and needs no explication. As to the complaint that analysis cannot be carried out, he must reply that it is a truth of the system that everything has every element in it, and that it is just a fact that the quantities of all the elements cannot be determined. (This does not rule out the possibility that for major components the quantities might be determined.) This situation may be disappointing, but that is how the world is organized.

In general, to both objections (a) and (b) Anaxagoras can reply that in a certain sense they miss the point. The systems that require elaborate analysis and definition are those which posit mysterious or recondite ele-

---

[13] Sec. 7.5.1.

[14] In Socrates' "dream" at *Theaetetus* 201dff., Plato examines and criticizes a similar account, one that could possibly be derived from Antisthenes (see Aristotle *Metaphysics*

ments. The beauty of my system, he might say, is that it requires no elements beyond those we are familiar with in everyday life. We already know what sorts of things there are in the world: wood, stone, air, water, flesh, blood, and so on. These things are the ultimate realities. If it is practically difficult to catalog them, nonetheless they are familiar to us; if it is vain to try to define the phenomenal substance, that is because there is no point in doing so. We already know everything we need to know about the elements from their phenomenal properties. What this system does is to show us how to stop asking fruitless questions and to help us appreciate the value of experience as a window to the world of basic realities.

The power of the system appears in its ability to explain how one thing appears to come out of another: water from clouds, fire from stones, and so on, flesh from bread. Since everything is in everything, the change is not miraculous, but is merely a change of concentrations of what is already there. What was concealed becomes manifest, what was manifest becomes concealed. But no new substances come to be, and no old ones perish. There is no secondary coming to be, we might say, but only phenomenal change. In that sense Anaxagoras's theory is more conservative and more Eleatic than Empedocles'.

8.4 Advantages of EST

8.4.1 Comparison of GST and EST

Concerning GST, we noted that it had several advantages: (1) physical explanation, (2) mechanical explanation, (3) appeal to empirical evidence, (4) comprehensive explanation, and (5) unified explanation. If we take for granted that GST marks a significant advance over mythological styles of explanation, the chief competitor of EST is GST. Fairly obviously, EST is a scheme of physical explanation. The elements are physical entities that by their proportions in a mixture or connections in a combination account for the properties of resulting substances. The only way in which EST is not thoroughly physical appears in the rule of forces. Are Mind, or Love and Strife physical entities? While cosmic Mind seems to have some properties that transcend the physical, such as its ability to start the vortex, and its knowledge of past, present, and future, mind seems to have location and extension:

1043b23). Note especially the criticism at 203b: one must know the letters in order to know the syllables (but they seem to be unknowable on the given account).

In everything there is a portion of everything except mind. And in some things mind too is present. (B11)

Everything else has a portion of everything, but mind is boundless and autonomous and mixed with no object, but it is alone all by itself. If it were not by itself, but were mixed with something else, it would have a portion of all objects, if it were mixed with any. . . . And the things mixed with it would hinder it from ruling any object in the way it does when it is alone by itself. For it is the finest of all objects and the purest, and it exercises complete oversight over everything and prevails above all. And all things that have soul, both the larger and the smaller, these does mind rule. (B12)

Mind, which always is, is very much present now where everything else is, in the vast surroundings and in both the things that have been aggregated and those that have been segregated. (B14)

There are some obvious problems here: how can mind be where everything else is and yet be all by itself? And how can it be in some things when it is not mixed with them? These are potentially serious problems. But for now I wish to simply point out that mind has location, and physical properties such as being fine. However much Anaxagoras may be groping toward a nonphysical concept of soul, a Cartesian dualism even, he has not achieved anything like a strict categorial distinction between mind and matter.

Empedocles' Love and Strife seem to act by occupying territory. This becomes clear in B35:

when Strife reached the innermost depth
of the vortex, and Love comes to be in the middle of the circle,
there all these things combine to be one thing alone,
not suddenly, but joining together willingly each from its own place.
When these things are mingled the myriad races of mortal things flow out,
and many things stay unmixed, alternating with the things that are being
    blended,
which Strife held still suspended; for not perfectly
yet had he withdrawn to the uttermost bounds of the circle,
but some of his limbs remained within while some had withdrawn.
As far as he would run ahead, so far would advance
the gentle immortal onset of blameless Love.
And suddenly those things grew mortal which before were wont to
    be immortal,
and what was before unblended became mixed, exchanging paths.
When these things are mingled the myriad races of mortal things flow out,
fitted with all sorts of shapes, a marvel to behold.

The most obvious interpretation is that, in Empedocles' spherical cosmos, Strife advances from the periphery toward the center as Love retreats; when it has reached almost all the way, Love begins to advance outward, as though a concentrated garrison in the acropolis of a city were to counterattack against a dispersed enemy force.[15] Strife is slowly pushed backward until Love completely unifies the sphere, leaving Strife at the extreme circumference:

> As they [the elements] were coming together, Strife was retiring to
> the extremity. (B36)

But finally Strife gains strength, presumably by being concentrated:

> But when great Strife was nourished in his limbs,
> and leapt up to lay hold of his office, as the time was fulfilled
> which had been fixed for them by a broad oath of mutual succession, (B30)

> All the limbs of the god shuddered in turn (B31).

Thus Love and Strife seem to act by physical contact with the elements. Though they are personified and endowed with intentionality, the forces control only the space they physically occupy.

As to (2) mechanistic explanation, while Anaxagoras in particular seems to provide room for intentional explanations, the testimony of Plato and Aristotle is that he does not make use of such explanations. Plato has Socrates look to Anaxagoras for a new kind of explanation:

> But when once I heard someone reading from a book of Anaxagoras, as he said, and saying that mind was what ordered and was the cause of everything, I rejoiced at this cause, and I approved of the saying that mind was the cause of all; and I considered that if this was right, that mind ordered all things, then mind must order and arrange each thing in whatever way was best. If, therefore, one wished to discover the cause by which anything came to be or perished or existed, this is what one would have to discover about it: how it was best for it that it should be or suffer or do anything whatsoever. And according to this approach there was nothing else for a person to inquire about concerning that or any object than the ultimate good. (Plato *Phaedo* 97b8–c3)

But Socrates is disappointed:

[15] Wright 1981, 55, makes the tactics of the constant strife between Love and Strife incomprehensible by saying "Strife strikes, as Love did, by rushing in to claim the center" (Inwood 2001, 52, concurs). This makes no military sense: why rush to face the front of an attacking phalanx when the rear is always the vulnerable part? It is best to take "Love comes to be in the middle of the circle" in B35.4 as meaning "comes to be *confined*" rather than "rushes in [from the outside] to occupy." See Graham 2005.

This marvelous hope of mine . . . was quickly crushed, when as I proceeded to read I found the man made no use at all of mind, nor referred to such causes to order the events, but he appealed to airs, aethers, waters, and all kinds of other strange causes. (98b7–c2)

Aristotle concurs:

When someone said that mind was present in nature as well as in living things, as the cause of the world-order and all its arrangement, he appeared as a sober man in comparison to his predecessors with their incoherent ramblings. We know that Anaxagoras arrived at this theory. . . . [Yet] it is plain that they hardly make any use of these causes, except ever so little. For instance, Anaxagoras uses mind merely as a *deus ex machina* for organizing the world, and whenever he is at a loss to say what cause necessitates an event, he drags it on stage, but in any other situation he appeals to anything but mind as a cause of events. (*Metaphysics* 984b15–19, 985a17–21)

Anaxagoras tends then to focus exclusively on mechanistic explanations, to the frustration of Plato and Aristotle, who prefer final and formal cause explanations.

As to (3), proponents of EST, like those of GST, are committed to an appeal to empirical evidence. There is, however, one area in which EST may be inferior: in that it requires more indirect evidence. We shall discuss this problem in the following section on disadvantages of EST.

As to (4), EST, like GST, attempts to give a comprehensive explanation of all phenomena.

The only obvious area in which GST has a prima facie advantage is in (5) unified explanation, for the simple reason that proponents of GST work from a single principle, the generating substance, whereas the proponents of EST have a plurality of principles—the elements plus the forces acting on them—as their starting points. We shall discuss this point also in the following section on disadvantages. In general, however, there is a large area of similarity in the aims of the two kinds of theory. The significant area of difference comes in the ability of EST to solve the problems of GST.

8.4.2 Solving the Problems of GST

We have seen (chapter 4) that GST is beset by significant problems. According to the Problem of Primacy, there is no systematic reason to recognize one of the basic substances as the generating substance. According to the Problem of Origination, there is no reason for the generation of the cosmos to take place if the generating substance is superior to the other basic substances and in a primeval state it alone existed. And according to the Problem of Being, we cannot say what any of the basic substances really is, given that one substance is transformed into another in a series. The strength of EST appears most clearly in its ability to deal with these problems.

8.4.2.1 The Problem of Primacy.

In EST every element is equal to every other: the theory is egalitarian rather than monarchial in its structure. This feature appears most clearly in Empedocles:

> For these things are all equal and of the same age,
> Each rules its own domain and has its own character,
> and they rule in succession as time rolls round. (B17.27–9)

"These things" include not only the four elements but also the two forces of Love and Strife. They are equal in power and even in age. They each rule in succession. The last point is an important feature of fifth-century democracy: any citizen might be called on to assume an office by lot, and to rule for his appointed term of office, as in the Prytany or executive committee of the Athenian senate. For democrats, even election could be undemocratic because it allowed for advantages of education, money, and influence for certain candidates. But any citizen had an equal chance of being chosen by lot to fill a magistracy.[16]

Equality of elements is built into EST. The elements all contribute to the composition of the cosmos. There is no need for one of them to have precedence over the others. We need only to have some power which makes them associate appropriately, as does Empedocles' Love.

In GST the Problem of Primacy is a major challenge: which of the basic substances is most real and most powerful? For EST, the problem simply disappears. With all the elements making equal contributions and being of equal power and importance, there is no need to make some invidious distinction as to which is the real substance or the real origin of things.

8.4.2.2 The Problem of Origination.

If the generating substance is the favored stuff, and if in the primeval state it alone existed, why did it change into anything else? Parmenides seems especially sensitive to this problem when he asks,

> And what need would have stirred it
> later or earlier, starting from nothing, to grow? (B8.9–10)

Among the several tacit objections in these words is the question, what does it need that it does not already have? Parmenides' property of completeness of what-is seems designed to block an appeal to need. Further,

---

[16] See Vlastos 1947.

why would it change at one time rather than another? The Principle of Sufficient Reason demands that there be some reason for a change happening when it does.[17] But if, in the primeval chaos, everything is one uniform stuff, and has been since time immemorial, why should it change now?

As long as the generating substance is viewed as the ultimate causal principle, the questions Parmenides raises will remain unanswered. But if we can separate the causal agency from the stuff of the world, we can reply to the challenge. And indeed, as we have seen, EST posits powers distinct from the elements: Love and Strife for Empedocles, Mind for Anaxagoras. It may be precisely the Eleatic demand that what-is be all alike that forces the differentiation. For if an element, as a representative of what-is, is all alike, it is too simple to account for change. Or, to be more precise, the element can account for change of the sort produced by contact with its properties—fire can heat, water can moisten—but not for any complex or directive change, such as producing order systematically.[18] An independent agency is needed for systematic or directive changes.

Anaxagoras endows his Mind with agency in the most straightforward sense: Mind directs the action of the universe, and it is aware of all things and orders all things. In other words, Mind acts intentionally on the basis of information:

> And of the whole revolution did mind take control, so that it revolved in the beginning. And first it began to revolve from a small revolution, now it is revolving more, and it will revolve still more. And things mixed together, things separated, and things distinguished, all these did mind comprehend. And the things that were to be—what things were but now are not, what things now are, and what things will be—all these did mind set in order, as well as this revolution, with which the stars, the sun, the moon, the air, and the aether which were being separated now revolve. For this revolution made things separate. And from the rare is separated the dense, from the cold the hot, from the dark the bright, and from the wet the dry. (B12, in part)

Now how precisely Mind orders things remains unclear, not only to us but also apparently to Plato and Aristotle as well.[19] Anaxagoras seems to have asserted the control of Mind, but not to have explained in detail how Mind asserts its will. The one means that it seems to employ is the vortex motion which it sets in motion: "this revolution made things separate." The vortex, as a great cosmic centrifuge, brings about physical effects resulting in the concentration of some elements in some areas. This

---

[17] See Barnes 1982, 187ff.
[18] E.g., Aristotle *Metaphysics* 1071b26ff.
[19] See above, sec. 8.4.1.

concentration ultimately articulates the cosmos and the things in it. The only obvious intervention of Mind is in starting the vortex in the first place. We get a kind of deistic scheme in which Mind accounts for the origination of the cosmos, after which it may take care of itself.

However incomplete his account of Mind's action is, Anaxagoras at least has an explanation of how cosmogony got started: Mind initiated the change, indeed an orderly kind of change, which organized the previously chaotic elements into a system. There is a first mover for the cosmos, and Parmenides' question has an answer.

Empedocles takes a very different route to account for cosmogony. His two opposing forces, Love and Strife, are always active. They alternately advance and recede, but they are always present in the universe, and always exerting themselves against each other. Love presides over a period in which the elements join to form compounds, including living things. She expands from the center of the world, and as she does, she draws more and more elements into compounds (B35). As the period continues, apparently, there is greater and greater unification of compounds, until finally everything becomes a single spherical unity in which the elements are mixed homogeneously. Strife, which has been pushed to the circumference of the Sphere, takes strength and suddenly invades the Sphere, causing it to tremble (B30). The elements begin to separate to their respective domains. "Whole-natured forms" are extruded from the fertile earth as it is emerging, and these forms become independent living creatures of a new generation. Eventually the elements become so stratified that no portions of one element are found in the domain of another; compounds are not possible, and Strife reins supreme. But at this point Love, which has retreated into the center of the circle, begins to expand again and the cycle repeats (B35).

Though the details of the cosmic cycle are controversial, there is little doubt that there is a repeating cycle.[20] What Empedocles gives us is not a linear cosmogony but a cyclical one. How does the cosmos originate? In a sense it does not: there is no one point at which differentiation begins, for there is an ongoing process of differentiation followed by unification. The process is "immobile in a circle."[21]

Are these satisfactory answers to the Problem of Origination? We can always ask for more. For instance, to Anaxagoras one can raise the question that was raised to Christians of Augustine's time: before God created the world, what was he doing?[22] Before Mind set the vortex in motion,

---

[20] See ch. 7, n. 25, on interpretations of the cosmic cycle.

[21] B17.13, B26.12.

[22] *Confessions* 11.10–12; the popular answer: preparing a hell for those who ask embarrassing questions about the creation.

what was it doing? What conditions made it right for Mind to start the motion at one time rather than another? To Empedocles one can raise the question, what evidence is there or could there possibly be that there are reiterations of cosmogony? The problems become increasingly abstract and increasingly metaphysical. It is not clear that either philosopher has the equipment to extend explanation much farther. But at least each offers a preliminary answer to a question the early Ionians may not have considered at all. And EST, with its articulation of causes separate from elements, provides the framework for an answer.

8.4.2.3 The Problem of Being.   The most disturbing problem for GST is to say what the basic substances are, given that they do not remain the same but undergo transformations into all the other substances. So troublesome is the ontological status of the substances that Heraclitus, a sympathetic critic of the tradition, sees the best move he can make is to abandon the substances as providing a foundation for explanation, in favor of the process itself. But if process philosophy is too radical a solution, one can find solace in EST. For EST assigns a permanent and unalterable character to every element: each one is what it is and will never be anything else. Yet by combining or mixing with other elements, it can produce other effects. Thus the elements can explain phenomenal changes while remaining unchanged at an ontological level.

The Problem of Being is the problem that seems to have especially exercised Parmenides, and it is in the solution to this problem that EST shows its superiority. The Problem of Primacy can be avoided by avoiding the claim that one substance is better than another; but the theorist who rejects the claim of primacy also gives up the elegance of a unitary explanation based on a single principle. The Problem of Origination is solved by EST, but only by separating powers and agencies from stuffs—only, that is, by complicating explanation again in such a way that no single principle can account for everything. In both these cases the theorist must make a trade-off. In ontology, too, one might argue that EST must posit a plurality where GST posits a single principle. But GST is really a pluralistic system too: although the other stuffs come from the generating substance, they are not ontologically reducible to it, as we have seen: each basic substance exists in its own right and is not identifiable with any other substance. GST is a pluralistic theory, but it cannot give any reliable account of its entities. They do not have a permanent character, and the temporary character they have is apt to disappear as they undergo transformation. EST can give a secure account of its entities as GST cannot.

8.5 Disadvantages of EST

Although EST has advantages relative to GST, it runs into its own problems. A first kind of problem is one that it, in fact, potentially shares with GST, but which becomes visible with the distinctions made in EST.

8.5.1 The Problem of Sortals

The basic substances familiar from both GST and EST are physical stuffs such as earth, water, fire. Anaxagoras may have a very broad conception of stuffs, including some properties as large and small, hot and cold; yet, from what we can see, he understands these properties on the analogy of stuffs. Stuffs are expressed by expressions called by philosophers and linguists "mass terms," substantive terms which do not admit of counting or pluralization, requiring auxiliary expressions to do that work; we cannot count or quantify, for instance, earth without specifying a measure, such as "two scoops of ice cream," "a shovelful of earth."[23] In contrast to mass terms are "sortal terms," which allow pluralization and counting by cardinal numerals: three oranges, five apples, two dogs.[24] The philosophical significance of this distinction is that while the Ionians seem to be good at talking about stuffs, they have difficulty dealing with discrete things. Many of the most important things in our world are those expressed by sortal terms: trees, tables, houses, dogs, cats, and, most importantly, people. What is the status of these kinds of things?

These are the objects we most tend to associate with "things," the ones Aristotle will be at pains to define and make central to ontology.[25] But the favorite theoretical entities of the Ionians are stuffs, things that are at best thinglike. Indeed, it seems such objects have some of the theoretical ambivalence captured by Heraclitus:

> Collections: wholes, not wholes; brought together, pulled apart; concordant, discordant; from all things one and from one all things. (B10)

That is, while there is a kind of unity in a stuff owing to its similar character, assuming some degree of consistency or homoeomereity, the stuff is easily scattered, mixed with other things, and adulterated. Discrete objects often, and especially biological individuals, Aristotle's favored class of things, have an ability to resist change and to maintain themselves in existence. They have a kind of unity and permanence not found in stuffs.

[23] Quine 1960, 91.
[24] Strawson 1963, 171.
[25] For his importance in making things central to ontology, see Mann 2000.

Ionian philosophy is good at accounting for stuffs. But how can it account for discrete things? Granted that flesh, bone, and blood arise from Empedocles' elements, and that each of them is an element for Anaxagoras, how do these stuffs constitute a human? The closest we come is Anaxagoras's B4a:

> Since these things are so, one must believe that there are many things of all kinds in all the composites, and [or: namely] seeds of all things having all sorts of figures, colors, and tastes; and that men too are compounded [sumpagênai], and every other kind of creature that has soul. . . . I have said these things concerning the separation, that it occurs not only among us, but elsewhere as well.

There is a cosmological question about what Anaxagoras means by saying the same thing happens elsewhere (other worlds? other parts of our world?),[26] but here we are interested in the question of how it happens at all. Apparently the cosmic separation drives the production of organisms out of stuffs. But are we to understand that the stuffs are further "compounded" into, as Aristotle would say, anhomoeomerous parts and organisms? Or are there latent in the "seeds," which contain diverse figures as well as colors and tastes, the biological patterns of organisms? Unfortunately, Anaxagoras does not expand.

Empedocles does, however, offer some hints at how stuffs and organisms are related. At one stage of the cosmic cycle, limbs move about and join in chance combinations, some of which (the viable ones) survive, while the monstrous ones perish.[27] Empedocles recognizes, then, a stage of organization in which limbs can exist by themselves at least for a brief time. The existence of limbs seems to occur as Love is increasing. In the part of the cosmic cycle in which Strife is increasing, the stuffs gradually differentiate themselves in living things as well as in the world as a whole:

> whole-natured kinds first rose up from earth,
> having a portion of both water and heat;
> fire sent them up in its desire to reach its like,
> forms which did not yet manifest any pleasant figure of limbs,
> nor voice, nor organ native to man. (B62.4–8)

---

[26] That there is only one world: Vlastos 1959, 199–203; Schofield 1980, 102, with the support of Aristotle *Physics* 250b21–7. That there are plural worlds: Burnet 1930, 269–79; Gigon 1936, 25–26; Aravantinou 1993. That there are plural human civilizations on the earth, Cornford 1934, 6–8, Guthrie 1962–1981, 2.313–15. That there are countless microscopic worlds similar to ours, Mansfeld 1980; that there are nested worlds, Sisko 2003. That there is an unseen world in addition to the visible, Simplicius *Physics* 157.16–24.

[27] B57–B61, Aetius 5.19.5 = A72, Aristotle *Physics* 198b23–32.

As fire rose out of earth, it distended pockets of earth, which had water and fire mingled with it. Gradually these swampy creatures seem to have been articulated into human beings. Mechanical forces of separation drove the process, as far as we can see. This kind of process could be called a compounding, to use Anaxagoras's term.

But there is much more to say. Why do just the species of animals and plants exist which we find? Could there have been others? Empedocles offers a nascent theory of natural selection to account for the absence of monsters, at least. But we may still ask why offspring resemble their parents so closely. What is the reason for such *lack* of diversity among natural kinds? Modern biology has learned to use natural selection to great effect to account for this. It is unclear how far Empedocles appreciates the problem. In any case, we can notice that in the fourth century BC philosophers will find Ionian philosophy inadequate just at the point where it tries to portray structure arising out of mixture, form being generated from matter. For, Plato and the Friends of the Forms will argue, order must be prior to disorder, and form to matter. Yet modern science will vindicate the Ionians in their attempt to see structure emerging from matter.

8.5.2 Eleatic Problems

As we have seen, most interpreters of the pluralists assume that they are replying to something like Strong Eleatic Theory, and that pluralism fails to reply to its challenges. I have argued that the pluralists interpret Parmenides as promoting instead Weak Eleatic Theory, as a result of their taking the *Doxa* as a constructive approach to cosmology. If the pluralists do read Parmenides as presenting only Weak Eleatic Theory, they are not obligated to rescue cosmology from the Eleatic challenge.

On the other hand, it seems likely that Parmenides does intend to argue for Strong Eleatic Theory, at least for the level of ontological explanation, and that the pluralists miss the strength of his critique. If this is so, they are guilty at least of misunderstanding his argument, of *ignoratio elenchi*. And that leaves open the problems raised by the Eleatic challenge. How can what-is come to be from what-is-not? And, in consequence of this general challenge, and of the consequences Parmenides draws from it, how can there be a plurality of beings? How is change possible? And how is knowledge possible?

8.5.2.1 The Problem of Plurality. One of the Eleatic properties is being all alike. If what-is is all alike, according to Strong Eleatic Theory there can only be one thing. For if there were any distinction within what-is, it could arise only from what-is-not; but there is no such thing as what-is-not. So there are no distinctions to be made within what-is. And what-is is one single thing. If so, plurality is ruled out a priori, and all pluralistic theories are ex hypothesi excluded.

This is not a problem for Weak Eleatic Theory, which is compatible with pluralism. But given that the pluralists start from Eleatic principles, their theory is in some sense responsible to those principles. If they get them wrong, and the principles are stronger than they supposed, their theories are doomed from the outset.

There is, indeed, some sort of fallacy in Parmenides' argument for uniformity. For instance, if Parmenides understands "be" in an existential sense, in his argument that there cannot be what-is-not to separate what-is from its like (B8.46–8), he is saying that existent stuff cannot be kept apart by nonexistent stuff. But, we might object, suppose what is F exists and what is G exists; then we might have two stretches of F separated by G. Parmenides would reply that if F is, then G is not. Why? Presumably because G is not F. But here we have gone from the existential "is" to the "is" of identity (or perhaps of predication). That G is not F does not entail that G is not, i.e., does not exist. Various interpretations can be given to Parmenides' verb "to be," but mutatis mutandis there will be a reply to the argument.

The problem for the pluralists, of course, is that the semantic distinctions necessary to reply to Parmenides had not been discovered yet. Plato (in the *Sophist*) and Aristotle (in *Physics* I) developed different adequate semantic-metaphysical schemes to answer Parmenides—but only in the late fourth century, at least a century too late.

The inadequacy of pluralist theory is well-known. But it is important first to recognize that the early pluralists were not even trying to address the problem because of their understanding of Parmenides' argument, and second to recognize the inherent difficulty in replying to the Eleatic challenge, as properly understood.

8.5.2.2 The Problem of Change.   Parmenides clearly rejects coming to be and perishing in developing his Eleatic properties, and Anaxagoras and Empedocles follow him in this. But in his third Eleatic property Parmenides seems to reject motion in place as well: what-is is *atremes* or *akinêton*.[28] Parmenides provides a basic typology of change when he either rejects or downgrades change in the form of coming to be and perishing, changing place, and alteration ("exchanging bright color").[29] Thus he

[28] B8.4, 26, 38. It is without starting or stopping, line 27.

[29] B8.38–41; how strong the criticism is depends on whether we read ὄνομ' ἔσται "are (mere) names" (most editors) or ὀνόμασται "have been named" (Woodbury 1958, followed by Mourelatos 1970, 180–85; Curd 1998, 89, n. 66). See ch. 6 n.60.

challenges not only coming to be and perishing, but other kinds of change as well.

The pluralists are perfectly willing to abandon coming to be and perishing for the ultimate realities: they are everlasting and always the same. But the possibility of explaining phenomenal change depends on there being some other kind of change that the elements participate in that can be manifested as phenomenal change. If all kinds of change are ruled out, the Eleatic criticism will block any chance of explaining observed changes, and cosmology will become impossible. It is not clear that Anaxagoras and Empedocles appreciate this threat to their program. As far as we can see from the fragments, neither philosopher gives a proper theoretical justification for change.[30] Anaxagoras posits a vortex motion that continues once it is initiated, causing changing concentrations of elements which approximate the different densities found in Anaximenes' theory.[31] Empedocles describes changing locations and interactions between the four elements as governed by the influence of the two forces. Both theories seem to presuppose change of place of elements as well as changing relationships among them. In terms of Aristotle's typology of change—coming to be and perishing (change of being), increase and decrease (change of quantity), alteration (change of quality), and locomotion (change of place)[32]—the pluralists' elements clearly have the last. Change of being is ruled out on Eleatic principles. Change of quantity can happen only locally: some amount of one element can migrate to another, increasing the amount at the given place; but since the element is not created or destroyed, the total quantity is fixed and unchanging. Change of quality arises for both philosophers by change in arrangement of elements—either the amount of one stuff present in the mixture, or the way they are configured. But this change of arrangement also seems to presuppose change of place, since the elements must change locations in order to come into new arrangements.

Thus it appears that the pluralists are committed to some sort of change, and most obviously that is change of place. If, however, Parmenides provides a convincing argument against change of place, the project of the pluralists cannot be realized.

8.5.2.3 The Problem of Knowledge.

At least since Xenophanes there had been a problem of knowledge: how can we know the truth of the

---

[30] See above, ch. 7, sec. 4.3, on Empedocles' interpretation of the elements as immobile; but Empedocles does not justify his interpretation or consider possible objections.

[31] Anaxagoras B16, echoing the stages and mechanisms of Anaximenes' cosmogony.

[32] Aristotle *Categories* 14, more strictly analyzed, *Physics* V.1. Parmenides recognizes a number of types of changes, if not yet a typology at B8.40–1.

philosophical or scientific claims? Xenophanes seems to abandon any claim of certain knowledge in favor of a fallible cognition that is capable of progressing, perhaps by trial and error.[33] But the Eleatic critique seems to raise even more substantial obstacles to scientific knowledge. In B7 Parmenides raises the problem of senses:

> Never shall this prevail, that things that are not are.
> But you, withhold your thought from this way of inquiry,
> nor let habit born of long experience force you along this way,
> to wield an unseeing eye and echoing ear
> and tongue. But judge by reasoning the contentious refutation
> spoken by me.

Taken by itself, this passage may suggest that Parmenides wishes to condemn the senses in their own right, and thus to use the inadequacy of the senses as a crucial premise in his argument. But his criticism of the senses comes only after his observation that they conflict with true understanding:

> From this first way of inquiry <I withhold> you,
> but then from this one, which mortals knowing nothing
> wander, two-headed. For incompetence in their
> breasts directs a wandering mind; and they are borne
> both deaf and blind, dazed, undiscerning tribes,
> by whom to be and not to be are thought to be the same
> and not the same, and the path of all is backward-turning. (B6.3–9)

The last two lines reveal that the fault of mortal thinking is to hold contradictory propositions, which make one "two-headed." If mortals are deaf, blind, and dazed, it is because they can embrace contradiction without demur. The point of B7 must be not that the senses are a priori unreliable and helpless, but that their reports are incompatible with the truth, for they entail contractions. If the inescapable truths of philosophy are incompatible with sense experience, the latter must go.

Hence the problem of the senses arises as a kind of dialectical objection to Eleatic philosophy: surely your theory cannot be right because it conflicts with the senses. The Eleatic answers the objector: surely the senses cannot be right because they produce contradictions in saying the same things are and are not; hence the senses are in no position to support everyday beliefs. Only one account of reality is self-consistent: the Eleatic one. The senses are disqualified by their contradictory results. The senses,

---

[33] B18, B34, B35.

then, are to be rejected, but only after reason has established the true account of things.

All of this is predicated on Strong Eleatic Theory. On Weak Eleatic Theory, the senses are only compromised when, or to the extent that, they report contradictory results. Thus if we think that air turns into every other basic substance, and what is one thing is now another, now another, we are guilty of thinking what could not be. But if we understand correctly that the senses report only what seemed to be one thing seeming to turn into something else, we need not err. The understanding makes a mistake in inferring that one thing is transformed into another, when reason could tell us only that a new state of affairs has arisen by a rearrangement of continuing elements. The senses are not ultimately at fault, but our understanding of the change.

On this reading the senses are potentially misleading, but they are necessary sources of information about the world. On the strong reading, the senses can never provide any information about the world. For on the strong reading change is not real, while on the weak reading change can be real even if the appearance of transformation among elements is not. The weak reading allows the pluralist to distinguish between a kind of naive realism and a critical realism. Let me hasten to add that this is not the Cartesian distinction based on the properties of consciousness of mind and the extension of matter, respectively. But it is a distinction between what appears, perhaps inevitably, to the subject, and what really is. A true understanding of the world would allow the subject to make correct inferences from sense experience, and to reconstruct from the data at least a plausible account of what underlies changes.

Because Parmenides criticizes the senses in the context of a reply to a potential objection, he does not rule them out a priori, nor is their rejection an axiom of his position. Thus a reply to him is possible. The reply would take the form of an argument to show that the senses do not produce contradictions by themselves: they produce contradictions only on certain assumptions. These assumptions, the defender would argue, arise from the incorrect ways we view the world, perhaps the uncritical ways we grasp reports from the senses. If we avoid these, we can save the phenomena. Of course, to defend the senses we must presuppose that there are changing phenomena for them to observe. But to report this falls within the competence of the senses, so a defense is possible.

One major concession the pluralist makes to the Eleatic is to accept the claim that what we see are appearances. To concede that point is to put oneself in the position of saying the real entities, in the case of the pluralists, the elements, are known only by inference. Philosophy drives a strong wedge between experience and reality by putting the real things irretrievably beyond the reach of the senses. They are to be known, if at all, by looking for entities which possess Eleatic properties and which can in principle generate the phenomena of experience. No longer is what we see what we get. There is now an epistemological gap between our experience and the world.

8.5.3 The Strong Eleatic Position

Taken together, the Problem of Plurality, the Problem of Motion, and the Problem of Knowledge constitute the familiar challenge to cosmology that we meet in the textbooks. But we see that this set of problems arises only to the extent that we recognize in Parmenides Strong Eleatic Theory rather than Weak Eleatic Theory. I have argued that the response of the early pluralists is consistent with the latter rather than the former position. By understanding Parmenides as advocating Weak Eleatic Theory, Anaxagoras and Empedocles construe him as providing a new and more powerful program for cosmology, one that escapes the problems of GST. They see him as not only raising the Problem of Being for Ionian philosophy, but also answering it. In short, they see Parmenides as the founder of EST, and themselves as following out the implications of a new paradigm. They are thus free to work constructively with the Eleatic bodies posited, so they think, by Parmenides: the elements.

On the other hand, it is plausible to think that Parmenides advocates the Strong Eleatic Theory and employs Weak Eleatic Theory only as a second-best strategy for dealing with appearances. Thus if he has followers who promote his strong agenda, they will come in conflict with the pluralists and raise anew the major problems surveyed here. Of course Parmenides did have such followers: Zeno and Melissus. And the remainder of the fifth century was occupied with the conflict between these defenders of Strong Eleatic Theory and the pluralists dedicated to Weak Eleatic Theory. At this point the debate looks very much like the textbook debate. But if the story I have told here is true, there is an important historical and dialectical difference between what happened and the way the debate is traditionally portrayed. The early pluralists did not see Parmenides as an advocate of Strong Eleatic Theory but of Weak Eleatic Theory. Hence they saw themselves as followers of Parmenides, not foes. They conceived of themselves as not as anti-Eleatic pluralists but as Eleatic pluralists. Meanwhile, the later Eleatics saw themselves as Eleatic purists or fundamentalists. The debate was carried out not as between two different schools but as between rival factions of the same school. And even if the concept of a school is anachronistic here, the point is that two traditions traced their roots and allegiances to the same master, much as rival Socratic schools would do in the fourth century and later.

There is no question that Parmenides stood at the head of all later philosophy. But his role was initially a positive one for natural philosophy. Only after the first blossoming of Eleatic pluralism did the specter of a complete Eleatic rejection of cosmology arise. The problems portrayed here were indeed philosophical problems of the most serious kind. But they remained latent problems, to be discovered or rediscovered as the pluralist program began to develop.

9

THE ATOMIST REFORM

T
HE ATOMISTS, Leucippus and Democritus, belong to a second generation of pluralists. Unlike Empedocles and Anaxagoras, who can be taken to be enthusiastic followers of Parmenides, they are aware of strong Eleatic criticisms of the scientific project. Aristotle reads Leucippus as responding to the arguments of Zeno and Melissus, and his interpretation on this point seems plausible.[1] Yet atomism retains a good deal of the Eleatic program as well as of the Ionian project. How are they able to reconcile the two apparently hostile theories? And do they have a good philosophical justification for the moves that they make? Although there is no question that their solution to problems of scientific explanation was fruitful to an unparalleled degree in the long run, there remains a serious question about how successful the atomists were in their immediate philosophical response to those problems.

9.1 The Challenge

Zeno invented dilemmas that seem aimed at producing absurdities for anyone accepting a plurality of real existences or the possibility of motion. Most relevant to the atomist project is his dilemma concerning divisibility.[2]

> [H]aving proved that unless what-is had size, it would not exist, Zeno adds, "if there exist [many things], each thing must have some size and solidity, and one part stands out from the other. And the same consideration applies

---

[1] But see Diller 1941, who gives interesting arguments to show that Melissus is answering Leucippus, as well as Diogenes of Apollonia. If this is the case, we gain evidence for the existence of the shadowy figure of Leucippus, and find Democritus replying to Melissus on behalf of Leucippus. Cf. also Klowski 1966, 32; Klowski 1971; and Bicknell 1967a, criticizing Kirk and Stokes 1960. In any case, there was a lively debate in the late fifth century, in which the atomists took part.

[2] B1; cf. B2, B3. According to Solmsen 1988, this kind of argument motivated Leucippus and Democritus more than the arguments of Melissus.

to what stands out from this. For it will have size and stand apart from it. And it is the same to say this once and always. For there will be nothing which can serve as a final part of this nor will one part be different from another. Thus if there are many things, they must be both small and great: so small as to have no size, so great as to be unlimited." (B1 = Simplicius *Physics* 140.34–141.8)

Suppose we divide a body and divide it again and again ad infinitum. Finally we will be left with parts which either have magnitude or do not. If they have no magnitude, when we reassemble them the body we reconstruct will have no magnitude either. If they have magnitude, when we reassemble the infinite number of parts, the body will have an infinite magnitude. Either limb produces an absurdity. Hence the assumption that a body is infinitely divisible into parts is mistaken. This argument provides the atomists with the motive for putting a limit to division. There must be some practical end to the division of a body, some lower threshold. We do not observe such a threshold with our senses, but reason tells us it must exist, if not at the level of the sensible, at some subsensible level. There must be ultimate particles of matter that are indivisible, *atoma*, the atoms.

Although we know a number of the dilemmas Zeno posed, we do not know how explicit he was in his defense of monism, or whether he was explicit at all.[3] But we do know that Melissus was a confirmed monist, possibly the first explicit defender of some strong monism in the Eleatic tradition. Unlike Parmenides, Melissus argues that what-is is infinite in its extent and also sempiternal in its existence. It is unique: there is only one type of thing, what-is, and only one thing, the infinite exemplification of what-is.[4] Hence Melissus embraces both a monism of type and a monism of token, intending to exclude all plurality of substance and even of qualities. Despite his differences from Parmenides, Melissus becomes the filter through which Parmenides and perhaps Zeno are understood in the later tradition: they are monists who exclude any real change and motion for what-is.[5]

The key argument for the atomists is found in Melissus B7:

(7) Nor is there any void, for the void is nothing, and what is nothing would not be. Nor does it move, for it does not have anywhere to retire, since it is full. For if there were void, it [what-is] would retire into the void. But since

---

[3] Solmsen 1971 challenges the orthodox view that follows Plato's interpretation in the *Parmenides* rather slavishly.

[4] B1–B6.

[5] And Parmenides' theory is distorted by Melissus's in the early history of philosophy; a point recently stressed by Palmer 2004.

there is no void, it does not have any place to retire. (8) And the dense and the rare would not be. For the rare could not be full like the dense, but already the rare is more empty than the dense. (9) This is the distinction we must make between the full and the not-full: if it yields in any way or accepts anything, it is not full; if it does not yield nor accept anything, it is full. (10) So it must be full if it is not empty. So if it is full it does not move.

Most important for our purposes are the second and third sentences taken in conjunction with the first. An argument may be reconstructed as follows:

M1

1.	Void is nothing.	
2.	What is nothing is not.	
3.	Hence, there is no void.	(1, 2)
4.	If there is void, there is a place to move to.	
[5.	If there is a place to move to, there is motion.]	
6.	If there is no void, there is no place to move to.	
[7.	If there is no place to move to, there is no motion.]	
[8.	Hence, if there is no void, there is no motion.]	(6, 7, HS)
9.	Hence, there is no motion.	(8, 3, MP)

Of course there is an ambiguity in "is not" in 2, but that is present already in Parmenides. Point 5, implicit in the second sentence of the quotation, is odd: it is not clear why what-is would *have* to move into what-is-not. However, the key part of the argument comes from 6 on and reveals that in the absence of void there is no motion. If there could be no motion without the void, or the empty (*to keneon*), void is a necessary condition of motion. If what-is is completely full, there is no place to move to, and hence no motion. If there is motion, there must be an empty place to move into, and hence a void. The dense and the rare, such as Anaximenes posited, depend for their existence on a distinction between full and empty. Only by admitting the empty into our ontology can we allow motion.

Objections can be raised to this argument. Why could not some sort of circulation be allowed?[6] Or even interpenetration?[7] But if one assumes

---

[6] As developed by Empedocles B100, with Aëtius 4.22.1 = A74, Plato *Timaeus* 79a–80c. Aristotle *Physics* 267a15–20 criticizes "antiperistasis" as not explaining anything. See also Simplicius *Physics* 1351.12–16, Philoponus *Physics* 639.3ff.

[7] The atomists at least were aware that ashes could absorb an equal volume of water: Aristotle *Physics* 213b21–2 = Leucippus A19. They of course attributed the phenomenon to the presence of void in the ashes.

that what is full is completely solid and impenetrable by nature, as Melissus seems to do, then the argument will stand. In fact, the full and the empty, as Melissus conceives them, are two sides of the same coin. If we understand what-is to be completely compact and impenetrable and what-is to be all alike, then the contrasting case is extension without any resistance at all, the absence of all physical properties. This sounds like an alleged something that is nothing at all, complete nonexistence or nothingness. Such an alleged entity would seem to fall victim of Parmenides' charge that it could not be thought of or expressed. It would be inconceivable and hence impossible. Nonetheless, if one accepted the characterization of what-is as completely full and all alike, then the only chance for motion would lie in the possibility that there is some state of being that is empty of such stuff.

One further argument by Melissus seems relevant to the atomists—that if there are many things, they must have Eleatic properties:

> (2) If there were many things, they would have to be such as I say the one is. For if there are earth and water and air and fire and iron and gold and the living and the dead, and black and white and the rest, which men say are real (*alêthês*)—if then these things exist, and we see and hear rightly, each thing must be such as it first seemed to us, and it must not change nor become different, but each thing must always be such as it is. Now we do say we see and hear and understand rightly. (3) But it seems to us that the hot becomes cold and the cold hot and the hard soft and the soft hard and the living dies and is born from what is not living, and all these things alter, and what was and what now is seem not to be alike at all, but iron which is hard seems to be worn away by contact with the finger, as well as gold and stone and whatever else seems to be completely solid. [So it turns out that we neither see nor know the things that are.][8] And from water, earth and stone seem to come to be. (4) Now these things do not agree with each other. For although we say that there are many things which are eternal and have characters and strength, all of them seem to us to alter and change from what they were seen to be on any occasion. (5) It is clear then that we have not seen rightly, nor do those things seem rightly to be many. For they would not change if they were real, but they would remain such as each seemed to be. For nothing is stronger than true being. (6) But if it changed, what-is would perish and what-is-not would come to be. So in this way, if there were many things, they would have to be such as the one. (B8)

The many things that allegedly exist seem to reflect at least in part the views of philosophical opponents. The first four items are Empedocles' four elements. Iron and gold are not included within Empedocles' ele-

[8] I athetize this sentence, following Barnes 1982, 622, n. 3.

ments but seem to satisfy Anaxagoras's requirements. And living and dead are favorite contraries of Heraclitus, and the interchange between the living and dead referred to in section (3) seems to be an allusion to Heraclitus's view of the sameness of opposites.[9] The claim that earth and stone seem to come from water reminds us of Anaximenes' series of transformations. Sensible substances seem to change into other substances or into opposites.

Allusions thus let us identify the targets of Melissus's argument, namely the great cosmologists, the Ionians and neo-Ionians. The argument seem to go as follows:

M2

1.	What is does not change.	
2.	Hence, if (a) the many exist and (b) we see and hear and understand rightly, the many must not change.	(2)
3.	Suppose (a).	
4.	We all suppose (b).	
[5.	Hence, the many do not change.]	(2, 3, 4, MP)
6.	But we see and hear and understand the many to change.	(fact)
7.	Hence, the many change.	(4, 6)
8.	Not both (a) and (b).	(3–7 RAA)
9.	Not (a) and not (b).	?

Point (1) is a product of previous argument. The proof clearly hinges on (2). Melissus is careless about propositional attitudes, e.g., in (4), but given the early date of composition we can hardly fault him for his failure. More troublesome is the move from (8) to (9), where he shows his ignorance of the logical truth we now call DeMorgan's theorem. In this case his argument seems strange as well as illogical, because we would have expected him to call attention to (4) in order to isolate (a) from (b) and to attack (a), presumably his main target in the argument. Then we could have

9a.	Not (a) or not (b).	(8, DM)
10.	Not (a).	(9a, 4, DS)

at least so long as we allow him to treat (4) not only as a fact about human beliefs but also as an actual assumption of the argument. Not-(a) would indeed follow from the argument as reconstructed. Of course, Melissus ultimately wants to overthrow (b) as well as (a). But the modified argument allows him a strong ad hominem reply to his opponents: because

---

[9] Heraclitus B88, cf. B62, B36, B48, B76, B26.

they are so confident in the reliability of sense experience, they are forced to accept monism. Now since Melissus tries to overthrow both (a) and (b) in the same breath, both the ontological and the epistemological assumptions, he fails to overthrow either. For he unwittingly leaves open the possibility that either one singly could be true. Now in fact, if there are not many things in the world, then it would turn out to be true that we do not perceive rightly because we perceive that there are many things. On the other hand, the fact that we do not perceive rightly does not entail that there are not many things, only that they are not such as we perceive them to be. This, then, is the potentially viable and philosophically interesting case, one that has not been precluded by the argument.

The later Eleatics have provided strong challenges to Eleatic pluralism. On the positive side, they have at least clarified certain positions and shown what would be required to build a satisfactory theory under the aegis of Eleatic metaphysics. Zeno provides the motive for positing particles of minimal size. Melissus provides the motive for positing a void, as well as a suggestive program: anyone who posits a plurality of beings must make them such as the Eleatic one.

One serious problem for understanding the atomist response to the second generation of Eleatics is that it is difficult to see how the atomists provide any satisfactory philosophical rejoinder. To posit atoms and the void provides a kind of model for how motion and change could be understood. But to be secure philosophically the atomists must do more than that: they must counter the arguments put forth by the critics of pluralism. Otherwise, they merely beg the question. They assume without justification what their critics have already denied. Such a move is, of course, dialectically indefensible. But if they have an argument, what is it, and how does it work? Rarely do modern interpreters provide an argument for the atomists, and even more rarely is it a good argument.[10]

A second problem grows out of the atomist response to the Eleatics. Whatever their full justification, the atomists say that there is void, and give as a partial explication that what-is-not exists no less than what-is. This simple claim is a very strong reply to the Eleatics, because it is one of the first premises of Parmenides' argument that what-is-not is not possible in some basic sense. Hence, if we overthrow this principle, we seem to undermine the Eleatic position so fundamentally that there is nothing left. Yet the atomists to a surprising degree adhere to the principles of Parmenides.[11] Their atoms have Eleatic properties insofar as it is possible

[10] Barnes 1982, 442, KRS 433, quoted above in ch. 7, sec. 2.

[11] Cf. Morel 1996, 47: "A défaut d'informations précises sur la patrie géographique de Leucippe, nous pouvons en effet lui attribuer au moins deux patries intellectuelles: Elée [et] Milet."

to assign them, and the overall system of explanation of the atomists adheres strictly to the theses of no becoming and no change (in the character of the basic entities). If they reject first principles of the Eleatic system, why do they not abandon the whole program? As far as we can see on most interpretations, they give a dogmatic and question-begging reply to the Eleatic challenge, and then go on to adhere to Eleatic principles as if they were unchallengeable dogmas. Why oppose the Eleatics only to embrace them? These problems remain, as far as I can see, unresolved and too often unappreciated.

In any case, we must recognize that the atomists have a complex relationship to the Eleatics. They accept some points of their program and reject others. They acknowledge a central role for what-is, endowed with the four Eleatic properties Parmenides had deduced for it: (1) without coming to be or perishing, (2) all alike, (3) unchanging (where for the atomists this does not preclude locomotion), and (4) complete. On the other hand, they allow void, which they understand as what-is-not, providing the boundaries for an infinite number of particles and hence supporting against Melissus a fundamental pluralism. And they posit everlasting motion of the particles. Whereas the Eleatics are either implicit or explicit type and token monists, the atomists are in different senses monists, dualists, and pluralists. They are material monists in allowing only one type of being as matter for the atoms. They are essential dualists in recognizing two basic kinds of entities, atoms and the void. And they are numerical pluralists in positing an irreducible plurality of substances, the atoms. Yet for all their differences from the Eleatics, they are deeply indebted to them and do not wish to abandon their basic principles, or some subset of them, nor some important part of their program, namely to explain things without appeal to coming to be and perishing.

9.2 Foundational Arguments

It is well to distinguish theories from the arguments that purport to establish those theories. It may turn out that even though the arguments given to establish a theory are inadequate, the theory may be interesting and valuable in its own right. For instance, the arguments given by Bentham and Mill for utilitarianism are dismal, yet the theory is of enduring importance for ethics, whether one accepts it or not. Thus, we need not throw up our hands at early atomism even if it does not have a good argument against previous positions, or indeed even if it does not try to argue against them. After all, Aristotle was fond of distinguishing between theories of nature and arguments against those that denied the possibility of a theory of nature. Nonetheless, to be a philosopher rather than a physicist is in part to be willing to defend the possibility of one's theory against potential critics. And so far as we know, the early atomists did not make any strong distinctions between dialectic, metaphysics, and physics or in any other way shelter themselves from the need for open debate on basic philosophical issues. It remains desirable, then, to see if the atomists have some arguments to establish their own principles, or, in default of an explicit argument, if one could be provided consistent with their own principles which could establish their position.

The one passage that seems to offer some evidence for an atomist defense of basic principles is found in Aristotle:

> Leucippus thought he had arguments (*logoi*) which stating what was in agreement with sensation would not deny coming to be or perishing or motion and the multitude of beings. He agreed with appearances in these things, but also with those who posit the one in holding that there is no motion without void, and he says the void is not-being and nothing of what-is is not-being. For what in the primary sense is, is fully. But such being is not one, but infinite in multitude and invisible because of the smallness of the masses. (*On Generation and Corruption* 325a23–30)

The first step in establishing the theory of atomism is to accept in some way the evidence of the senses. To appeal to the senses does not by itself beg the question against the Eleatics. For neither Parmenides nor Melissus had used a denial of the senses as an early premise in his argument. In Parmenides, it comes at a relatively late point in the argument (B7.3–5) to forestall possible objections: the senses cannot contradict the arguments of reason. In Melissus it appears late (B8) to show that the evidence of the senses is inconsistent with the theorem that what-is does not change. To appeal to the senses, then, does not immediately contradict Eleatic principles, although it is inconsistent with results deduced from those principles. The atomist will somehow have to neutralize certain principles which entail the inadequacy of the senses, but of course, he will have to challenge some Eleatic principles anyway.

Suppose, then, that we simply accept the evidence of the senses that there is change in the world. From that we may argue to a plurality of beings:

A1

1.	There is change.	(empirical fact)
2.	If there is change, there is a plurality of beings.	
3.	Hence, there is a plurality of beings.	

Point 2, though not self-evident, is at least an obvious move to make for a philosopher steeped in the Eleatic tradition, and influenced by the views of the Eleatic pluralists. If there is only one thing, namely what-is, and if what-is is all alike, homogeneous or homoeomerous, then there is no internal change or differentiation in it. And there is change and difference only if there is a plurality of beings. The Eleatic opponent can try to block the deduction by attacking the senses, but the attack will succeed only if he can maintain the argument against the senses in face of the atomist attack.

Aristotle supplies an intelligent argument of his own for the common sense view concerning the value of appearances:

> We have in fact already argued that it is not possible for all things to be at rest, but we shall argue the point again now. For if things are truly like this, as some claim, and what-is is infinite and motionless, at least this is not at all how it appears to the senses, but many beings *seem* to move. If then there is false opinion, or opinion in general, there is also motion, and even if there is only imagination, or even if opinion varies, still there is motion. For imagination and opinion are considered to be kinds of motion. (*Physics* 254a23–30)

Even the *appearance* of motion is a fact that must be explained. We can explain that fact only by supposing that there is some kind of motion, even if it is confined to the realm of the mental. For even if we only imagine a change took place, our own perception of the change constitutes a change of belief on our part. Now the critic of motion must grant that there is an appearance of change, if he is to get his own argument started. But to concede an appearance of change is to concede a change of appearance, which entails that change exists in the world (for the perceiver and his perceptions are assumed to be part of the world). Aristotle does not attribute this argument to the atomists (it is his own), but one sees how it could be developed by anyone committed to the data of sensation.

Furthermore, there can be change only if the many change, at least on the model suggested by Melissus, in which the each of the many is itself possessed of Eleatic being and hence internally unchanging.

A1 (continued)

4.	There is (phenomenal) change only if the many change.	
5.	The many can change only in place.	
6.	Hence, the many change in place.	(1, 4, 5)

But if we take Melissus's argument **M1** seriously, we must deal with point 8 of that argument: if there is no void there is no motion. Now, by contraposition (transposition), **M1.8** entails

7.	If there is motion (change), there is void.	
8.	Hence, there is void.	(1, 7)

Precisely this argument is identified by John Philoponus.[12] We have now deduced a plurality of existents and the existence of void from the fact of phenomenal change and from Eleatic principles.[13] Of course it is now evident that what the conclusions of **A1** are incompatible with Melissus's theory. And we must now show how we can block Melissus's conclusions.

But before proceeding, let us take a look at the overall structure of **A1**. We have moved from the fact of motion to the existence of a plurality of beings, and to the existence of void. Our starting point is drawn from experience, and so we have one important empirical premise. But note that this is not an empirical argument in the usual sense of that term. The argument moves from what is given in experience (in general) to what would have to be the case to have an experience in general. In this it is much like a Kantian transcendental argument. In the transcendental argument we consider what exists universally and necessarily, and hence (according to Kant) a priori; we note that this could exist a priori only if some structure were present; and we conclude that this structure must be present. In the case of the atomists no a priori validity is advanced. But we are arguing from the fact of a certain kind of universal experience to the conditions which make that experience possible. Like Kant against Hume, we take some experience not as a dubitandum but as a starting point for argument. The strategy seems especially strong when the choice (the *krisis*, in Eleatic terms) is between either accepting experience or rejecting it *as a whole*. If we reject change and differentiation, we reject the whole of sense experience, whose point is to help us make discriminations between, e.g., what is edible and what is not, what will save our lives and what will kill us. If, moreover, we reject experience, what will we have to explain? If philosophy is an attempt to explain our experience, but there

---

[12] "Now Democritus, taking principles testified of by sensation, that there is division and plurality in things, as well as motion, on the basis of these he introduces the void, constructing the obscure from the evident. For if there is division, he says, and plurality in things, there is void; but there is division and multiplicity; hence there is void. He takes the conditional premise just like Parmenides, but not the minor premise. For Parmenides maintained that there is no void, which is obscure, but Democritus that there is multiplicity and division. And similarly with motion: if there is motion, there is void; but there is motion; hence there is void. Here too he takes the same conditional premise as Parmenides, but not the same minor premise." (John Philoponus *On Generation and Corruption* 155.10–19). The only problem with this exposition is the fact that Philoponus seems to put Melissus's remarks in the mouth of Parmenides; see esp. ibid., lines 19–22, where he attributes to Parmenides the view that what-is is unlimited.

[13] Furley 1987, 121 and n. 7, "corrects" his acceptance of this argument (which he now attributes only to the Epicureans) to: "[i] no motion without void; but [ii] there is void; therefore [iii] there can be [*sic*] motion." Let M = "There is motion"; V = "there is void." Then [i] translates to "$\sim V \supset \sim M$," which is equivalent to "$M \supset V$"; [ii]–[iii] translate to "V, therefore M," committing the fallacy of affirming the consequent. Furley's interpretation seems to agree with that of Gomperz 1939, 1: 347.

is no experience, what need is there of philosophy? If you and I do not exist, why should I try to convince you that my philosophy is right and yours is wrong? As Aristotle puts it:

> Now to claim that all things are at rest, and to defend this thesis disregarding sense perception is a case of intellectual failure—indeed it calls into question the whole of experience rather than some part of it, and not only in relation to the natural scientist, but in relation to virtually all the sciences and all judgments, since they all make use of motion. (*Physics* 253a32—b2)

These observations no doubt go far beyond anything Leucippus said in defense of his atomism, but they show that the appeal to experience is neither naive nor wrongheaded in a case such as he confronted. At some point it makes perfectly good sense to take one's stand with sense experience and to challenge anyone who would deny it to say anything meaningful at all. Indeed, we are reminded of the famous retort Democritus puts in the mouth of the senses:

> Wretched mind! After taking your evidence from us, do you overthrow us? Our overthrow is your downfall. (B125)

Though Democritus is well aware of the limits of sensation, he recognizes the indispensability of sense experience. Whatever one's views about the details of epistemology, experience is the beginning and end of philosophy; without experience there would be nothing to philosophize about and nothing to test theories against. Democritus seems to appreciate this fact as Melissus does not.[14]

Thus far, then, Leucippus's strategy seems defensible and he seems capable of some sort of sound argument for atomism, however obscure it may have been in his actual writings. But now that we have argued to a plurality of beings and to the existence of void, we must see how Leucippus could maintain these posits against a strong Eleatic counterargument. Our task will be somewhat simpler if we recognize that the real ontological question here is not about plurality but about the void. For the possibility of a plurality is, at least for the atomists, in turn founded on the existence of void. Empedocles and Anaxagoras posit qualitatively different kinds of matter in a continuum. But the atomists reject qualitative differences at the ontological level and accept the uniformity of matter. The only possible ground for plurality, then, is spatial separation, breaks in the continuity of matter. In fact, a problem arises as to whether any two atoms can be in actual physical contact with one another.[15] Thus,

---

[14] This argument is echoed in Lucretius 1.422–5, 693–700; cp. Epicurus *Letter to Herodotus* 38, *Principle Doctrines* 23.

[15] Philoponus *Physics* 494.19–25, *On Generation and Corruption* 158.26–159.7 = Leucippus A7; Aristotle *Metaphysics* 1039a9–11, with Taylor 1999b, 186–88; Taylor 1999a, 184–

there is a plurality of beings only if there is void to separate them. So the major ontological disagreement between the Eleatics and the atomists turns out to be the acceptance by the latter of void.

And precisely here the atomists seem to have been most boldly innovative. According to Aristotle,

> Leucippus and his companion Democritus say the elements are the full and the empty, saying they are being and not-being, respectively, and of these the full and solid are being, the empty not-being. For this reason they say that being is no more than not-being: because void is no less than body. (*Metaphysics* 985b4–9)

> [Democritus] calls place by these names: the empty [or: the void], not-thing, [*ouden*] and the boundless; and each of the substances by these: thing [*den*], the compact, and being. (*On Democritus* from Simplicius *On the Heavens* 295.3–5)

> Thing (*den*) is no more than not-thing (*mêden*). (Democritus B156)

The atomists call atoms and the void *to on* and *to mê on*, respectively, being and not-being, or what-is and what-is-not. Here, of course, they fly in the face of Eleatic theory, according to which what-is-not cannot be thought or expressed.

We find in Democritus B156 a remarkable argument—or a fragment of one—which defends the atomist principle of void. What are we to make of this argument? It is an "indifference argument" of the sort that goes back at least to Anaximander. Formally, the argument is quite simple:

1. Thing is no more than not-thing.
2. Thing exists.
3. Hence, not-thing exists.

The real question is why one should accept the first premise, which flies in the face of strong arguments to the contrary from Parmenides and his followers.[16] Stephen Makin, who has studied indifference arguments in

---

85, 188–89. This account is problematic: in other accounts motion is caused "by impact" (Democritus A47). If we do not recognize impact, we must allow some sort of action at a distance, as Taylor notes; but the atomists do not expressly acknowledge such an action.

[16] Matson 1963 stresses the fact that Democritus puts the accent on the negative side: "Hing exists no more than Nothing" (using Kirk's nonsense syllable as a translation of nonword *den*—see next two notes), which is intended to show that being will be not be any more real than not-being; if then not-being does not exist, neither does (Eleatic) being. But why should the Eleatic accept this claim? According to Matson, the argument goes: "If not-*x* is necessarily not-*y*, then *y* is necessarily *x*. So if (as Melissus insists) μη-δέν, not-hing, necessarily is not-existent, it follows that what necessarily *is* existent is the δέν, hing. But δέν, hing, so far from being a necessary existent, *is not even a word*!" (29, italics in original). This whole argument seems rather *recherché*, and is not echoed in ancient testimonies. The view is criticized by McGibbon 1964, who, however, hints at a missing argument he fails to

early Greek philosophy in detail, surprisingly has almost nothing to say about this particular one, which has far-reaching consequences for the history of thought.[17] Why does Democritus find it compelling? And what are the consequences he draws from it?

The challenge is to justify (1). Consider the following argument, which I shall call an ontological argument:

O

1. For every F, if F is, we may understand F.
2. To understand F, we must understand not-F.
3. Hence, not-F is conceivable (meaningful, intelligible).
4. If something is conceivable, it is.
5. Hence, not-F is.

If we look for some presuppositions which would lead us from accepting the existence of F to recognizing that of not-F, we find some quasi-Eleatic assumptions. Parmenides holds that what-is is available as an object of thought and expression. We point out that to understand F is to know how it is different from other things which are not-F, implying that we understand not-F, etc. But the argument is elaborate, the presuppositions extravagant, and the whole argument sounds like an anachronistic anticipation of Aristotle's *Metaphysics* IV.

We might try a simpler argument I shall call the semantic argument:

S

1. For all F, if F makes sense, not-F makes sense.
2. F makes sense.
3. Thus, not-F makes sense.

This argument is at least simpler and more direct than the previous one. But it suffers from a fatal problem. It starts by positing a term F; but in Democritus's argument that term is filled by *den*, a strictly nonsensical syllable.[18] The argument gets its force and meaning from the negative pronoun, *ouden*. Somehow, the argument starts from a derivation of the positive pronoun from the negative. The argument is, indeed, based on a wordplay by which we derive a nonexistent positive term from an existing negative one.[19] How are we to account for that? Surely not on the hypothesis that the negative term gets its meaning from the positive.

---

articulate (254–55). See further Klowski 1971, who criticizes both Matson and McGibbon and sees in the void Leucippus's criticism of Parmenides' limited being.

[17] Makin 1993, 98. Similarly Taylor 1999b, 161–62, seems to gloss over the problem.

[18] It is formed by ignoring the formative elements *oud(e) hen*. See next note.

[19] See Moorhouse 1962, who argues that even the one antecedent of Democritus, Alcaeus fr. 320 Lobel-Page, does not differ in formation or meaning from Democritus's term. The positive *den* is dependent on negative *ouden* or *mêden* for its sense.

The wordplay effectively accomplishes the isolation of the negative particle, the morpheme *ou*, which marks the difference between the two terms. What the particle negates here is not a verb (though *ou* is the standard negative particle for verbs, "not"), but a substantive—or what Democritus by juxtaposition shows to function as a substantive. Formally, the term *ouden* is taken to signify "not-thing." The move is crucial because it shows Democritus using the negative to negate a predicate rather than a verb (see Furley 1987, 119–20). Now "nothing" does not mean "nonexistent" or more generally "not-being," but "not-thing." To be nothing is not necessarily to be nonexistent.

The argument must be an attempt to rehabilitate "nothing" as an potential reality, since its present opposite, *den*, is a virtual unknown. But toward whom is the argument directed? Presumably toward someone who had attacked the reality of nothingness. Parmenides is not a likely candidate, for his argument does not turn on the use of "nothing."[20] Now in Melissus B7 (cited above) we do have an argument that turns on the concept of nothing. The argument, captured in M1.1–3, identifies the void with nothing and declares that nothing is not (does not exist) to reject void. What Democritus's argument accomplishes is to disprove statement (M1.2), thus rendering the argument unsound.

If something like Melissus's argument is the target of Democritus's argument, we may see that the latter does not need an elaborate background theory to make his argument work. He merely needs to block the inference from "nothing" to "nonexistent." And it appears that the analysis forced on us by the playful juxtaposition of "thing" and "not-thing" is sufficient to show us a sense in which nothing could exist. Not-thing is that which contrasts with whatever it is that we identify as a thing. From this we could perhaps generate something of a background theory. Whenever we perceive or think of a thing, we implicitly contrast it with what it is not, not-thing. We identify the thing precisely by distinguishing it from everything else around it, including its *periechon* or surroundings. Far from thought excluding what-is-not, in this sense the possibility of thought presupposes what-is-not as a contrasting item. True, we may not know not-thing directly or primarily, but we must in some sense be aware of it in the very act of knowing thing.

Barnes (1982, 402) translates the last sentence of the *Metaphysics* passage, "That is why they also say that what exists exists no more than what does not exist—because the empty [exists no less] than body," and glosses: "The void is non-existent; the void exists: hence the non-existent exists."

---

[20] The term μηδέν appears at B6.2 in an ambiguous statement; later it appears in B8.10 as an apparent synonym for τὸ μὴ ἐόν, but as a kind of afterthought. Nowhere does the argument focus on nothingness per se.

He goes on (403–4) to develop two senses of existence, revived from Frege, by which to disambiguate the sentence "the non-existent exists" and to dissolve the paradox. But this is to miss the whole point of the atomist reform. For it is a reform not only of metaphysics but also of language. Before Parmenides the Presocratics tended to analyze reality in terms of opposites; Heraclitus had emphasized change as an alternation between opposite states. But Parmenides attacked such thinking by arguing that being and not-being cannot be treated as alternatives: there is a fundamental asymmetry between what-is and what-is-not. The former is intelligible, the latter is not. But all other contraries, insofar as they exemplify a contrast, embody a distinction between what-is and what-is-not. Contrariety itself proves to be unintelligible and impossible. There is only qualitative uniformity and (by parity of reasoning, since the dense and the rare are contraries) quantitative consistency. All this is achieved (on one plausible interpretation) by reducing being to existence and not-being to nonexistence. Ontologically speaking, existence and nonexistence are not equal but opposite states: existence is everything, nonexistence is nothing.

What the atomists have done is to go back to thinking in terms of contrasts. They line up atoms and the void as contraries and then make the ultimate leap: they identify one of them with being and one with not-being. The obvious inference is that they are not thinking in terms of existence and nonexistence at all, but in terms of some other more subtle contrast. And Democritus brings out that contrast in terms of his opposition between *den* and *ouden*, "thing" and "not-thing." By isolating *ou* as a morpheme he shows it as negating a noun (or noun substitute), i.e., as negating a predicate. Moreover, the contrast shows that the negation is of a certain type of predicate, a predicate representing the subject as a thing. "Nothing" does not mean pure nonexistence but the absence of substantiality. It is to be understood as "not-F," or rather, given a meta-variable Φ ranging over all substantive predicates, "not-Φ."[21] So the negative particle, properly understood, attaches not to the verb "to be" but to a substantive pronoun. We can safely deny substantiality without talking nonsense.

What has so far not been properly appreciated is that the analysis implicit, at least, in Democritus's contrast provides a far-reaching correction to Eleatic reasoning. It is generally recognized that no fully adequate reply

[21] Cf. Furley 1987, 120: "[P]erhaps [Democritus] meant to indicate that 'nothing' is not always a negation, but sometimes a negative predicate, of the form 'not-x.' " This comes close, but the problem is not with negation, a syntactic feature of language, but with nonexistence, and the alternative is not just any negated predicate, but one that negates thinghood or substantiality. The general point is seen also by Schofield (KRS 415). Sedley 1981, 182–83, though he is wary of Barnes's formulation, favors two senses of existence.

to the Eleatics appears until Plato's *Sophist* and Aristotle's *Physics* I. Both Plato and Aristotle employ a sharpened awareness of semantics linked to a powerful metaphysical background theory to explore the sense or senses of being and particularly not-being. But the atomists are not given proper credit for the insight that breaks the Eleatic argument: that there is some sense of being/not-being apart from the existential sense.[22] Yet clearly Democritus's wordplay indicates that he has grasped the fundamental truth, that one can utter a negative sentence without denying existence. Once we see that point, the stranglehold of Eleatic reasoning is broken. We can have our cake and eat it too: being and not-being, full and empty, a plurality of beings separated by not-being. It is likely that neither Leucippus nor Democritus was equipped to provide an articulate defense of the insight he had achieved.[23] Linguistic science was still in its infancy.[24] But the atomists at least showed how it would make sense to allow not-being back into philosophical theory. They had discovered, if not articulated, semantic pluralism as a basis for asserting ontological pluralism.

At this point the atomists have shown how the notion of void is a coherent one, and have defended conclusion A1.8, there is void, against their Eleatic opponents. For they have showed tacitly that the argument against void is flawed by a failure to make the relevant semantic distinctions between non-being and not-being-something.[25] The remaining problem is to show that by their defense of atomism the atomists have not totally

---

[22] Furley 1987, 122, observes, "[I]t becomes clear that although the Atomists admitted the being of what *is not*, they had not yet progressed as far as Plato in the *Sophist*. They had not reached the notion of 'is not' as equal to 'is other than,' and so they were not yet ready to admit that a subject that *is* may nevertheless bear innumerable predicates containing 'is not.' " It is of course true that the atomists do not launch a general semantic theory on the basis of their insight. But the key point is not that they fail to see the full significance of "is not," but that they see that "not" can attach to terms other than "is."

[23] Schofield 2003a has recently argued that the *ou mallon* argument for the void should be attributed only to Democritus, not to Leucippus, on the authority of Theophrastus.

[24] Prodicus was interested in lexicography, while Protagoras made elementary distinctions between the gender of substantives (Aristotle *Rhetoric* 1407b6–9 = A27) and moods or illocutionary acts (Diogenes Laertius 9.53–4 = A1). The basic distinction between the parts of speech of noun and verb is not found until Plato.

[25] Curd 1998, 188ff., offers a good review and criticism of previous explications of the atomist argument for the void. Her own view (198–206) is that the void is rescued from not-being by being knowable to the understanding, stressing Plutarch's point (*Against Colotes* 1108f) that the void has a nature and existence of its own (204). But there are problems with this claim. First, Simplicius points out that Democritus identifies *phusis* with atoms, not the void (*Physics* 1318.32–1319.1, a point stressed by Salem 1996, 65). Further, the *den* argument stresses the negative character of the void, as does its role as *to mē on*. Also, if we take Parmenides' four properties as the criteria for essence, even if we can accommodate being without generation and destruction, being everywhere alike, and changeless, we cannot make void complete in the sense of having limits. Finally, there is no evidence that Democritus gives a positive account of the nature of void.

overthrown the Eleatic project. Once we establish the multivocity of being, why should we fear the Eleatic elenchus? And why should we adhere at all to the restrictions placed on inquiry by an exploded theory? Yet the atomists adhere almost slavishly to as much of the Eleatic program as they can accommodate.[26]

In B2 Parmenides argues that there are only two ways of inquiry, the way that it is and the way that it is not. The latter way is not thinkable or expressible, so it is excluded in favor of the former. In B6 we learn that the previous argument has ontological implications: what-is *is*, while nothing is not, *mêden d' ouk estin*. This last point is what the atomists are committed to denying: in some sense nothing, that is, not-thing, is. But it seems possible at this point to challenge the argument. Parmenides is right to say that we can conduct inquiries into the world only in terms of what-is, for only that is knowable directly. But it does not follow from this that what-is-not does not exist at all. The ontological claim does not follow from the epistemological or heuristic. To think that it does is just to confuse two senses of "be," which should be kept distinct. So let us reject the ontological claim that what-is-not does not exist. Can we keep the Eleatic program intact? We might wonder whether we can. For Parmenides' argument that what-is does not come from what-is-not in B8 depends directly on the argument of B2 under the ontological interpretation (B8.7–9). And that point becomes the basis of further argument.

Here, however, the atomist can settle for a more modest version of the argument. We have not shown that what-is-not does not exist. But the fact that what-is is directly knowable while what-is-not is not indicates that they have different properties. (I shall assume here that the fact that these are intentional properties or properties relative to external subjects would not count adversely for either the Eleatics or the atomists.) By an intuitive application of Leibniz's law, we can infer that they are different and, because the properties are taken to be fundamental, fundamentally different. We have not shown that one entity or set of entities exists, the other not, but rather that one entity is radically different from another. For it is basic to what-is that it is knowable, to what-is-not that it is not. Further reflection will show that the former is limited in extent, the latter unlimited; the former is solid, the latter rare, etc. What we have established is a radical difference in the basic categories of reality. In a similar way Descartes argues that we can conceive of our own existence without a body, but we cannot conceive of it without thought, and hence we are essentially thinking things irreducible to extended things. The fact that we can conceive of mind without matter and matter without mind counts as a compelling argument for their essential distinctness. So the atomists

---

[26] This problem is often overlooked, but it is appreciated by Furley 1987, 122.

may infer the essential difference between atoms and the void. What follows from the argument of Parmenides B2 is not, then, a thoroughgoing monism, but a radical dualism. The argument against coming to be in Parmenides B8 can now be founded not on the impossibility of there being not-being, but on the impossibility of not-being being identical with being. What-is cannot come from what-is-not because they are radically heterogeneous.

Now we can preserve the set of Eleatic properties with only minor modifications. There will be no coming to be or perishing of what-is (or of what-is-not, for that matter). What-is will be all alike qualitatively, like separate pieces of gold, in Aristotle's analogy.[27] Parmenides seems to want to say more than this: he wants no gaps in what-is. Of course, the atomists introduce gaps, but not internally into the individual atom, which is uniform. Furthermore, the atom will be without change. The major qualification we must make here is to allow the atoms to undergo locomotion. They do not, however, change in any essential way, in their internal constitution or their shape or size, and to this extent they are changeless. Finally, the atoms are complete in themselves, requiring no external source of fulfilment, and they do meet with limits (though not necessarily with spherical boundaries). Thus we can provide an interpretation of each Eleatic property in light of our reevaluation of not-being. The result is, no doubt, a modified or reformed Eleaticism, but it is recognizable as an Eleatic theory. From the atomists' point of view, they have not destroyed Eleatic theory but reformed it by synthesizing it with the Ionian program. We concede that there must be not-being to allow for phenomenal change. But we explain change strictly in terms of unchanging essences in unchanging individuals, founding our explanation on what-is. The confusions of mortals that Parmenides rails on in B6 are thus avoided: we have kept from saying that to be and not to be are the same and not the same, for we have maintained a strict ontological distinction, a far cry from the generating substance theories of the early Ionians. Methodologically, the Eleatics are right, but they must be careful not to throw the baby out with the bathwater.

Let us return briefly to argument M2, in which Melissus argued against the existence of the many on the grounds that if we see and hear correctly and there are many, they both change and do not change. To this the atomists now have a good reply. The many that we perceive as changing are not the ultimate realities. We can defeat the argument formally by accepting the conclusions up to M2.8. Then we must accept 9a rather than 9. And finally we must reject (10) "Not (a)" in favor of

---

[27] *On the Heavens* 275b32–276a1, cf. Plato *Timaeus* 50a–b.

10a. Not (b): We do *not* see and hear and understand rightly.

We do not perceive rightly because the many that we perceive are not the many that do not change. But there are entities which do not change, invisible to us but knowable through reason and through their effects on us. That there is change is an undeniable fact. That the ultimate realities are unchangeable is a truth of reason. In order for change to be possible, there must be many such realities moving in a void, and since we do not perceive such things, they must be below the threshold of our perception. Thus our perception is not reliable insofar as it does not reveal the ultimate realities to us. But here reason can take over and tell us what we need to know. In Democritus's words,

> Of understanding [*gnômê*] there are two kinds, one legitimate, one bastard. Of the bastard kind are these: sight, hearing, smell, taste, touch. And there is the legitimate kind, which is distinct from this. . . . Whenever the bastard kind is no longer able to see anything smaller or hear or smell or taste or perceive by touch . . . [sc. the legitimate kind takes over]. (B11)

We can, then, accept Melissus's argument so long as we identify the contradiction as resulting from a failure of perception, not of ontology. Our senses are unreliable in reporting what really exists, but a many does really exist. Alternatively, the atomists can point out that "the many" is ambiguous in **M2** between the many realities and the many appearances. Until the ambiguity is clarified, the argument will be at best misleading.

We may now see the strategy of the atomists in overview. Being challenged by the Eleatics' arguments that seem to entail that what-is is one and unchanging, they begin with an argument from the world of experience to the presuppositions of that experience, namely atoms and the void. Confronted with Melissus's argument that the void is nothing and what is nothing is not, Democritus produces a clever argument to show that nothing, qua not-thing, is just as intelligible as thing and indeed presupposed by our understanding of substantial beings. There is, then, no reason to dismiss the notion of the void, and every reason to accept it as a foundation for the possibility of our experience. The argument not only answers the Eleatic objection, but also tacitly shows how to overthrow the Eleatic position altogether: what-is-not is just as defensible as what-is, and hence there is no reason to rule out the way of Is-not at all. The latter point, however, is not exploited by the atomists so far as we can see. For them the possibility of not-being allows them to turn Eleatic monism into a dualism of principle: what-is, or atoms, and what-is-not, or void. They remain committed to an Eleatic program like that of the *Doxa* in which phenomena are explained in terms of unchanging principles. By admitting what-is-not as a principle, they have reformed the Eleatic program; by maintaining explanation as a derivation from changeless elements they adhere to Eleatic principles.

In the end we see that the atomists do have an intelligent and even compelling case against the Eleatic critics of pluralism. It is not clear how articulate they are in presenting the case, and to what extent the argument is fully spelled out. But the things they say can be fleshed out to provide a cogent reply to criticisms and a firm foundation for further theory construction. It seems consistent with the principle of charity to attribute to them something like the argument that makes sense of their responses to the Eleatics. We see also that the atomists do not seek to dismantle Eleatic theory but simply to neutralize its most paradoxical conclusions so as to accommodate it to the Ionian project of scientific explanation. They, like the earlier pluralists, see in Eleatic substantialism the solution to the problem of change.

9.3 Atomism and EST

Like the early pluralists, the atomists are committed to explaining the cosmos as a set of natural phenomena, and of doing so within the framework of Eleatic assumptions. Let us recall the structure of the Elemental Substance Theory:

EST

1. There is a set of substances $\{E_i\}$ which are the basic substances.
2. The E_i are permanent existences.
   a. The E_i are (i) without coming to be and perishing, (ii) homogeneous, (iii) unchanging, and (iv) complete. (Eleatic Substantialism)
   b. The set has a plurality of members. (Pluralism)
3. Derivative substances S_j are a product of relation R_k of E_i.
   a. (Definition) The E_i are *elements*.
4. There is a mechanism M that controls the production of R_k.
5. There is a set of forces $\{F_l\}$ that governs M.
6. (a) The world comes to be through the orderly application of the F_l to E_i, and (b) continues to exist through a balance of forces.

Does atomism instantiate EST? Clearly the ontological requirements (1–2) are satisfied. There is a plurality of substances making up the basic set, namely the atoms. And those atoms are Eleatic substances. As we noted, they do change in respect of place, but then so do the substances of Empedocles and Anaxagoras. Otherwise they are changeless. The atoms differ from the substances of Empedocles and Anaxagoras in that they are all qualitatively alike, being individuated only by their place or by their boundaries with void. Yet this difference does not affect the status of the

atoms as a plurality of Eleatic substances, and in this there are no differences with the early pluralists.

The macroscopic objects we experience are made up of microscopic
atoms. Democritus specifies relationships which enter into the construction of a complex body:

> And just as those who make the underlying substance one generate other
> things from the affections of this (positing the rare and the dense as sources
> of these affections), in the same way [the atomists] say the differences [of
> their atoms] are the causes of all other things. Now they say these are three,
> shape and order and position. For they say what-is differs by "contour" and
> "contact" and "rotation." Of these contour is shape, contact is order, and
> rotation is position. Thus A differs from N in shape, AN from NA in order,
> and Z from N in position. (Aristotle *Metaphysics* 985b10–19)

If all atoms are alike qualitatively, they can differ only in their shapes
and relations to other atoms. Bodies compounded out of atoms get their
properties from the shapes of component atoms and the arrangement in
which they are linked together. Furthermore, we are told that when bodies
are more densely packed, i.e., have less void, they are heavier. In principle
all the properties of perceptible bodies are derivable from the properties
of the arrangement of atoms—in relation to a potential perceiver. Now
in practice the analysis cannot be carried out, for we have no direct access
to the atoms. But for all Democritus's skepticism about firm knowledge
of the world, he is no less convinced that in principle whatever properties
a body has is the product of its atomic composition. Hence, a compound
substance S is the product of some relation R of some set of elements E
(EST-3).

The atoms interact, according to the atomists, simply by traveling
about in whatever path their last collision impelled them, until they have
another collision with an atom. In some cases they become entangled, and
if enough of them join together they can generate a perceptible body.
There are processes by which like atoms get sorted with like atoms, as
when pebbles of similar sizes are collected together by the action of the
waves on a beach.[28] But all of this is the result of atoms in motion coming
into contact with other atoms in motion under the same conditions. The
processes of nature are purely mechanical processes without the intervention of any higher will or purpose. We find in Leucippus and Democritus
no such personified forces as the god of Xenophanes or the Love and
Strife of Empedocles or the Mind of Anaxagoras. Such forces at least
bring the trappings of external agencies controlling the world as had the
deities of Homer and Hesiod. Furthermore, we find no internal agency

---

[28] Democritus B164.

either, such as had been hinted at by the Milesians with their generating substance that steers all things. Even Epicurus would later endow his atoms with an ability to swerve spontaneously. But Democritus's atoms do not swerve.

Finally, there seems to be no action at a distance in the early atomists.[29] It is not entirely clear how personified forces act among the Eleatic pluralists. They may operate by contact, as Empedocles B35 seems to suggest. Anaxagoras's Mind is a physical body, even if it is not like other bodies, and it too may act only where it is physically present.[30] Xenophanes' god, by contrast, practices a kind of telekinesis without physical effort.[31] In the case of the atomists, there can be no causal connection except by collision and possible subsequent entangling. There is only matter in motion, with one being communicating with another only by direct physical contact. Though the atomists develop no mathematical physics, they would be comfortable with the conservation of matter and energy and with Newton's first two laws.

In sum, atomism appeals to mechanical means to account for the relationship of atoms. It is more purely and resolutely mechanistic than any other theory we have examined, dispensing even with the appearance of foresight or intervention in the processes of nature.

9.4 Birth of the Cosmos

Thus far we have been dealing with first principles. But one of the most important features of the atomist system is its ability to explain the sensible universe in terms of the invisible world of atoms. Unfortunately, we have only one reliable account of cosmogony, and that is brief and rather obscure. Yet it is the basis on which we must reconstruct the atomist cosmogony:

> [Leucippus says] the worlds come to be in this way: "by being cut off from the boundless," many bodies having all sorts of shapes travel into a "great void," which being gathered together produce a single vortex, in which, as they strike each other and circle every which way, the like are separated apart

---

[29] C.C.W. Taylor supports action at a distance, citing Philoponus (see above, n. 15). But in our early sources and in the doxographic tradition we get no report of such action, but rather claims that atoms strike each other and sometimes grab hold of each other: Aristotle *On Generation and Corruption* 325a32–6, Simplicius *On the Heavens* 242.23–6, Democritus A47; the sources, including Aristotle, do not seem to notice any contradiction between separate atoms being in contact. See Konstan 2000, 130–32.

[30] Anaxagoras B11, B12, B14.

[31] Xenophanes B25, B24, B26.

to the like. When they are no longer able to travel around in equilibrium because of their number, the light ones move to the void outside as if they were sifted, while the rest "stick together" and becoming entangled run around together with each other and make a first spherical structure. And this like a membrane stands apart, containing in itself all sorts of bodies. While the bodies spin around in response to the resistance of the middle, the surrounding membrane becomes attenuated as the inner bodies are eroded by contact with the vortex. And so the Earth came to be, when the things borne to the middle stuck together. And again the membrane-like container grows by an influx of bodies from outside, and as it is borne by the vortex, it captures whatever things it touches. Some of these becoming entangled form a structure, at first moist and muddy, but then being dried out as they are carried about with the vortex of the whole, they are ignited and produce the substance of the stars. (Diogenes Laertius 9. 31–2 = Leucippus A1)

A general picture emerges from this description, specifically attributed to Leucippus. It has been noted that the puzzling detail of "being cut off from the boundless" echoes a thought pattern going back to Anaximander and tends to confirm the antiquity of the report, even as it suggests Leucippus had difficulty separating his type of story from that of his predecessors.[32] The "great void" where the atoms collect to create a world is likewise reminiscent of the *chaos* required from Hesiod on to provide room for a cosmos.[33] But other puzzling features remain. What is the point of sending the light atoms off into the void if they are not going to be part of the world-making process? Why does the vortex produce a spherical shell when a vortex motion seems to be cylindrical or conical? Why does the center resist the motion of the vortex? Why do heavy things collect there?

The whole question of weight and the distribution of bodies has been a vexed one for scholarship. While the evidence tends to point to atoms having weight for the early atomists,[34] it is difficult to know how to make

[32] Cf. Bailey 1928, 92.

[33] Hesiod *Theogony* 116. For etymology and explanations, see West 1966, 192–93. Competing theories have it that chaos is either the space below earth separating earth and Tartarus, or above earth separating earth and heaven. But we should not overlook the possibility that chaos is the whole space in which the cosmos comes to be, separated only by the birth of Gaia into an upper and a lower space; on this interpretation it is a sort of womb of creation.

[34] Of the texts collected at 68A47, two seem to say Democritus's atoms do not have weight, Aëtius 1.3.18, 1.12.6. But these statements can be reconciled with texts implying that they do have weight on the assumption that Aëtius is speaking of a lack of downward motion outside the dynamic forces of a cosmic vortex. See Taylor 1999b, 179–84; Chalmers 1997. O'Brien 1981, 1: 345, thinks the atoms perhaps were assigned weight. Furley 1983 thinks this account cannot account for the phenomena of falling bodies.

sense of weight. For there is no theory of gravity per se, nor do the early atomists claim there is a universal tendency for atoms to move downward, as do the Epicureans later. What then accounts for the phenomenon of heavy bodies falling? Without going into great detail, we may identify a possible solution. Atoms have something like momentum, which is determined by both the motion (velocity, in modern terms) and the size: larger atoms have more momentum because they have more being and strike with more force against other atoms. The property atoms have is resistance to change of motion, or mass. Mass can be turned into weight in the proper physical or meteorological environment. If motion in the center of a vortex is minimal, as in the eye of a storm, heavy bodies will tend to collect there. Lighter bodies will tend to be picked up and carried, as well as maintained, aloft. One significant difficulty is that the center of a vortex is a vertical axis, not a flat plane. Why does the material around the central axis settle into the flat plane of the earth? One possibility is that there is some downward pressure from the top of the vortex. If heavy bodies were susceptible to the pressure of a cosmic downdraft, they would tend to squeeze more rare bodies upward. Some such picture is at least suggested in our other main report of atomist cosmogony:

(1) The world arose with a curved[35] shape in this way. As the atomic bodies with an unplanned and random motion were moving continuously with high speed many of them gathered to the same place and thus manifested an abundance of figures and sizes. (2) As they gathered in the same place, those that were larger and heavier settled down everywhere, while those that were small, round, smooth, and slippery were squeezed out with the confluence of atoms and carried into the upper regions. Now as the pressure decreased with altitude it could no longer push them higher, but it did prevent them from falling down, and it pushed them into the regions that could absorb them. These were the outer regions, where the multitude of bodies circulated. They became entangled with each other in the revolution to produce the heaven. (3) The atomic substances being of all sorts, as has been said, and ever maintaining their own peculiar character, those that were pushed upward formed the nature of the stars. The mass of bodies that were evaporated pressed on the air and squeezed it out. When this became wind by reason of its motion, it surrounded the heavenly bodies and carried them around with it so that they maintain their present orbits above. Then from the atoms settling down the earth was generated, while from those aloft were produced heaven, fire, and air. (4) Since a great deal of matter was still contained in the earth, which was compacted by the force of the winds and breezes from the stars, all the fine-shaped formations of this were squeezed

[35] περικεκλασμένωι the root meaning of the term is "twist around" or "bend" (LSJ).

together to produce the moist substance. Since this had a fluid character it flowed down to subterranean hollows, which were able to hold and contain them; or as it sank down the water was able to hollow out the subterranean regions by itself. So the principal zones of the world were generated in this way. (Aëtius 1.4.1–4 = Leucippus A24)

This account is more detailed and in some ways perhaps more coherent than the passage from Diogenes Laertius. But it has some features found in Epicurean accounts and for that reason is suspected of having been confounded with later theory.[36] Of course, it is possible that some common source influences both the present passage and Epicurus. The present passage also differs in that it makes the light bodies squeezed out by the vortex to become the heavens rather than to be expelled from the cosmos altogether. It is not clear what relationship the present account has to the account preserved by Diogenes. But it does at least provide us with some additional possibilities where we have so little reliable evidence. In any case, on this account it appears that some pressure is exerted on heavy bodies to force them to the center. There is also some pressure from evaporation, which tends to drive light bodies upward.

One possible way to reconstruct the theory is to suppose that the vortex does not simply create a closed rotation of bodies in the upper atmosphere, but that it draws in new material, e.g., from the earth and its peripheries. In the vortex air spirals ever upward as it circles the earth until it reaches the apex of the whirl, at which point it flows downward toward the earth, pressing bodies down with its motion. Massive bodies, being resistant to motion, tend to move down but to stay put once they have reached the equator of the vortex, where motion is most sluggish. Resistance grows and only lighter bodies are carried away from the center plane, forming a massive body that becomes the earth. The theory would be strengthened if we supposed that from the celestial equator downward the air spirals down, producing a mirror effect with pressure up to the central plane from the south celestial pole. Yet what little information we get tends to indicate that the earth was supposed to ride on a cushion of air as in Anaximenes, where the most obvious reading is that the air is trapped below the earth, not pressing it upward. The present reconstruction must remain highly speculative, given the scarcity of evidence. But here is at least one way a theory such as that of the atomists might account for the phenomenon of gravity within the cosmos.

Whatever the details of cosmogony, the atomist theory is remarkable for presenting a cosmogony that is purely mechanistic. Or, to put it an-

---

[36] E.g., Bailey 1928, 143, n. 2, following Liepmann 1885, 19–30. Zeller 1919–1920, 1104–5 and 1105, n. 1, takes Pseudo-Plutarch as providing valuable evidence, even if filtered through an Epicurean source.

other way, the mechanism that the atomists use to account for the cosmos operates purely and explicitly without a mechanic. Even as the earlier pluralists had moved in the direction of mechanistic explanations, they had kept the vestiges of the voluntary or rational agent, the Love and Strife of Empedocles, the Mind of Anaxagoras. They are agents capable of purely consistent action, but still endowed with a heart or a mind. For the atomists, by contrast, there is only a space bombarded by randomly moving atoms, which by chance interact synergistically to produce a cosmic storm. The storm produces its own shell which will preserve it intact for an extended period of time, while in the calm at the eye of the storm heavy bodies collect and flatten out to form an earth, while light bodies ignite to form heavenly bodies illumining the earth. The occasion for the cosmogony is purely chance meetings of atoms. The process is in principle repeatable, and, given the vastness of space and matter, it will be repeated, though not with precisely the same results in each cosmos.[37]

On the model produced by the atomists, a chance meeting of atoms can produce an arrangement that is self-sustaining. When the motion of an assemblage of atoms becomes a vortex, it sifts atoms in such a way that like atoms congregate with like. A membrane of lighter atoms surrounds the central storm, while heavy atoms settle into the center of the vortex. Further atoms are taken into the system and arranged by the dynamic interactions so as to reinforce the structures already present. The cosmos becomes differentiated into a dynamic but stable system. The atomists go on to explain how living things emerged in our cosmos, then how humans emerged and elaborated their culture. But the general principles are the same: chance arrangements of atoms which tend to preserve themselves and to survive in the environment become stabilized and dominate in their sphere. Those that do not preserve themselves disappear. In the biosphere a process of natural selection takes over to weed out the poorly adapted in favor of the well adapted. The strong survive, the weak perish.

In general atomism builds order out of disorder by a reverse entropy process. Sometimes by chance unstable conglomerations of atoms produce stable systems. Those systems expand to order the disorderly processes around them. In the case of the cosmos, increasing order gives rise to ever more orderly subsystems, ultimately to plants and animals and human beings. No external guidance is needed: order supervenes on disorder. The vortex grows out of random collisions; the cosmos evolves out of the vortex; the biosphere evolves from natural substances sorted by the vortex; humans evolve from animals; humans develop speech and characteristic behaviors that give rise to society and civilization. Finally humans come to understand the cosmos. Yet in all this no external agency is re-

---

[37] Hippolytus *Refutation* 1.13.2 = DK 68A40.

quired to introduce or maintain order. Increasing degrees of organization result from a world stabilized by its own natural processes. Order arises naturally out of disorder, organization out of chance, life out of inorganic processes, reason out of blind force.

Atomism is an instance of EST in which the elements, together with their motions, do all the explaining. No external agents or internal powers are needed. Nature is just matter in motion, and explanation is just showing how in principle particles of matter in motion can account for the existence of a cosmos and for all the phenomena of the cosmos. Nature has been thoroughly demythologized and explanation exclusively founded on mechanical interactions. Ultimately, nature has become completely autonomous.

10

DIOGENES OF APOLLONIA AND

MATERIAL MONISM

WE HAVE COME almost to the end of the story, and yet there remains one problem that we have not addressed. Where did the notion of Material Monism that is so prominent in Aristotle and the doxographical tradition arise, if not from the early Ionians? How did it come to be so prominent in the historiography of early philosophy without its being an important feature of that philosophy? Could it really be only a mirage? Certainly Aristotle did not invent Material Monism. By taking a closer look at the appearance of MM, we can form a clearer picture of how it came to be thought of as the original approach of Ionian philosophy. In this chapter I shall argue that the source of MM is Diogenes of Apollonia, a figure who has occupied an anomalous role in accounts of Presocratic philosophy. Though most assessments of his contribution have assigned him a minimal significance at best, we shall find reason to see him as an important figure.

10.1 Diogenes in Modern Accounts

Diogenes was a member of the last generation of natural philosophers to have an impact before Socrates. His activity can be dated by the fact that his cosmology is parodied by Aristophanes in the *Clouds*, performed in 423. Diogenes portrays air as the single source of all things. Thus he is obviously influenced by Anaximenes' theory. He clearly holds that all things not only arise out of air but *are now* air, that is, forms or manifestations of air as it appears in different guises. Thus he holds Material Monism. According to orthodox interpretations, he is asserting the same theory that Anaximenes introduced in the previous century—though it is now obsolete. Thus Diogenes can be seen as a throwback to an earlier time, someone who has missed the point of Eleatic criticisms and who does not appreciate the need for a pluralistic theory to answer them.

Another approach to Diogenes is to see him as attempting to combine several points of view. Theophrastus seems to support this reading:

Diogenes of Apollonia, who was one of the last to study this kind of thing, wrote mostly in a composite manner, following either Anaxagoras or Leucippus. He too said the nature of the totality is air, which is boundless and everlasting, from which, by condensation and rarefaction and change of affections, the form of everything else comes to be. Theophrastus reports these views of Diogenes, and the treatise that has reached me, entitled *On Nature*, clearly makes air that from which everything else comes to be. (Simplicius *Physics* 25.1-8 = A5)

The word I have translated "composite," *sumpephôrêmenôs*, is often rendered "eclectic," and consequently Diogenes is treated as an eclectic philosopher. The Greek term does not quite mean "eclectic"—it is typically used of things jumbled together[1]—yet it at least suggests a lack of unity and cohesiveness. According to Simplicius, he was dependent on Anaxagoras and Leucippus at different points—not to mention Anaximenes for his ultimate insights. While Anaxagoras and Leucippus are more recent figures and hence more modern or up-to-date, they also have very distinct theories, both from each other's and from Anaximenes' theory. Hence the interpretation of eclecticism raises the question of how Diogenes could possibly combine such different viewpoints in a meaningful unity. He may be broad-minded and versatile, but Diogenes can hardly be a consistent philosopher if he draws on such disparate sources, at least if he combines them in an indiscriminate way. "By common scholarly consent, he was least as well as last" (Barnes 1982, 567).

The one positive notice that Diogenes has enjoyed in recent times has been that of Willy Theiler. In a dissertation originally published in 1925, Theiler looked for the sources of Xenophon's argument for God from design.[2] The trail led to Diogenes, so that Theiler could declare Diogenes the founder of teleological thinking in the Greek tradition. Theiler's study has had a major influence on subsequent scholarship, and has raised Diogenes' stock in the eyes of most scholars.[3] Yet his thesis has serious problems. What should be most obvious is the lack of reference to Diogenes in Plato's *Phaedo*—set in 399 when Diogenes' work was well-known, or in Aristotle, in the context of teleological explanation.[4] Plato and Aristotle consider Anaxagoras to be like a sober man among his contemporaries for saying that the world is arranged for the best. Yet they do not credit

---

[1] See Laks 1983, 93.
[2] Theiler 1965.
[3] Laks 1983, xxvii–xxviii.
[4] Plato *Phaedo* 97bff., Aristotle *Metaphysics* 984b8–22, 985a10–21; Hüffmeier 1963.

Diogenes for carrying the insight forward. Evidently they do not see him as making significant progress in advancing an idea they are sympathetic with. While there can be no doubt that Diogenes accepted teleological ordering in principle, it remains unclear how novel his use of it was.[5]

Finally, André Laks has studied Diogenes in his own right, attempting to view him as the creator of a unified theory, wherever he derived its elements.[6] Laks has clarified many features of Diogenes' theory and at last taken him seriously as a thinker. But he remains for Laks, as for most others, the inventor of the last Presocratic cosmology, and hence the end of the line.[7]

10.2 Diogenes in a New Light

Diogenes' shortcomings according to modern historiography have resulted largely from his failure to present any new ideas; at most he appears to be recycling concepts and principles invented by others. If, however, as I have argued, Material Monism is not to be found in sixth-century cosmological theories, then Diogenes, by introducing the theory, has made a significant theoretical innovation. This point has been seen already by M. C. Stokes, who in many ways pioneered the viewpoint I have been arguing for.[8] But it was buried in a chapter on "Miscellaneous Presocratic Contexts" in a section on "The Eclectics," and it seems to have gone mostly unnoticed.[9] Parenthetically, we should note that Hippon may have been another fifth-century Material Monist, who like Thales took water to be the source of all things—though our sources are ambiguous on this point.[10] In any case, Aristotle judges him to be hardly worth considering, and we do not have sufficient texts by which to judge for

[5] Cf. Barnes 1982, 567. On the other hand, perhaps Diogenes assigns a much bigger role to teleology than Anaxagoras; according to Laks 1993, 30, Anaxagoras's "νοῦς is a teleological power not in the sense that it wants everything to be the best . . . but in the sense that it is capable of handling the data in a sovereign manner, so that a complex cosmic order emerges from the least possible expenditure of νοῦς' energy."

[6] Laks 1983, xxiii–xxiv.

[7] The subtitle of Laks's book is La dernière cosmologie présocratique; the title of Barnes's chapter on Diogenes, "The Last of the Line." Laks's sense is somewhat equivocal: he rules out Democritus from consideration since Democritus is contemporary with Socrates (xix, n. 1).

[8] Stokes 1971, 238–44.

[9] Barnes 1982, 567, dismisses the view without referring to its author or arguing against it; Laks 1983, xxxii, n. 1, takes only passing notice; KRS ignores it.

[10] DK 38; Hippolytus Refutation 1.16 = A3 identifies two principles, water and fire, and says fire is begotten (γεννώμενον) by water, suggesting an instance of GST. On two principles cf. A5.

ourselves;[11] hence Diogenes remains the only figure to whom we can confidently ascribe Material Monism, and the only figure of his age to leave a significant mark on his contemporaries, resulting in a significant trail of evidence for modern students.

In the present chapter I have two different objectives: first, to show that *if* the argument of the preceding chapter is correct, Diogenes becomes a significant thinker with a new and important theory, one which became so influential that it eclipsed and obscured another similar theory, namely GST; and second, to provide some confirmation of the argument of the preceding chapters. It would be ideal if we could find in Diogenes evidence that he was consciously reshaping GST for his own purposes; unfortunately the texts are too exiguous to allow any such direct confirmation. What I hope to accomplish is the more modest task of making plausible connections among contemporary theories to show that Diogenes may reasonably thought to be a reviser of GST rather than a reviver of MM. In the present section I will explain the role of Diogenes entailed by taking GST as the standard early Ionian model of explanation. In the following section I will explore connections between Diogenes' theory and those of other Presocratics in an attempt to provide historical evidence for the account arrived at ex hypothesi.

Suppose, then, that Ionian philosophy of the sixth century BC instantiates, more or less, GST. There was no instance of MM in pre-Parmenidean theories. In the wake of Parmenides' criticisms of GST, there was no instance of MM: the cosmologies of the early or mid-fifth century are pluralistic theories positing elemental stuffs of some kind. Yet we find MM in Diogenes:

> B2. My view, in general,[12] is that all existing things are altered from the same thing and are the same thing. And this is manifest: for if the things presently existing in this world-order: earth, water, air, fire, and the rest, which plainly exist in this world-order, if any of these was different the one from the other, being other in its own nature [*phusis*] and not the same as it changed often and altered, in no way would it have been able to mix with another, neither would benefit nor harm <come to one from the other>, nor would any plant grow from earth nor any animal nor anything else come to be, unless they were so constituted as to be the same. But insofar as they are all altered from the same thing, they become different at different times and turn back into the same thing.

Here Diogenes gives us an argument that the existing things (*eonta*) of the world, including the four elements (recently identified by Emped-

[11] *Metaphysics* 984a3–5 = A7.
[12] The phrase could mean something like "to describe the totality" (as Laks takes it).

ocles), are really all the same thing. Often the existing things are thought of as the basic realities; but here they cannot be, for they (or three of the four elements named) share a more basic nature. If they did not, Diogenes argues, they could not interact with one another for good or ill; there could be no causal interaction between them. We shall come back to his argument presently. For now the main point is his claim that all the different things of the world have a common nature.[13]

But what is that nature? It is air:

B4. Moreover in addition to these there are the following compelling proofs: men and the other animals live by breathing air. And this is their soul (life) and intelligence, as will be convincingly explained in this treatise; and if this departs they die and their intelligence comes to an end.

Air is the source of life and intelligence, as we may see by considering that when living things cease to breathe, they die and cease to be sentient. This same substance is the source of intelligence and control in the world:

B5 (beginning). And it seems to me that the source of intelligence is what men call air, and by this all things[14] are steered and it controls all things. For this seems to me to be God, and to reach everywhere, to arrange all things, and to be present in everything. And there is nothing which does not partake of this.

Thus we can generalize the principle of life to be the power which controls and arranges the whole world.[15]

Diogenes makes two claims: (1) there is a single substance in the world, (2) which is to be identified with air, one of the existing things. Thus he is a monist whose one is a material reality, namely air. He is, then, a Material Monist. Assuming that he has no predecessor, we may recognize him as the first Material Monist, the inventor of an important and influential theory.

Clearly, Diogenes borrows from Anaximenes in identifying air as the principle of all things. But if Anaximenes is a pluralist rather than a mo-

[13] Although McDiarmid 1953, 102–6, sees the connection Theophrastus, following Aristotle, makes between Anaximenes and Diogenes, he argues that "the air of Diogenes is no more an Aristotelian substrate than is that of Anaximenes" (105). True, it is not an Aristotelian substrate in the sense of prime matter, but it does in fact serve as the kind of concrete substratum for other properties that Aristotle envisages as being what some Presocratics posit as their ultimate reality. In other words, Diogenes does satisfy the conditions of MM.

[14] Reading πάντα for πάντας. See Perilli 1988–1989.

[15] Barnes 1982, 574–76, argues that B4 and B5 concern psychology rather than cosmology; the ultimate reality is simply matter or stuff. But Simplicius has Diogenes' book before him and is in a position to satisfy his own doubts about the matter (Physics 1528–9, 153.16–17). Barnes's interpretation would allow Diogenes to make a great conceptual leap (of the

nist, Diogenes is an innovator precisely in his first claim. Now whether Diogenes is aware of his theoretical difference from Anaximenes is not clear and, for present purposes, irrelevant to us. Whether he thinks of himself as simply following Anaximenes in his monism or whether he is conscious of his innovation, Diogenes is going beyond Anaximenes' theory in his own construction. He is probably using the same considerations for the primacy of air as Anaximenes used, but his air is not just the original and generating substance, but the only substance.

What makes Diogenes think that there is only one substance in the world? The closest we can come to answering that question is to examine B2. There Diogenes argues that if two different existing things were not really the same, they could not causally interact with one another. The force of his argument may be seen by contrasting his position with that of a famous dualist whose main classes of existents are not the same. Descartes explains the real difference between mind and body:

> And first of all, because I know that all things which I apprehend clearly and distinctly can be created by God as I apprehend them, it suffices that I am able to apprehend one thing apart from another clearly and distinctly in order to be certain that the one is different from the other . . . and, therefore, just because I know certainly that I exist, and that meanwhile I do not remark that any other thing necessarily pertains to my nature or essence . . . I rightly conclude that my essence consists solely in the fact that I am a thinking thing. . . . And although possibly . . . I possess a body . . . yet because, on the one side, I have a clear and distinct idea of myself inasmuch as I am only a thinking and unextended thing, and as, on the other, I possess a distinct idea of body, inasmuch as it is only an extended and unthinking thing, it is certain that this I . . . is entirely and absolutely distinct from my body, and can exist without it. (Meditation VI, trans. Haldane and Ross)

Here the fact that I can conceive of one thing without another, backed by God's omnipotence, shows that two things with different natures or essences are really or metaphysically distinct. Mind is thinking, unextended thing while body is extended, unthinking thing. The two kinds of substance have nothing in common. Hence arise the two great problems of modern philosophy: how can mind know body, and how can mind and body interact? Ultimately, Descartes must have recourse to the omnipotence of God to assure an answer to the former. The difficulties of several centuries of wrestling with these problems attests to the difficulty of bridging the gulf between two kinds of substance different in essence. The problem seems to be not just that their essences are different, but that they are

---

sort that seems to me anachronistic); on this interpretation he should be a much more significant thinker than Barnes himself is willing to concede ("a judicious eclectic," 583).

mutually exclusive and have nothing in common: one is unextended, one extended; one is thinking, one nonthinking.

The parallel with Descartes' argument shows that Diogenes' argument is perceptive and powerful. Yet, one might complain, the alleged existents that Diogenes has in mind—for instance, the four elements—have a good deal in common even if they are different in some properties: they are all extended, for instance. Thus we find a metaphysical basis for causal connection among the several elements. There is room here for a healthy debate. But the whole question of causal interaction seems to be new in Diogenes. It is not found, for instance, in Melissus.[16] There is one Eleatic property Diogenes seems to have in mind: homogeneity, which Diogenes takes as a necessary condition for causal interaction.[17] If two beings were completely other, they could not interact. Diogenes does not seem to be able to distinguish between two things that are different in part of their essence but the same in some respect. In any case, Eleatic thinking about natures seems to play an important role in his argument. We might think of his argument as a kind of argument that causal connection presupposes ontological homogeneity. Without a common likeness in the existents, they cannot interact. And the only kind of likeness that will support causal connection is sameness of nature or essence.

Diogenes' argument in B2 (quoted above) seems to go as follows:[18]

D1

1. If two things interact they must have a common nature.
2. Existing things (e.g., the four elements) interact.
3. Hence, they have a common nature.

Diogenes argues for D1.1 by pointing out that if the existing things did not have a common nature, they would not be able to mix or affect each other. But we can observe that they do interact (2), so they have the same nature. From Descartes' perspective, one billiard ball can move another because they have a common makeup, and more generally both are extended physical objects. Diogenes tacitly requires that they have a common matter. Of course there could be much controversy about how much is to be read into "common nature" in the argument. But at least Diogenes articulates a specific requirement for causation interaction, one that in some sense is vindicated by modern science: Newtonian bodies have mass and velocity and interact in ways determined by these properties.

---

[16] I bracket the question of whether Melissus or Diogenes wrote first; they are in any case roughly contemporary thinkers.

[17] Melissus reasserts this Parmenidean property, but does not connect it with causal efficacy per se: Pseudo-Aristotle On Melissus, Xenophanes, Gorgias 974a12–14 = A5.

[18] Cf. Barnes 1982, 572–74.

The standard view of Diogenes is that he is reviving MM in the wake of Eleatic objections to cosmology. On this view, he is going back to an obsolete system and can offer his modern interpreter no fundamental insights.[19] Yet precisely how Eleatic objections defeated MM we are usually not told; in fact, MM seems to offer resources for replying to the Eleatics. On the present view, MM never existed before Diogenes wrote, and hence he alone is responsible for this important theory—important at least in the later tradition of interpreting the Presocratics. Diogenes could not be reviving MM, but at most revising a superficially similar theory, GST. He is, then, introducing an innovation, whether it is timely or untimely, progressive or vain. Whether Diogenes is consciously revising GST, or whether he thinks he is reviving MM with new and improved arguments, we cannot say on the basis of the evidence we have. But in either case, if the present analysis is correct, he is an innovator who changed the terms in which the original substance was understood and defended, drawing on an Eleatic conception based on a fixed and invariable nature or essence.

10.3 Diogenes in Historical Context

Thus far my interpretation of Diogenes is based on a hypothesis, namely that the early Ionians are followers of GST rather than MM. Now we must examine the interpretation in light of what we know about its historical context. Does it make historical sense? The first point to notice is that, according to Barnes (as discussed above),[20] a strong point in favor of MM as the correct interpretation of early Ionian philosophy was the fact that the condensation-rarefaction entailed or presupposed MM. I have already argued against that claim. But we may point out that Diogenes himself does not make use of condensation in his account of air in the fragments, which clearly embodies MM. Instead, he uses another feature:

> B5. And it seems to me that the source of intelligence is what men call air, and by this all things are steered and it controls all things. For this seems to me to be God, and to reach everywhere, to arrange all things, and to be present in everything. And there is nothing which does not partake of this. But nothing partakes of it in the same way as anything else does, for air itself and intelligence have many forms; for it is mani-

---

[19] Curiously, Cherniss, who rejected MM for the early Ionians, fails to see Diogenes as responding to Parmenides: "Hippo of Samos and Diogenes of Apollonia, who sought to derive the articulate world, the one from water, the other from air, are our chief witnesses to the fact that there were still men who could talk as if Parmenides never lived" (1951, 344).

[20] Ch. 3, sec. 3.

fold, warmer or colder, drier or moister, more stable or more lively in motion, and many other differentiations are present in it, and countless differentiations of flavor and color. Furthermore, the soul of all animals is the same: air that is hotter than that which surrounds us, yet much colder than that around the sun. The heat of no two animals is alike (since not even the heat of different men is the same), but it differs—not greatly, but in such a way as to be similar. Nothing, however, of those things that are differentiated one from another is able to become exactly like the other without becoming the same. Thus since differentiation is manifold, animals too are manifold and many and unlike each other in form, behavior, and intelligence because of the multitude of differentiations. Nevertheless, all things live, see, and hear by means of the same thing, and all have the rest of their intelligence from the same source.

Here it is not the scale of density but the scale of temperature that determines the phenomenal character of air. Density is irrelevant, or perhaps only a parameter of heat. The relative temperature determines what the character of a thing is, apparently what species of thing it is, and even the individual character of a thing. Thus heat controls both the specific differentiation of a thing and the individuation of one particular from another. Diogenes does see air as undergoing variations on a single scale while determining its character, but the scale is not that of density. In this respect Diogenes is independent of Anaximenes and has a different conception of the mechanism of change from his predecessor. Diogenes Laertius reports that Diogenes of Apollonia uses density in his cosmology, but we cannot confirm this from the fragments.[21]

Air has special properties that make it causally efficacious. Speaking of air, Diogenes opines:

B7. But this seems plain to me, that it is great, powerful, everlasting, immortal, and much-knowing.

Air is great insofar as it is boundless; it is powerful apparently as controlling other things; it is everlasting and deathless because it remains through all changes; and it is much-knowing as the source of all intelligence. Some features imply the control or agency of air, and bring us back to the question of teleology.

The statement about teleology is found in B3:

B3. For it would not be possible without intelligence for this distribution to come about in such a way that it had a measure of all things, of winter

---

[21] "These are his views: air is the element, there are numberless worlds, and the void is boundless. Air by being condensed and rarified generates the worlds" (Diogenes Laertius 9.57 = A1).

and summer, of night and day, of rain, wind, and fair weather. And everything else, if one would take notice, one would find arranged in the finest
way possible.

The world results from a balanced distribution of qualities. But this distribution presupposes intelligence. Diogenes seems to lead us from particular examples of balance in cycles of seasons, days, and weather, inductively to generalize a pattern in all cosmic events: there is order, implying
intelligence. But where does the argument go from there? The most obvious step is to say that intelligence is found only in its bearer, air. Hence
air is the source of order and direction in the cosmos.

If that is how the reasoning proceeds, it does not instantiate an argument from design, which argues from cosmic order to a divine source.
That argument would go as follows (from Barnes 1982, 577):

D2
1. The world is organized in the best possible way.
2. Hence, there is an intelligent arranger of all things.

But we do not find evidence for a *personal* agent in Diogenes. His argument (as reflected in B4) moves from cosmic order to a particular physical
source, air. Now it is true that for Diogenes the source is divine; but on
the present reconstruction, that consequence would be a concomitant of
the source's being air. The argument seems to go as follows:

D3
1. The world is organized in the best possible way.
2. Something can be organized in the best possible way only by intelligence
3. Hence, the world was organized by intelligence.
4. Intelligence requires a physical principle.
5. The physical principle of intelligence is air.
6. Hence, air is the principle of the world.

D2 moves from an observation of order in the world to an individual
agent, a personal deity who organizes it, in an argument for the existence
of God from design. Diogenes' argument D3 moves rather to some impersonal stuff that presumably distributes itself appropriately; the argument
aims at demonstrating the superiority of air to other stuffs—it answers
something like the Problem of Primacy. The argument is not an argument
for theism, that is, an argument from order to a benevolent designer, but
rather from order to a physical principle. Hence the explication of Diogenes' argument as the prototype of arguments from design would fail.[22]

---

[22] Contra Theiler 1965, 15ff., cf. Barnes 1982, 576–79, who in fact recognizes the distinct
emphasis of the argument but does not distinguish it from arguments from design; Hüffmeier 1963, 134.

The present interpretation seems to be confirmed by the opening sentences of B5: the first sentence identifies the source of intelligence as air, while the second uses the divine attribute as a confirmation, rather than a conclusion of the argument. There is also a significant difference between Diogenes' argument and Xenophon's argument, that according to Theiler, is based on it, in the fact that Xenophon sees God as arranging the world for man's sake (Hüffmeier 1963, 136). Accordingly, we see Diogenes not as an innovative theologian so much as a cosmologist arguing for a certain physical principle as the *arche*.

Yet there is an important teleological dimension to Diogenes' thought. Air is something that brings order to the world in an intelligent, apparently purposive way. The pluralistic systems, following Parmenides' cosmology, had driven a wedge between matter and the forces which direct it. Anaxagoras's Mind was distinct from his numberless elements, while Empedocles' Love and Strife were distinct from his four elements. Although the separation of force and matter brought conceptual clarification, it raised new questions. What is a force apart from matter? How does it interact with matter? These questions are analogous to the problem of causal connection raised in B2. Diogenes could reasonably argue that by locating causal efficacy in matter, he was avoiding insoluble problems. His identification of air with the moving agency of the world was a much-needed integration that avoided a disastrous schism. He could also claim that his theory integrated thought and being in a way suggested by Parmenides but not realized in pluralistic theories.

Overall, Diogenes could claim that while pluralistic theories attempted to make sense of the world in terms of compounds or mixtures of elements, they in fact ignored the monistic principles that are the consequence of Parmenides' *Aletheia*. What-is must, among other things, be homogeneous; pluralistic theories are by hypothesis heterogeneous. Diogenes makes room for plural manifestations of what-is, but these remain mere manifestations or phenomenal determinations—just as light and night might be thought of as mere appearances in Parmenides' cosmology. In the present context, Diogenes' theory makes remarkably good sense as a post-Parmenidean system, responding to the challenges raised by the Eleatics while accommodating the cosmological speculations of the Ionians. We are justified from a historical viewpoint in seeing his Material Monism as a robust theory that makes good sense as a response to the concerns of the later fifth century.

Diogenes' ontology makes distinctions necessary to responding to the Eleatic challenge. On the one hand, he recognizes plural existents, *eonta*, including the four elements of Empedocles (B2). On the other hand, he argues that they must all have a single nature (*phusis*) in order to be able to affect one another. Although the concept of *phusis* is clearly older than

Diogenes, it does not always have the connotations it does in Diogenes. Empedocles uses the term and its cognates to denote birth or coming to be.[23] But clearly for Diogenes it has the sense of a fixed and unalterable nature, and essence in Aristotelian terminology. More significant, perhaps, is the fact that Diogenes associates existents with a special kind of change, *heteroiousthai*, apparently "alteration."[24] The same term appears in Melissus, leading Diller to think that Melissus is responding in part to Diogenes.[25] Whatever the precise dialectical order, the terminology shared by the philosophers seems to witness to a more subtle appreciation of issues concerning change. The kind of change that existents undergo is not coming to be and perishing, but alteration or differentiation in properties. The one historical point we can make is that there is no sign of an analysis of change in the sixth century. Parmenides lists a series of suspect changes, perhaps for the first time making some sort of preliminary analysis (B8.40-41). As we have seen, reports of Anaximenes simply attribute coming to be to the basic substances, and Heraclitus explicitly views elemental change as coming to be and perishing—at the same time—since the coming to be of one element is the perishing of another.[26] From a historical perspective, it is plausible to think that the attribution to existents of alteration arose only in the fifth century in response to the Eleatic challenge. We can remark in Anaxagoras and Empedocles a rejection of coming to be and perishing. There is no need for them to talk about alteration, since for them some sort of mixture is the mechanism behind phenomenal change. The only cosmologist in whom we can attest the language of alteration is Diogenes—giving some support to Diller's interpretation. Whether he invented the distinction or not, he exploited it in his argument. And in that we can track an advance in discriminations over the sixth century.

One further distinction between Anaximenes appears at this point. If Aristotle is right (as I think he is not), Anaximenes accounts for (phenomenal) coming to be and perishing in terms of changes in density. Hence, he reduces coming to be and perishing to quantitative change, or, in Aristotle's terminology, increase and decrease. But from what we have seen, we can observe that Diogenes accounts for phenomenal coming to be and perishing in terms of changes in temperature, a qualitative distinction. Hence, in Aristotle's terminology, he reduces coming to be and perishing to alteration. Diogenes has some discussion of a void, and seems to have

---

[23] Esp. B8.1, 4.

[24] Cf. Aristotle's term for alteration, *alloiôsis*, e.g., *Categories* 15a14.

[25] Diller 1941, 366–67. Terminology provides only part of Diller's evidence. The common view is that Melissus is earlier, supported, e.g., by Jouanna 1965. On the relation between Leucippus and Melissus, see also Klowski 1971.

[26] B36, B76.

at least made room for an analysis of density.[27] But he does not pursue this approach. Why not? I doubt that we can give any certain answer, but one possibility is that Diogenes saw Anaximenes' theory of density as already compromised by Anaximenes' belief that the elements are generated and destroyed. Furthermore, Parmenides had attacked Anaximenes' strategy as incoherent.[28] In any case, we see that Diogenes does not follow Anaximenes slavishly, and that he makes some significant reconsiderations of the Milesian's theory.

There is no doubt that Diogenes is a Material Monist of the sort that Aristotle envisages. He gives a new and elegant argument for MM, what I shall call the argument from causal coherence: existent things could not interact unless they shared the same nature. In an era in which arguments are few, we should appreciate the ingenuity of this one. According to Diogenes, the connectedness of matter is evidence of its underlying unity. The phenomenal stuffs of the world result from alterations of a basic kind of matter, namely air. The causal argument and the mechanism of alteration are clearly inventions of Diogenes. He does not give a list of basic substances, other than to cite Empedocles' four elements. He does not seem particularly concerned with enumerating the phenomenal stuffs, which are in any case only contingent and derivative beings. His basic concern seems to be to establish the primacy of air and to argue that all phenomenal existents are really air. In the pursuit of his argument, he points out that the world is arranged for the best, that this presupposes intelligence, and that air is the cause or principle of intelligence. If he enunciates less than an argument from design, he at least provides a coherent argument for Material Monism with air as the primary reality.

Does any of this prove that Diogenes invented MM? No. Yet we do see him as a perceptive respondent to the debates of the fifth century. He is not slavishly dependent on Anaximenes; he makes distinctions that seem impossible for the sixth century but appropriate for the fifth. If they are appropriate, they are in no way inevitable, and they show considerable insight in their own setting. They could, of course, be seen as refinements needed to bolster MM in the theoretical environment of the fifth century. But we may still ask if, in the absence of clear distinctions between nature or essence and attributes, generation and alteration, even reality and phenomenon, MM was possible. Perhaps in some intuitive form, but it is difficult to imagine a well-articulated theory.

I come back to Plato's testimony of Anaximenes.[29] He describes a theory of change according to which seven elements change into one another

---

[27] Diogenes Laertius 9.57 = A1; Aristotle *On Respiration* 471a4 = A31.

[28] Parmenides B8.22–24, 44–45, 47–48.

[29] *Timaeus* 49b-c, with sec. 3.4.4 above.

by generation, based on condensation and rarefaction, seeing it as providing a helpful description of our sensory observations. He invokes Anaximenes' theory. We must assume he knew Diogenes' theory that was so prominent in Aristophanes' parody of his master. Yet he does not assimilate Anaximenes' theory to Diogenes'. I must suppose that Plato saw significant differences in the two theories, and found the former useful as a preliminary account of how transformations occur in the sensible world. That he does not use the latter even in his account of teleological insights, in the *Phaedo* discussion, shows that he did not take Diogenes to have much to add to Anaxagoras. But clearly Plato could distinguish between the two theories on the basis of their ontological foundations, and I think we should too.

10.4 A New Theory of Matter

The reports that ascribe Material Monism to the early Ionians are suspect. On the present interpretation, the early Ionians adhere to GST. There is one original stuff, which is transformed by a process such as condensation into other basic substances which make up the cosmos. These stuffs can be transformed by an inverse process back into the original stuff. There is a kind of evolution of matter from one form to another which, when stabilized, supports the complexity of the world. In GST types of matter come to be and perish in such a way that the birth of one stuff is the death of another. Heraclitus identified this relationship as a basic principle of cosmology, and used it to suggest that only process philosophy can capture the insights of Ionian cosmology. GST provides the pattern of change in all or most sixth-century Ionian cosmology.

In the early fifth century, Parmenides criticizes cosmology following the pattern of GST. He criticizes the notion of coming to be and perishing, rendering the foundations of GST problematic. In criticizing traditional cosmologies he develops an alternative style of cosmology which does not rely on coming to be and perishing, but rather on the mixture of enduring stuffs. Picking up on the new style of cosmology, the pluralists develop a theory in which a plurality of stuffs, or elements, mix and separate in such a way as to account for the changes of the world. The new style of cosmology puts a strong emphasis on the distinction between phenomenal appearances and ultimate realities. Coming to be and perishing turn out to be mere appearances derived from changing relationships of invariable elements. The resulting theory is EST. The new theory distinguishes between the fixed natures or essences of real things and the changing properties of phenomena.

Finally, as a further consideration of problems of change and cosmology, Diogenes develops a theory according to which there is only one substance, air, which by altering or changing its states, can appear as a plurality of substances. His theory fits the characterization of Material Monism identified by Aristotle. It draws on the distinctions of essence and attribute, coming to be / perishing and alteration, reality and appearance developed in EST. But it draws also on the Eleatic property of homogeneity and the Eleatic insight that only one substance could satisfy the formal demands logic places on being. MM is thus a post-Parmenidean theory that seems possible only in light of Eleatic criticisms of matter and change. Hence MM is a neo-Ionian theory using Eleatic principles to modify and reconceive the principles of cosmology. The order of historical development is not MM, then Eleatic criticism, then EST, but rather GST, then Eleatic criticism, then EST, then MM.

On this account, Diogenes is an important innovator. He sees that the techniques of pluralism can be used to render pluralism obsolete: if phenomenal changes result from rearrangements of real but unchanging elements, why not make do with one element that undergoes complex changes of state? The only thing missing from this account would be an external cause to initiate change; but if we can locate the cause in the original stuff itself, then there will be no reason in principle why differentiation cannot be self-caused. Thus we can go beyond pluralism to a simpler theory. The resulting theory will be more true to its Eleatic roots than pluralism because it will posit only one kind of being; it will remain Eleatic in distinguishing between real unvarying being and phenomenal change, which will be only phenomenal and derivative. In this sense, Diogenes can claim to be more Eleatic than his pluralist predecessors, while yet preserving a robust account of cosmology. Diogenes emerges as an innovative philosopher with an interesting response to contemporary problems.

That Diogenes was seen as an important philosopher in his own time can be seen from Aristophanes' response to him. In the *Clouds* of 423, air is taken to be the ultimate reality. A vortex motion accounts for many phenomena, and the gods themselves are rendered obsolete in the new theory. Euripides seems to react to Diogenes,[30] and the Hippocratic treatise *On Breaths* also reflects the theory of Diogenes, as do other Hippocratic treatises.[31] In his own time Diogenes seems to have been promi-

---

[30] *Trojan Women* 884–9 = C2.

[31] For other allusions in the Hippocratic literature, see DK 64C3, 3a, 3b; Jouanna 1965 shows how the Hippocratic *Nature of Man* criticizes Diogenes by using arguments inspired by Melissus.

nent—perhaps the most prominent philosopher after Anaxagoras.[32] There is certainly no evidence to indicate that his contemporaries looked at him as a mere copycat or imitator of earlier theories. On the contrary, his popularity suggests that he was seen as offering the latest and most up-to-date theory of cosmology.

The modern judgment that Diogenes has little to offer results from a certain interpretation of the Presocratic tradition in which he appears merely to revive a theory refuted well before he wrote. This interpretation is, of course, directly dependent on Aristotle's reading of Ionian philosophy, in which the earliest philosophers were Material Monists. If, then, the last Ionian philosopher was a Material Monist, he could have nothing interesting to offer the tradition by way of theoretical novelty. But if the account of the early Ionians I have offered is correct, then Diogenes was not reiterating a hackneyed and time-worn theory. Indeed, Aristotle's very assumption that he was reiterating such a theory provides evidence for Diogenes' influence: so completely had Diogenes' theory dominated later thought that Aristotle could not read the early Ionians without projecting Material Monism onto them. The fact that Plato could distinguish the theories shows that the conflation of them was not inevitable. Yet that so astute an observer as Aristotle could conflate them shows how cogent the latter theory could appear. In a manner documented by McDiarmid (1953, 102ff.), Aristotle expanded a common theoretical principle, the primacy of air, into a common philosophy.

Let us return to the point where most histories of Presocratic philosophy begin. After introducing the Milesians as Material Monists, Aristotle makes specific attributions:

> Anaximenes and Diogenes [of Apollonia] posit air as prior to water as the simple body that is most properly the source. (Aristotle *Metaphysics* 984a5-7)

Aristotle mentions Anaximenes of the sixth century and Diogenes of the fifth in the same breath. He sees no essential difference between the two, but merely an original monistic theory on the primacy of air and a re-statement. On my interpretation, the original statement is that of Diogenes, and Aristotle's attribution of the theory to Anaximenes conceals a major sea change in one-source theories. How did Aristotle get Anaximenes and his generation so wrong? He simply assimilated their theories to a much more familiar and, no doubt, well-articulated theory of the

---

[32] Contrast Barnes, who concedes the fact that Diogenes is influential: "Such a reputation implies not stature and novelty but rather the reverse; it is unoriginal men who are thus representative" (Barnes 1982, 567–68). This seems to me tendentious.

recent past.[33] It would not be the first or the last time an able thinker had missed the historical difference between two superficially similar theories—and it may not have been the first time this particular identification was made.[34] We must remember that the awareness of a Zeitgeist or, less portentously, a historical framework, was not even clearly articulated until the nineteenth century. We need not fault Aristotle for failing to appreciate changing historical perspectives in the infancy of history and the very beginning of the historiography of philosophy.

Nothing I have said in this chapter definitively proves that Diogenes was not a reviver of an old theory. But at least his complex response to philosophical questions on the one hand and the reaction of his peers to him on the other suggests that he played a different role: that he ingeniously revised and adapted an old theory to respond to the challenges of the fifth-century debate. The very success of Diogenes' theory offers a diagnosis of how Aristotle could have misunderstood the early Ionians: he mistook the familiar theory for the unfamiliar, the articulate for the inchoate, the recent for the remote. If he did, his identification of the two theories is a conflation, and we are justified in departing from it. Diogenes turns out to be an interesting and even creative player in the debates of the fifth century.

---

[33] Cf. Alt 1973, esp. 158: "Dabei kann kein Zweifel bestehen, dass man im 4. Jahrhundert hinreichend Kenntnis von ihm [sc. Diogenes] besass, weit mehr als von Anaximenes, und dass man sich im Peripatos für ihn interessierte." Her reference is to a time after Aristotle's death, but the remark is true of Aristotle's time also.

[34] Hippocrates Nature of Man 1 refers to anonymous philosophers who derive the world from a single principle, whether air, fire, water, or earth. While we can assign Diogenes to air and possibly Hippon to water (n. 10 above), presumably we would have to connect early Ionians with the other principles: Heraclitus with fire and perhaps Xenophanes with earth; and if we have license to include early Ionians, Thales can be associated with water, Anaximenes with air. However, the Hippocratic author provides no real doxography, nor does he even name the philosophers he has in mind.

11

THE IONIAN LEGACY

WE HAVE COME to the end of a long story. We have seen that there are two major systems of explanation to be found among the Presocratics: the Generating Substance Theory of the sixth century BC, and the Elemental Substance Theory of the fifth. But what implications does this fact have for our understanding of Presocratic philosophy as a whole, and for the origins of critical and scientific thinking in the West?

11.1 Paradigms of Explanation

According to GST, a single stuff, the generating substance, changes into all other stuffs. As it changes, it ceases to be what it was and becomes something new: it perishes and a new substance comes to be. The change of stuffs is taken to be a cosmic event that generates not only a new series of stuffs but also an ordered distribution of them that constitutes a world. To turn a cosmogony into a quasi-scientific theory, we need some sort of account of how it is that one stuff turns into another—in an orderly and regular way. Anaximenes supplies this theory, in the form of a pair of reciprocal processes, condensation and rarefaction, that mechanically determine the succession of stuffs. With Anaximenes we have a more or less complete theory that can account for cosmogony and cosmology in terms of something like chemical changes. On the present account Anaximander supplies the general content of Ionian science: cosmogony and cosmology, while Anaximenes supplies the mechanism that makes the story into a rational theory. After Anaximenes, thinkers like Xenophanes and Heraclitus can draw on a general scheme of explanation to understand the world—or can criticize it for its inadequacies. In GST perception is direct acquaintance with the stuffs of the world in a more or less transparent process.

After GST we find EST. According to this type of theory, there is a plurality of stuffs, which may include all those of GST, but which instead of being related to each other genetically, are all coexistent. The primeval chaos does not consist of only one substance, the generating substance, but a mixture of all stuffs. These stuffs are fixed and unchangeable in their natures, and are everlasting: they do not come to be or perish. They are like modern chemical elements, which are unalterable by normal chemical processes. The world arises as like elements are separated out of the chaos into the same places and the great masses of the world are articulated. In this scheme the forces that were tacitly contained in the generating substance of GST become distinct realities in their own right. They act on the elements to bring them together or separate them, so as to produce the substances we experience. Phenomenal substances are the result of a mixture of elements, and are either manifestations or compounds of those elements. In EST sense perception meets with limits: we perceive either only some of the elements present (Anaxagoras) or only appearances supervening on combinations of them (Empedocles) or only ideas of secondary qualities utterly unlike them (Democritus).

If the preceding account has been even roughly correct, it accomplishes two things. First, it defines the major explanatory schemes of Ionian philosophy in a philosophically defensible manner, and second, it allows for the possibility of a historical and dialectical linkage between them. According to the Standard Interpretation, some at least of the pre-Parmenidean Ionians subscribe to Material Monism, according to which the original stuff of the universe is the only stuff there is; it changes into successor stuffs—for instance, air into wind into cloud into water—but the changes it undergoes are only what Aristotle calls alterations: the original stuff changes appearances, but it retains its essential identity. Thus Thales' water is always water, even when it looks like air, and Anaximenes' air is always air even when it looks like water. This interpretation makes it impossible to understand Parmenides historically, because MM already has a ready answer to his criticisms of change: it rules out the most objectionable kind of change, coming to be and perishing, and posits an everlasting changeless being as the foundation of all reality. Further, according to the Standard Interpretation, the pluralists desperately try to save cosmology from Parmenides' devastating attacks, but they only succeed in begging the question against him. Indeed, they violate one of the most basic principles of Parmenidean metaphysics by positing a plurality of beings in the first place. If the Standard Interpretation were right, the neo-Ionians should have emphasized the Eleatic features of MM, which already embodied an answer to Parmenides' criticisms, rather than positing a plurality of beings for which they could produce no theoretical justification.

If, instead of MM, we take GST as the original explanatory scheme, we can supply a dialectical opponent for Parmenides, and even propose a missing link: Heraclitus, who criticizes GST from the standpoint of a sympathetic successor, one who wants to modify the theory but not totally reject it. Heraclitus's modification focuses attention on the paradoxical implications of GST, and suggests that it is radically incoherent. Parmenides misses the irony in Heraclitus and thus finds in him a target for radical criticisms of GST. Heraclitus awakens Parmenides from his dogmatic slumbers to reorient philosophy. From this perspective we are able to appreciate why Parmenides appends a revisionary cosmology and cosmogony to his critical theory: his target is the Ionian cosmologists, and he is aware of the need to compete with them on their own turf. His aim is either to invent an anti-cosmology to expose the flaws of cosmology, or to offer a second-best cosmology to apply to a realm in which demonstrative knowledge is not possible. But with only a shift of emphasis—reading the *Aletheia* in light of the *Doxa* rather than vice versa, that is, seeing the cosmology as the fruits of the ontology—his successors can see him as offering a constructive alternative to GST, with no disclaimers needed for the limits of physical theory. And so the early pluralists read Parmenides. Thus EST appears as the stepchild of Parmenides. And the absence of an argument against Parmenides in the Eleatic pluralists makes historical sense.

The present interpretation allows us to make plausible historical connections between the protagonists of Presocratic philosophy. Or rather, it allow us to supply appropriate dialectical connections rather than superficial historical ones. With the dialectical connections, we may see that GST and EST emerge as something like Kuhnian paradigms of problem solving. Each explanatory scheme provides a kind of model for accounting for the phenomena of the natural world. A paradigm is more powerful than an abstract theory precisely because it provides concrete examples of how to deal with difficult problems. In this light, Anaximenes' own theory provides a concrete expression of the features abstracted in our formalization of GST. The paradigm is also prescriptive in indicating how things should be explained. But appropriate explanations within the scheme can run up against serious problems, which if not resolved, eventually lead to challenges to and abandonment of the paradigm. This, we can say, happened to GST.

The kinds of problems that led to the demise of GST are not empirical failures, but conceptual ones reflecting theoretical inadequacies: how can something that is always changing be the ultimate reality? Of course, many of the conditions of Kuhn's theory of scientific change are missing: there is no professional scientific community, there is no program of experimental testing per se, and no institutional setting. Yet there is a real, if rudimentary and nonprofessional, scientific community, and some sort of feedback. Most important, members of the inchoate community seem to have constructed theories within the framework of successful explanatory models. The cosmologists conducted their research in a kind of intellectual community which shared a basic set of assumptions and recognized theoretical and methodological constraints. They constituted a proto-scientific community of individuals in far-flung cities that learned from and criticized one another's speculations.

The most important consequence of this interpretation is that it allows us to see the Ionian tradition as a unified movement, both through time and at the same time. Through time the cosmologists adopted, criticized, modified, and adapted theories of their predecessors. At any given time, despite the differences in content, style, and scope of different theories, there was a sense of participating in a common program and addressing a community of intellectuals engaged in the same enterprise. Adherents of the Ionian tradition sought to explain the phenomena of the world in natural terms on the basis of their natural constituents. The paradigms expressed by GST and EST allowed these adherents to grasp what counted as an explanation and what did not, what sorts of constituents might qualify as explanantia and what not.

The recognition of the Ionian tradition as a powerful, vibrant, and long-lived intellectual endeavor suggests that it was *the* major movement of early philosophy. Other schools or movements were forced to confront the Ionian tradition in some way or other to establish their own credibility. It is when the Pythagorean tradition advanced a competing cosmology that it became visible as part of early Greek philosophy. Empedocles introduced Pythagorean psychology into his Ionian cosmology. Philolaus developed a cosmology in which Pythagorean principles such as limit, unlimited, number, and harmony played a key role. It is unclear how much of Philolaus's theory antedated him in an oral tradition. But Philolaus in a certain sense founded the Pythagorean tradition anew by expressing his principles in a cosmology of the sort that could be recognized as part of the cosmological discourse originating in Ionia. His effort to express his views in the format of a cosmology and cosmogony, in a prose treatise, broke with the mystical silence of his school, and brought it into the dominant scientific conversation by giving it a voice in public space. To that extent, the Pythagoreans owe their later influence to their effort to engage the Ionians.

The Eleatics, as we have seen, likewise owed their existence to the Ionians. Parmenides occupies an anomalous position insofar as he attempts to overthrow GST and to expose its weaknesses while devising a substitute cosmology of his own. His substitute cosmology becomes the inspiration for a new paradigm, EST. He would, in my view, repudiate EST and reject its foundations, at least insofar as they claim to provide secure knowledge of the world; but he cannot escape the charge of trying to play, however deceptively, the Ionian game. Meanwhile it seems likely that Zeno and

Melissus had to argue against the Eleatic Pluralists that they themselves were the genuine disciples of Parmenides and the Pluralists a misguided heresy. The Ionian tradition provides the sine qua non of the Eleatic school, and not vice versa.

We also have reason to reevaluate the role of Diogenes of Apollonia. Typically treated as an eclectic whose combination of Milesian cosmology, Eleatic metaphysics, and Anaxagorean theology allows him to make modest contributions in the area of teleology, he becomes an afterthought. But if the present argument is right, he is innovative in a surprising way. For it is Diogenes who combines Eleatic monism with Ionian cosmology to come up with a monistic physical theory. Diogenes holds the theory that Aristotle and almost everyone after him up to the present day have attributed to the Milesians: Material Monism.[1] So natural and elegant is the theory that it seems as if he is simply repeating Anaximenes' account in light of Anaxagoras's Mind. Yet his argument presupposes the unity of Eleatic being and has no antecedents in Miletus. Diogenes offers a powerful and appealing theory that is responsive to the philosophical problems of the fifth century and seems to some readers to offer the only natural interpretation of sixth-century Ionian thought. Accordingly, Diogenes' synthesis displaces Anaximenes' theory so as to completely efface it in the doxographical tradition. Plato, meanwhile, never conflates Diogenes with Anaximenes, preserving the essence of the Milesian theory. But modern commentators, when they read Plato's meditation on Anaximenes, cannot recognize it as expressing a Milesian theory because they have been conditioned by the doxographers not to take seriously any departure from Aristotelian historiographical orthodoxy.

11.2 Explanatory Progress

Thus far I have argued that the Ionian tradition is a unified movement. It is unified historically as a series of connected explorations, and pragmatically as a set of contributions to a common program. To that degree it shows some of the traits of modern scientific research. But is there something like the feature that is so important to modern science, a record of progress?

Here we should notice a theoretical difficulty. In Kuhn's model of science that has proved so helpful for understanding long-term developments, one can recognize progress within a paradigm as one answers the questions posed and works out the problems suggested by the paradigm. But when questions arise which do not seem to find adequate

---

[1] Aristotle *Metaphysics* 984a5–7, with 983b6ff.; see above, ch. 3, sec. 2.

answers within the paradigm, the framework itself is called into question and sometimes overthrown in a revolutionary episode, which gives rise to another paradigm that at least promises to solve the previously insoluble problems. Kuhn cannot assure us that there is progress from one paradigm to another.[2] Yet it seems clear in the course of modern science that new paradigms have been more successful. Ultimately scientists do not seem to embrace a new paradigm unless it works at least as well as the old on old problems and promises to deal with the unsolved problems of science better than the previous paradigm.[3] Over the course of history, modern natural science seems to have made consistent progress as it abandoned one paradigm in favor of another—objective progress, as seen in the ability to manipulate nature to the point of splitting atoms, traveling into outer space, and altering genes. Thus it does not seem naive to ask if the Ionian tradition is like modern science in having an ability to make progress.

The present study seems to show that there was significant progress in the Ionian tradition. The most obvious kind of progress took place in the theoretical coherence of the explanations advanced. The criticisms Heraclitus and Parmenides directed against GST were intelligent and incisive: GST was open to the Problem of Primacy, the Problem of Origination, and the Problem of Being. Heraclitus's response was radical: he advocated a process philosophy to replace the emphasis on explanation by stuffs. It was, though difficult to comprehend—indeed ahead of its time—theoretically more defensible than GST, and ontologically more advanced. Parmenides' criticism demanded strict identity in the explanans, which Parmenides could achieve only by severely limiting the scope of explanation so as to abandon the whole Ionian project. His second-best cosmology—or perhaps anti-cosmology—marked a new and innovative model, even if it did not satisfy the strict standards its author demanded. EST, built on Parmenides' model, avoided at least the most obvious failings of GST, and provided identity conditions for the explanantia. A second wave of objections offered by Zeno and Melissus provoked a modified Eleaticism among the atomists, in which what-is-not was allowed a positive status in the ontology, and what-is and what-is-not became opposite con-

---

[2] Since competing paradigms construe the world in mutually incommensurable ways, change from one to another is a kind of conversion that brings people to see the world in a new way: Kuhn 1996, chs. 10, 12. Since there is no single way of understanding the world, "We may . . . have to relinquish the notion, explicit or implicit, that changes of paradigm carry scientists and those who learn from them closer and closer to the truth" (170).

[3] On a positive note, Kuhn observes, "The decision to reject one paradigm is always simultaneously the decision to accept another, and the judgment leading to that decision involves the comparison of both paradigms with nature *and* with each other" (77, Kuhn's italics).

ditions in a thoroughgoing dualism replacing the monism of what-is. The Eleatic pluralists are guilty of misunderstanding the scope of Parmenides' attack on cosmology, and the atomists are less than clear in their rehabilitation of what-is-not. But there is a coherent dialogue between the Ionians and their critics in which the Ionians responded to criticisms and strengthened their positions accordingly.

There is even empirical progress in the Ionian tradition. The Ionians identified the water cycle, long-term climatic trends, and ecological balance. Some of their meteorological theories, such as the explanation of hail, turned out to be substantially correct.[4] Other theories which did not succeed, such as the explanation of the sun's heat and of lightning, had to await modern discoveries; in fact, both the phenomena mentioned were not adequately explained until the twentieth century. But given the assumptions of their time, the Ionians provided at least reasonable conjectures. More significant is the explanation of the moon's light. Once Parmenides saw that the sun lights the moon, he and his successors, most notably Anaxagoras, were able to deduce the shape of heavenly bodies, their continued existence after they set, their circular path around the earth, and the cause of eclipses.[5] No earlier culture seems to have arrived at this knowledge. The Ionians put astronomy on the secure path of a science, and did so by using only reason and the observations of the naked eye. Although there were several competing theories of the heavens in the fifth century, most were informed by the core astronomical discoveries of Parmenides and Anaxagoras.

Finally, there is conceptual progress. Here the anticipation of modern theories is striking. GST, with its account of changes in a basic substance, anticipates the modern theory of the phase states of matter, which depending on temperature and pressure may take the form of solid, liquid, gas, or plasma. EST in different versions anticipates several modern theories. The mixture without emergence of Anaxagoras is reminiscent of the chemical theories of mixtures. The mixture with emergence of Empedocles anticipates the theory of chemical combination, in which discrete elements combine by chemical processes to produce a countless number of compounds, whose properties are the result of relations of the elements. The ancient atomic theory is the forerunner and inspiration of the modern atomic theory of matter. The modern theory has gone far beyond the picture of unbreakable solid particles, but without the basic insight of that theory there would have been no incentive to develop the first models of atomic matter. More exotic examples can be found: Anaxagoras, with his recursive account of matter and his notion of infinite repeatability,

[4] Anaxagoras from Aristotle *Meteorology* 348a14–20, b8–15 = A85.
[5] See above, ch. 6, sec. 5.

seems to have foreshadowed fractal geometry and chaos theory.[6] Anaximander and Empedocles anticipated, however crudely, the theory of evolution including natural selection, while Xenophanes recognized the value of fossil evidence for natural history.

Why were the Ionians so fertile in their theorizing? One reason was that in their search for explanations, they severally tried out all possible explanations. Either matter is changeable or it is stable. Either matter is continuous or it is discrete. Either mixtures produce new combinations or they produce only concentrations and dilutions. Either forces are inherent in matter or they are separate. The Ionians covered all possible alternatives in a kind of survey of the conceptual space. They thought through each alternative and provided arguments or at least a theoretical framework for any alternative that seemed promising. Wherever modern scientists searched for an answer, the Ionians had left their tracks, twenty or more centuries earlier. Though the Ionians lacked the moderns' powerful apparatus for empirical testing, they shared their successors' conceptual acuity. The Ionians were, in a sense, conceptual pioneers, exploring the range of concepts that would be relevant to a naturalistic explanation of the world. After they had mapped the territory, it was easier for their successors to find their way in the wilderness.

Besides the conceptual advances made by the Ionians, they perhaps invented a practice that was more important than the sum total of their concepts: that of natural science. Among the concepts they invented, or almost assumed, were three interrelated ones that made the practice of science possible: the world, nature, and science. The world, *kosmos*, appears first as a concept in Heraclitus B30, where the philosopher uses a term whose root meaning is "order" and which can also mean "adornment." It is possible that some Ionian before Heraclitus had used the term, but at least by the end of the sixth century BC it had appeared.[7] The connection with the notion of order, especially in Heraclitus, assures that in a sense the world is not just a fact or a conglomeration of things, but a systematic organization of them.

The term "nature" (*phusis*) is prominent also in Heraclitus as denoting the constitution of things found in the world. Indeed, the world exists as an organization of natures, each of which acts or reacts according to its nature. Eventually, at least, the idea would arise of Nature as the sum total of all natures, the *natura rerum*.[8] The behavior of things in the world was accessible because things had natures which could be studied. Fire would do what fire does, water would do what water does, and so on.

[6] See Graham 1994.
[7] See Kerschensteiner 1962; Vlastos 1975, ch. 1.
[8] See Heinimann 1945, 89–109.

Furthermore, dogs would do what dogs do, and lions would do what lions do. Thus natural things were available for study. And the totality of things would result from the interactions of individual things of determinate kinds, each acting according to its nature. Hence the totality of things, the world, was indeed a system, and the system itself was accessible to study.

The Ionians, including historians and physicians as well as cosmologists, believed in *historiê*, investigation. Investigation could at its worse be no more than collecting facts—as Heraclitus warned—which would not confer understanding on the investigator. On the other hand, it could fruitfully be turned toward important issues, even toward the self, and it could uncover true relationships, including the discovery of natures that usually lie hidden. It could also reveal the structure of the world and its principles of operation. This amounts to a secure knowledge of the world. It may, according to Xenophanes, be only an all too human construct, but it is valuable nonetheless. It may, according to Empedocles, confer on the researcher almost superhuman powers. In any case, it can tell us how the world functions, how it began, and how (or if) it will end. The researcher replaces the seer and the bard as the messenger of hidden knowledge. The Ionians do not produce a distinctive word for science such as Plato's *epistêmê*, but they understand how research can produce a new and transcendent kind of knowledge.

11.3 The Primacy of Ionian Research

In this study I have emphasized the Ionian tradition and the advances it made in a scientific conception of the world. What, it may be asked, about the many other practices that contributed to science? Certainly many other activities influenced the development of Greek science. I have chosen to concentrate on one important practice, but that does not of itself preclude other practices from contributing to the story of science. My claim, however, has been more than just that Ionian speculation was one among many scientific tendencies. My claim is that the Ionian tradition provides the impetus that gives meaning and unity to the other related practices.

One of these is history. Of the originators of history, Hecataeus of Miletus set a pattern of informed observation with his history, geography, and ethnography. Herodotus, while critical of Hecataeus, at least has much in common with him in his style of peripatetic storytelling. Yet Hecataeus is himself influenced by the Milesians at least in his use of the map, inspired by Anaximander. For his part, Herodotus reacts to Ionian theories attempting to account for the Nile floods. Indeed, he refutes theories proposed by Thales, Hecataeus, and Anaxagoras, and proposes his own theory in their place—a theory very much dependent on the assumptions of the Ionians concerning natural regularities and meteorological cycles. In effect, he enters into the conversation initiated by the early Ionians and thus shows himself as a devotee of Ionian science.[9] His own usual brand of *historie* is, of course, much different from that of the cosmologists: he is concerned with historical developments, ethnographical tendencies, and geographical influences. He is unapologetically engaged in *polymathiê* rather than in attaining to some unifying insight into reality. Yet the conceptual world he lives in is at least partly shaped by the Ionian cosmologists. It is impossible to know how much the intellectual world was altered in the century before Herodotus lived, but we can be certain that the critical methods and attitudes he developed in history were influenced by the cosmologists.

Certain schools of physicians practiced extensive observation of the course of diseases and the occasions on which victims fell ill. They have been hailed by some as the real founders of natural science. In the medical treatises attributed to Hippocrates we find criticisms of the "hypotheses" of theorists who propose one or two principles to explain sicknesses.[10] One author criticizes Empedocles and those like him who require a general knowledge of man prior to an understanding of medicine.[11] Yet another author sees a knowledge of astronomy or meteorology—in the province of natural philosophers—as necessary for medicine,[12] while some authors pattern their explanations after philosophers such as Heraclitus and Diogenes of Apollonia.[13] Indeed, even the critics of the cosmologists give explanations that draw on the types of explanantia developed by the cosmologists: powers, humors, types of matter, and configurations of organs.[14] Thus even when they are most critical of the philosophers the medical writers show their indebtedness to the kinds of explanation invented by the Ionian cosmologists.[15] At least since the time of Alcmaeon of Croton[16] (early fifth century), some physicians had been both natural

---

[9] See Graham 2003c.

[10] *Ancient Medicine* 1, cf. *The Nature of Man* 1. For the convenience of nonspecialists I shall use the English titles of the Loeb Classical Library edition, W.H.S. Jones, translator.

[11] *Ancient Medicine* 20.

[12] *Airs, Waters, Places* 2.

[13] *Regimen* I, *Breaths*.

[14] E.g., *Ancient Medicine* 22, *The Nature of Man* 4. These authors seem to be more concerned with false identifications of natural powers than with methodology per se, though the first-named treatise does criticize a priori theorizing in general.

[15] One recent study of the Hippocratics, Longrigg 1993, is sensitive to this relationship.

[16] DK 24.

philosophers and medical practitioners, and medical science, presumably as opposed to folk and magical practices, had drawn on natural philosophy.[17] Thus, while there is no question physicians had much to teach philosophers about close observation, record keeping, and the like, their own understanding of medical science was in some fundamental sense derivative of the Ionian program.

Mathematics is a field that developed significantly during the time of Presocratic philosophy. Unfortunately, it is very difficult to track the early development of mathematics. Thales is reputed to be the founder of Greek mathematics as well as of science,[18] while Pythagoras is traditionally credited with important discoveries.[19] But in the absence of written records for either, it is impossible to validate these claims. It is possible that Thales brought back from Egypt practical mathematics relating to land surveying, and it is possible that Pythagoras was interested in the ratios of strings associated with musical intervals. In any case, there is evidence of independent research in mathematics in the fifth century BC.[20] Earlier, Anaximander had speculated that the heavenly bodies were spaced at regular whole-number intervals from the earth. Heraclitus recognized fixed mathematical relationships in the transformation of basic stuffs. Empedocles saw the elements as combining in whole-number ratios in chemical compounds. Philolaus posited the principles with mathematical applications as principles of the world. At this early period mathematics provided only a priori schemes for explanation, not a basis for empirical research. Yet the fact that philosophers in the Ionian tradition speculated about mathematical relationships at least raised the question of mathematical foundations of physical phenomena. In the context of Ionian science, mathematics could be glimpsed as a foundational study.

During the sixth and fifth centuries BC, technology made great progress in the Greek world. By the end of this period, the Greeks were building some of the finest buildings, turning out some of the most elegant pottery, and constructing some of the most effective warships in the world. Anaximander seems to have been influenced by architectural technology of his time in his conception of the world as like a column drum.[21] Various arti-

---

[17] In particular Empedocles and Diogenes of Apollonia are said to be physicians: Diogenes Laertius 8.58 with Empedocles B111, B112.10–12; Diogenes B6 and Pseudo-Galen On Humors 19 = A29a. Recent discoveries also indicate that Parmenides was a physician, or at least thought to be so by later citizens of Elea: "Parmenides, son of Pyres, member of the association of [Apollo] the Healer, naturalist" (inscription on herm found in Velia [Elea] excavations, Ebner 1962).

[18] Proclus Commentary on Euclid 157.10–11, 250.20–251.2, 299.1–4, 352.14–18 = Thales A20, drawing in part on Eudemus's authoritative treatise Researches in Geometry.

[19] Burkert 1972, ch. 6.

[20] Burkert 1972, 420ff.

[21] Hahn 2001, 194ff.

facts and gadgets such as the column drum, the lyre, and the clepsydra provided models of cosmic bodies and processes for the Ionians. Meanwhile, practices such as medicine and rhetoric made claims to being *technai* or crafts with reliable methods for producing predictable results. But Ionian explanation was not driven by technology in the way modern science sometimes would be, nor was improved technology seen as the inevitable outcome of improved science. The point of scientific research was to understand the world, how it came to be and how it worked, for its own sake. At most, improved technology might be a sign of improved science.

In the fifth century rhetoric came to the fore as the most desirable knowledge to have. As democracy developed and spread to more cities, the ability to speak convincingly became the political skill par excellence, and those who could claim to teach it effectively could command generous compensation. Rhetoricians could claim that their art gave them the ability to learn and communicate effectively any kind of knowledge. We see the rhetorical approach even in Hippocratic medical treatises, and hear of its advantage for physicians. Rhetoricians sometimes claimed to be competent in the field of cosmology; and even those who avoided the subject as abstruse or irrelevant sometimes claimed to be able to debate effectively about natural philosophy.[22] One of the leading rhetoricians, Gorgias, provided an elaborate counterargument against philosophical ontology by imitating (or parodying) the arguments of Parmenides and Zeno.[23] Whatever their specific orientation toward cosmology, the rhetoricians were deeply aware of Ionian theory and generally influenced by it, even if sometimes negatively.[24]

Although many practices contributed to the scientific conception of the Greeks, it was the Ionian tradition that provided a comprehensive framework of explanation to which they *could* contribute. The Ionians saw the world as a complex system of natural processes which was in principle accessible to human understanding. Indeed, even human understanding itself was a natural process mediated by sense perception and made possible by causal interactions between the human agent and the world. If nature as a whole was knowable, so were particular manifestations of nature, such as sickness, human culture, and history, including technology and language, and ultimately even the art persuasion, based on the interaction of language and emotion. We cannot, of course, say how far prac-

---

[22] Plato *Protagoras* 318d–319a has Protagoras reject Hippias's eclectic style of education.

[23] Gorgias *On What Is Not*, B1, Pseudo-Aristotle *On Melissus, Xenophanes, Gorgias* 979a12ff.

[24] Recently, Bett 2002 has pointed out that what the sophists had in common was a "naturalistic" or "social-scientific" approach to human affairs (254–58). They seem to see the social world in terms of the kinds of categories developed by natural science. The anthropological approach was pioneered in part by Democritus; see Cole 1967.

titioners of various arts saw themselves as part of the Ionian tradition. But the new attitude of intellectual freedom and control that was part of the fifth-century Enlightenment seems to owe much to the pioneering work of the Ionian thinkers. Even the thinkers most critical of the Ionian tradition, the Eleatics, shared with their opponents a faith in the sovereignty of reason as capable of attaining ultimate truth about reality.

For all the limitations of Ionian science, it presented a conception of the world as a natural system of phenomena no different in kind from everyday processes we meet in experience, which allowed its followers to explain the world. If they were far from arriving at the satisfactory explanation they claimed to provide, they at least showed what a scientific explanation of the world should look like, and in many cases their explanations prefigured empirically supported theories of the modern era. Xenophanes' observations have come to seem wiser with the passing years:

> Now the plain truth no man has seen nor will any
> know concerning the gods and what I have said concerning all things.
> For even if he should completely succeed in describing things as they come
>   to pass,
> nonetheless he himself does not know: opinion is wrought over all. (B34)

> From the beginning the gods have not revealed all things to mortals,
> but in time by seeking they come upon what is better. (B18)

Precisely by believing in a natural world order accessible to human cognition we have, in time, come to find out what is better. The gift of the Ionian tradition was not a final explanation of phenomena—indeed, we still have no final explanation to this day. Rather it was the ability to believe there was an explanation that we by searching could find out. And somehow, they implied, the knowledge of the search would make us better:

> Sound thinking is the greatest virtue and wisdom: to speak the truth and to
> act on the basis of an understanding of the nature of things. (Heraclitus B112)

Perhaps we could not state our commitment to the value of science more eloquently than did Heraclitus so many centuries ago. Although no theory of the Ionians has survived the assessment of time without paying restitution, the program they invented has transformed the world. The program embodies the rational investigation of the world, taken as a knowable set of events occurring in accordance with discernible laws. To this program researchers of the early modern period added the experimental method, mathematical metrics, and the rigorous testing of hypotheses, and in the process transformed it into a powerful method of discovering truth about the world. Yet for all the modern advances in methodology and application, the Ionian tradition has not vanished, but lives on in the science of today; for contemporary science is a logical extension and a historical continuation of the Ionian program. The intellectual and scientific world we live in is built on Ionian foundations. Without our Ionian forebears, the world we live in would be so different we do not know what it would be like. But probably it would not be a world built on advanced science and technology. Even if the Ionian tradition produced no single theory that correctly explained the world, it provided the basic conceptual tools of science, and more fundamentally the habit of seeing the world as a set of substances and events governed by scientific laws. And it taught us to believe, however quixotically, that we could discover those laws. The ultimate legacy of the Ionian tradition is the intellectual world we inhabit. Today, whether we recognize it or not, we are all Ionians.

REFERENCES

Aall, Anathon. 1896–1899. *Geschichte der Logosidee in der griechischen Philosophie.* 2 vols. Leipzig. Reprint, Frankfurt: Minerva, 1968.

Adkins, A.W.H. 1960. *Merit and Responsibility.* Oxford: Clarendon Press.

Algra, Keimpe. 1999. "The Beginnings of Cosmology." In Long 1999, 45–65.

Alt, Karin. 1973. "Zum Satz des Anaximenes über die Seele: Untersuchung von Aetios Περὶ Ἀρχῶν." *Hermes* 101: 129–64.

Aravantinou, A. 1993. "Anaxagoras' Many Worlds." *Diotima* 21: 97–99.

Arnim, Hans von. 1902. "Die Weltperioden bei Empedocles." In *Festschrift Theodor Gomperz.* Vienna: Alfred Hölder.

Austin, Scott. 1986. *Parmenides: Being, Bounds, and Logic.* New Haven, CT: Yale University Press.

Baeumker, Clemens. 1890. *Das Problem der Materie in der griechischen Philosophie: Eine historisch-kritische Untersuchung.* Munster: Aschendorff.

Bailey, Cyril. 1928. *The Greek Atomists and Epicurus.* Oxford: Clarendon Press.

Baldry, H. C. 1928. "Embryological Analogies in Pre-Socratic Cosmogony." *Classical Quarterly* 26: 27–34.

Barnes, Jonathan. 1979. "Parmenides and the Eleatic One." *Archiv für Geschichte der Philosophie* 61: 1–21.

———. 1982. *The Presocratic Philosophers.* Rev. ed. London: Routledge, 1979.

———. 1983. "Aphorism and Argument." In Robb 1983a, 91–109.

Beatty, Mario. 1997–1998. "On the Source of the Moon's Light in Ancient Egypt." *ANKH: Révue d'Égyptologie et des Civilisations Africaines* 6/7: 162–77.

Bernays, Jacob. 1885 [1850]. "Heraclitea." In *Gesammelte Abhandlungen.* Edited by H. Usener. Vol. 1, 1–108. Berlin: Wilhelm Herz.

Bett, Richard. 2002. "Is There a Sophistic Ethics?" *Ancient Philosophy* 22: 235–62.

Bicknell, Peter J. 1966a. "Anaximenes' *Pilion* Simile." *Apeiron* 1: 17–18.

———. 1966b. "ΤΟ ΑΠΕΙΡΟΝ, ΑΠΕΙΡΟΣ, and ΠΕΡΙΕΧΟΝ." *Acta Classica* 9: 27–48.

———. 1967a. "Parmenides' Refutation of Motion and an Implication." *Phronesis* 12: 1–5.

———. 1967b. "Xenophanes' Account of Solar Eclipses." *Eranos* 65: 73–77.

———. 1967c. "A Note on Xenophanes' Astrophysics." *Acta Classica* 10: 135–36.

———. 1969. "Anaximenes' Astronomy." *Acta Classica* 12: 53–85.

Bignone, Ettore. 1916. *Empedocle: Studio Critico.* Turin: Bocca.

Boeder, Heribert. 1959. "Der frühgriechische Wortgebrauch von *Logos* und *Aletheia.*" *Archiv für Begriffsgeschichte* 4: 82–112.

Boeder, Heribert. 1966–1967. "Parmenides und der Verfall des kosmologischen Wissens." *Philosophisches Jahrbuch* 74: 30–77.

Böhme, Robert. 1986. *Die verkannte Muse: Dichtersprache und geistige Tradition des Parmenides*. Bern: Francke.

Bollack, Jean. 1965–1969. *Empédocle*. 3 vols. Paris: Les Éditions de Minuit.

———. 1990. "La cosmologie parménidéenne de Parménide." In *Herméneutique et ontologie: Hommage à Pierre Aubenque*. Edited by R. Brague and J.-F. Coustine, 19–53. Paris: Presses Universitaires de France.

Bonitz, Hermann. 1870. *Index Aristotelicus*. Berlin: G. Reimer.

Bormann, Karl. 1971. *Parmenides: Untersuchungen zu den Fragmenten*. Hamburg: Felix Meiner Verlag.

Bowra, C. M. 1937. "The Proem of Parmenides." *Classical Quarterly* 32: 97–112.

Broad, C. D. 1925. *The Mind and Its Place in Nature*. London: Routledge & Kegan Paul.

Brown, Geoffrey. 1984. "The Cosmological Theory of Empedocles." *Apeiron* 18: 97–101.

Burch, George Bosworth. 1949. "Anaximander, the First Metaphysician." *Review of Metaphysics* 3: 137–60.

Burkert, Walter. 1972 [1962]. *Lore and Science in Early Pythagoreanism*. Translated by E. L. Jr. Minar Jr. Cambridge, MA: Harvard University Press.

———. 1992. *The Orientalizing Revolution*. Cambridge: Cambridge University Press.

Burnet, John. 1892. *Early Greek Philosophy*. London.

———. 1930 [1892]. *Early Greek Philosophy*. 4th ed. London: Adam & Charles Black.

Calogero, Guido. 1977 [1932]. *Studi sull'Eleatismo*. New ed. Florence: "La Nuova Italia" Editrice.

Cappelletti, Angel J. 1979. "Notas para una biografía de Anaxágoras." *Diálogos* 14: 7–28.

Carteron, Henri. 1923. *La notion de force dans le philosophie d'Aristote*. Paris: J. Vrin.

Cary, M., and E. H. Warmington. 1929. *The Ancient Explorers*. London: Methuen & Co.

Caston, Victor, and Daniel W. Graham, eds. 2002. *Presocratic Philosophy: Essays in Honour of Alexander Mourelatos*. Aldershot: Ashgate.

Chalmers, Alan. 1997. "Did Democritus Ascribe Weight to Atoms?" *Australasian Journal of Philosophy* 75: 279–87.

Cherniss, Harold. 1935. *Aristotle's Criticism of Presocratic Philosophy*. Baltimore: Johns Hopkins University Press.

———. 1951. "The Characteristics and Effects of Presocratic Philosophy." *Journal of the History of Ideas* 12: 319–45.

Cherubin, Rose. 2001. "Λέγειν Νοεῖν, and Τὸ 'Εόν in Parmenides." *Ancient Philosophy* 21: 277–303.

Clarke, Michael. 1995. "The Wisdom of Thales and the Problem of the Word ἱερός." *Classical Quarterly* 45: 296–317.

Classen, C. Joachim. 1965. "Bemerkungen zu zwei griechischen 'Philosophiehistorikern.' " *Philologus* 109: 175–81.

————. 1977. "Anaximander and Anaximenes: The Earliest Greek Theories of Change?" *Phronesis* 22: 89–102.

Cole, Thomas. 1967. *Democritus and the Sources of Greek Anthropology.* Chapel Hill, NC. Reprint, Atlanta: Scholars Press, 1990.

Collobert, Catherine. 1993. *L'être de Parménide ou le refus du temps.* Paris: Éditions Kimé.

————. 2002. "Aristotle's Review of the Presocratics: Is Aristotle Finally a Historian of Philosophy?" *Journal of the History of Philosophy* 40: 281–95.

Cordero, Nestor-Luis. 1979. "Les deux chemins de Parménide dans les fragments 6 et 7." *Phronesis* 24: 1–32.

Cornford, F. M. 1912. *From Religion to Philosophy.* London: Edward Arnold.

————. 1930. "Anaxagoras' Theory of Matter." *Classical Quarterly* 24: 14–30, 83–95.

————. 1934. "Innumerable Worlds in Presocratic Cosmogony." *Classical Quarterly* 28: 1–16.

————. 1935. *Plato's Theory of Knowldedge.* London: Routledge & Kegan Paul.

————. 1937. *Plato's Cosmology.* London: Routledge & Kegan Paul.

————. 1939. *Plato and Parmenides.* London: Routledge & Kegan Paul.

————. 1942. "Was the Ionian Philosophy Scientific?" *Journal of Hellenic Studies* 62: 1–7.

————. 1950. *The Unwritten Philosophy and Other Essays.* Cambridge: Cambridge University Press.

————. 1952. *Principium Sapientiae.* Cambridge: Cambridge University Press.

Couprie, Dirk L. 1995. "The Visualization of Anaximander's Astronomy." *Apeiron* 28: 159–81.

Couprie, Dirk L., Robert Hahn, and Gerard Naddaf. 2003. *Anaximander in Context.* Albany: State University of New York Press.

Coxon, A. H. 1986. *The Fragments of Parmenides.* Assen: Van Gorcum.

Cunliffe, Richard John. 1924. *A Lexicon of the Homeric Dialect.* London: Blackie & Son.

Curd, Patricia. 1991. "Parmenidean Monism." *Phronesis* 36: 241–64.

————. 1998. *The Legacy of Parmenides.* Princeton, NJ: Princeton University Press.

————. 2002. "The Metaphysics of Physics: Mixture and Separation in Empedocles and Anaxagoras." In Caston and Graham 2002, 139–58.

Dancy, Russell M. 1989. "Thales, Anaximander, and Infinity." *Apeiron* 22: 149–90.

Davison, J. A. 1953. "Protagoras, Democritus, and Anaxagoras." *Classical Quarterly* N.S. 3: 33–45.

Decher, Friedhelm. 1998. "Sein—Werden—Vergehen: Vorsokratische Erklärungsmodelle der Veränderung und Stoffumwandlung." *Prima Philosophia* 11: 307–24.

Deichgräber, Karl. 1933. "Hymnische Elemente in der philosophischen Prosa der Vorsokratiker." *Philologus* 42: 347–61.

————. 1938. "Xenophanes φύσεως." *Rheinisches Museum* 87: 1–31.

————. 1958. *Parmenides' Auffahrt zur Göttin des Rechts.* Mainz: Akademie der Wissenschaften und der Literatur.

Diels, Hermann. 1897. *Parmenides Lehrgedicht.* Berlin: Georg Reimer.

Diels, Hermann. 1899. *Elementum*. Leipzig: B. G. Teubner.

Diller, Hans. 1941. "Die philosophiegeschichtliche Stellung des Diogenes von Apollonia." *Hermes* 76: 359–81.

———. 1946. "Hesiod und die Anfänge der griechischen Philosophie." *Antike und Abendland* 2: 140–51.

———. 1956. "Der vorphilosophische Gebrauch von Κόσμος und Κοσμεῖν." In *Festschrift Bruno Snell*, Munich, 47–60. Reprinted in *Kleine Schriften zur antiken Literatur*. Edited by H.-J. Newiger and H. Seyffert, 73–87. Munich: C. H. Beck, 1971.

Dirlmeier, Franz. 1938. "Der Satz von Anaximandros von Milet." *Rheinisches Museum* 87: 376–82.

Dodds, E. R. 1951. *The Greeks and the Irrational*. Berkeley: University of California Press.

Dolin, Edwin F., Jr. 1962. "Parmenides and Hesiod." *Harvard Studies in Classical Philology* 66: 93–98.

Ebner, Pietro. 1962. "Scuole di Medecina a Velia e a Salerno." *Apollo: Bolletino dei Musei Provinciali del Salernitano* 2: 125–36.

Emlyn-Jones, C. J. 1976. "Heraclitus and the Unity of Opposites." *Phronesis* 21: 89–114.

———. 1980. *The Ionians and Hellenism*. London: Routledge & Kegan Paul.

Engmann, Joyce. 1991. "Cosmic Justice in Anaximander." *Phronesis* 36: 1–25.

Fehling, Detlev. 1994. *Materie und Weltbau in der Zeit der frühen Vorsokratiker*. Innsbruck: Institut für Sprachwissenschaft der Universität Innsbruck.

Finkelberg, Aryeh. 1986. "The Cosmology of Parmenides." *American Journal of Philology* 107: 303–17.

———. 1988. "Parmenides: Between Material and Logical Monism." *Archiv für Geschichte der Philosophie* 70: 1–14.

———. 1993. "Anaximander's Conception of the *Apeiron*." *Phronesis* 38: 229–56.

———. 1997. "Xenophanes' Physics, Parmenides' Doxa and Empedocles' Theory of Cosmogonical Mixture." *Hermes* 125: 1–16.

———. 1998a. "On the History of the Greek Κόσμος." *Harvard Studies in Classical Philology* 98: 103–36.

———. 1998b. "On Cosmogony and *Ecpyrosis* in Heraclitus." *American Journal of Philology* 119: 195–222.

Frank, Erich. 1923. *Plato und die sogenannten Pythagoreer: Ein Kapitel aus der Geschichte der griechischen Geistes*. Halle: Max Niemeyer.

Fränkel, Hermann. 1973 [1962]. *Early Greek Poetry and Philosophy*. Translated by Moses Hadas and James Willis. New York: Harcourt Brace Jovanovich.

Freudenthal, Gad. 1986. "The Theory of Opposites in an Ordered Universe: Physics and Metaphysics in Anaximander." *Phronesis* 31: 197–228.

Friedländer, Paul. 1914. "Das Proömium der Theogonie." *Hermes* 49: 1–16.

Fritz, Kurt von. 1938. *Philosophie und sprachlicher Ausdruck bei Demokrit, Plato und Aristoteles*. New York: G. E. Stechert & Co.

———. 1971. *Grundprobleme der Geschichte der antiken Wissenschaft*. Berlin: Walter de Gruyter.

Furley, David J. 1967. *Two Studies in the Greek Atomists*. Princeton, NJ: Princeton University Press.

―――. 1973. "Notes on Parmenides." In *Exegesis and Argument*. Edited by E. N. Lee, A.P.D. Mourelatos, and R. Rorty, 1–15. Assen: Van Gorcum.

―――. 1976. "Anaxagoras in Response to Parmenides." *Canadian Journal of Philosophy*, suppl. vol. 2: 61–85.

―――. 1983. "Weight and Motion in Democritus' Theory." *Oxford Studies in Ancient Philosophy* 1: 193–209.

―――. 1987. *The Greek Cosmologists*. Vol. 1, *The Formation of the Atomic Theory and Its Earliest Critics*. Cambridge: Cambridge University Press.

―――. 1989. *Cosmic Problems*. Cambridge: Cambridge University Press.

Furth, Montgomery. 1968. "Elements of Eleatic Ontology." *Journal of the History of Philosophy* 6: 111–32.

Gallop, David. 1979. " 'Is' or 'is Not.' " *Monist* 62: 61–80.

Gemelli Marciano, Laura. 1991. "L' 'Atomismo' e il Corpuscolarismo Empedocleo: Frammenti di Interpretazioni nel Mondo Antico." *Elenchos* 12: 5–37.

Giannantoni, Gabriele. 1988. "Le Due 'Vie' di Parmenide." *La Parola del Passato* 43: 207–21.

Gigon, Olof. 1935. *Untersuchungen zu Heraklit*. Leipzig.

―――. 1936. "Zu Anaxagoras." *Philologus* 91: 1–41.

―――. 1968 [1945]. *Der Ursprung der griechischen Philosophie*. 2nd ed. Basel: Schwabe.

Gilbert, Otto. 1907. *Die meteorologischen Theorien des griechischen Altertums*. Leipzig: B. G. Teubner.

―――. 1909. "Ioner und Eleaten." *Rheinisches Museum* 64: 185–201.

Giussani, Carlo. 1896. *Lucreti Cari De rerum natura libri sex*. Turin: Ermanno Loescher.

Gomperz, Theodor. 1939 [1896]. *Greek Thinkers*. Vol. 1. Translated by Laurie Magnus. London: John Murray.

Görgemanns, Herwig. 1970. *Untersuchungen zu Plutarchs Dialog De facie in orbe lunae*. Heidelberg: Carl Winter.

Gorman, Vanessa B. 2001. *Miletos, the Ornament of Ionia*. Ann Arbor: University of Michigan Press.

Gottschalk, H. B. 1965. "Anaximander's *Apeiron*." *Phronesis* 10: 37–53.

Graham, Daniel W. 1984. "Aristotle's Discovery of Matter." *Archiv für Geschichte der Philosophie* 66: 37–51.

―――. 1987a. *Aristotle's Two Systems*. Oxford: Clarendon Press.

―――. 1987b. "The Paradox of Prime Matter." *Journal of the History of Philosophy* 25: 475–90.

―――. 1988. "Symmetry in the Empedoclean Cycle." *Classical Quarterly* 38: 297–312.

―――. 1994. "The Postulates of Anaxagoras." *Apeiron* 27: 77–121.

―――. 1997. "Heraclitus' Criticism of Ionian Philosophy." *Oxford Studies in Ancient Philosophy* 15: 1–50.

―――. 1999. "Empedocles and Anaxagoras: Responses to Parmenides." In Long 1999, 159–80.

Graham, Daniel W. 2002a. "Heraclitus and Parmenides." In Caston and Graham 2002, 27–44.

———. 2002b. "La lumière de la lune dans la pensée grecque archaïque." In Laks and Louguet 2002, 351–80.

———. 2003a. "A Testimony of Anaximenes in Plato." *Classical Quarterly* 53: 327–37.

———. 2003b. "A New Look at Anaximenes." *History of Philosophy Quarterly* 20: 1–20.

———. 2003c. "Philosophy on the Nile: Herodotus and Ionian Research." *Apeiron* 36: 291–310.

———. 2004a. "Was Anaxagoras a Reductionist?" *Ancient Philosophy* 24: 1–18.

———. 2004b. "Thales on the Halys?" *Ancient Philosophy* 24: 259–66.

———. 2005. "The Topology and Dynamics of Empedocles' Cycle." In *The Empedoclean Kósmos: Structure, Process and the Question of Cyclicity*. Edited by Apostolos Pierris, 225–244. Patras: Institute for Philosophical Research.

Grote, George. 1881. *A History of Greece*. 4 vols. New York: American Book Exchange.

Guthrie, W.K.C. 1957. "Aristotle as a Historian of Philosophy: Some Preliminaries." *Journal of Hellenic Studies* 77: 35–41.

———. 1962–1981. *A History of Greek Philosophy*. 6 vols. Cambridge: Cambridge University Press.

Hahn, Robert. 2001. *Anaximander and the Architects*. Albany: State University of New York Press.

Hankinson, R. J. 1998. *Cause and Explanation in Ancient Greek Thought*. Oxford: Clarendon Press.

Harris, William V. 1989. *Ancient Literacy*. Cambridge, MA: Harvard University Press.

Havelock, Eric A. 1966. "Pre-Literacy and the Pre-Socratics." *Bulletin of the Institute of Classical Studies* 13: 44–67.

Hegel, G.W.F. 1971. *Werke*. Edited by Eva Moldenhauer and Karl Markus Michel. Vol. 18, *Vorlesungen über die Geschichte der Philosophie*. Frankfurt am Main: Suhrkamp.

Heidel, W. A. 1906. "Qualitative Change in Presocratic Philosophy." *Archiv für Geschichte der Philosophie* 19: 333–79.

———. 1911. "Antecedents of Greek Corpuscular Theories." *Harvard Studies in Classical Philology* 22: 111–72.

———. 1913. "On Certain Fragments of the Pre-Socratics." *Proceedings of the American Academy of Arts and Sciences* 48: 681–734.

———. 1940. "The Pythagoreans and Greek Mathematics." *American Journal of Philology* 61: 1–33.

———. 1941. *Hippocratic Medicine: Its Spirit and Method*. New York: Columbia University Press.

———. 1943. "Hecataeus and Xenophanes." *American Journal of Philology* 64: 257–77.

Heinimann, Felix. 1945. *Nomos und Physis*. Basel: Friedrich Reinhardt Verlag.

Heitsch, Ernst. 1974. *Parmenides: Die Anfänge der Ontologie, Logik und Naturwissenschaft*. Munich: Heimeran.

Hölscher, Uvo. 1944. "Die milesische Philosophie und die Lehre von den Gegensätzen." *Philologus* 96: 183–92.

———. 1953. "Anaximander und die Anfänge der Philosophie." *Hermes* 81:
257–77; 358–418.

———. 1965. "Weltzeiten und Lebenskyklus." *Hermes* 93: 7–33.

———. 1968. *Anfängliches Fragen*. Göttingen: Vandenhoek & Ruprecht.

How, W. W., and J. Wells. 1912. *A Commentary on Herodotus*. 2 vols. Oxford:
Clarendon Press.

Huffman, Carl A. 1993. *Philolaus of Croton*. Cambridge: Cambridge University
Press.

Hüffmeier, Friedrich. 1963. "Teleologische Weltbetrachtung bei Diogenes von
Apollonia?" *Philologus* 107: 131–38.

Hussey, Edward. 1972. *The Presocratics*. London: Duckworth.

———. 1995. "Ionian Inquiries: On Understanding the Presocratic Beginnings
of Science." In *The Greek World*. Edited by Anton Powell, 530–49. London:
Routledge.

Huxley, G. L. 1966. *The Early Ionians*. London: Faber & Faber.

Inwood, Brad. 1986. "Anaxagoras and Infinite Divisibility." *Illinois Classical
Studies* 11: 17–33.

———, trans. 2001 [1992]. *The Poem of Empedocles*. 2nd ed. Toronto: University of Toronto Press.

Jacoby, F. 1909. "Euthymenes von Massilia." In *Paulys Real-Encyclopädie der
classischen Altertumswissenschaft*. Edited by G. Wissowa. Vol. 6, cols. 1509–
11. Stuttgart: J. B. Metzler.

Jaeger, Werner. 1947. *The Theology of the Early Greek Philosophers*. London:
Oxford University Press.

Jöhrens, Otto. 1939. *Die Fragmente des Anaxagoras*. Bochum-Langendreer:
Heinrich Pöppinghaus.

Jones, H. 1972. "Heraclitus—Fragment 31." *Phronesis* 17: 193–97.

Jones, W.H.S., trans. 1923. *Hippocrates*. Cambridge, MA: Harvard University
Press.

———. 1979 [1946]. *Philosophy and Medicine*. Baltimore. Reprint, Chicago: Ares.

Jouanna, Jacques. 1965. "Rapports entre Mélissos de Samos et Diogène d'Apollonie." *Revue des Études Anciennes* 67: 306–23.

Kahn, Charles H. 1960. *Anaximander and the Origins of Greek Cosmology*. New
York: Columbia University Press.

———. 1966. "The Greek Verb 'to Be' and the Concept of Being." *Foundations
of Language* 2: 245–65.

———. 1969. "The Thesis of Parmenides." *Review of Metaphysics* 22: 700–24.

———. 1973. *The Verb 'Be' in Ancient Greek*. Dordrecht: Reidel.

———. 1974. "Pythagorean Philosophy before Plato." In Mourelatos 1974,
161–185.

———. 1979. *The Art and Thought of Heraclitus*. Cambridge: Cambridge University Press.

———. 1991. "Some Remarks on the Origins of Greek Science and Philosophy."
In *Science and Philosophy in Classical Greece*. Edited by Alan C. Bowen, 1–10.
New York: Garland.

Kahn, Charles H. 2001. *Pythagoras and the Pythagoreans*. Indianapolis: Hackett.

Kerschensteiner, Jula. 1962. *Kosmos: Quellenkritische Untersuchungen zu den Vorsokratikern*. Munich: C. H. Beck.

Keyser, Paul T. 1992. "Xenophanes' Sun (Frr. A32, 33.3, 40 DK⁶) on Trojan Ida (Lucr. 5.660–5, D.S. 17.7.5–7, Mela 1.94–5)." *Mnemosyne* Ser. 4, 45: 299–311.

Kingsley, Peter. 1995. *Ancient Philosophy, Mystery, and Magic: Empedocles and the Pythagorean Tradition*. Oxford: Clarendon Press.

———. 2002. "Empedocles for the New Millennium." *Ancient Philosophy* 22: 333–413.

Kirk, G. S. 1951. "Natural Change in Heraclitus." *Mind* 60: 35–42.

———. 1954. *Heraclitus: The Cosmic Fragments*. Cambridge: Cambridge University Press.

———. 1959. "Ἐκπύρωσις in Heraclitus: Some Comments." *Phronesis* 4: 73–76.

———. 1960. "Popper on Science and the Presocratics." *Mind* 69: 318–39.

Kirk, G. S., and Michael C. Stokes. 1960. "Parmenides' Refutation of Motion." *Phronesis* 5: 1–4.

Klowski, Joachim. 1966. "Das Entstehen der Begriffe Substanz und Materie." *Archiv für Geschichte der Philosophie* 48: 2–42.

———. 1971. "Antwortete Leukipp Melissos oder Melissos Leukipp?" *Museum Helveticum* 28: 65–71.

———. 1972. "Ist der Aer des Anaximenes als eine Substanz konzipiert?" *Hermes* 100: 131–42.

Konstan, David. 2000. "Democritus the Physicist." *Apeiron* 33: 125–44.

Kranz, Walther. 1912. "Empedokles und die Atomistik." *Hermes* 47:18–42.

———. 1916. "Über Aufbau und Bedeutung des parmenideischen Gedichtes." *Sitzungsberichte der königlichen preussischen Akademie der Wissenschaften* 1148–76.

———. 1934. "Vorsokratisches." *Hermes* 69: 114–28.

———. 1938. "Kosmos als philosophischer Begriff frühgriechischer Zeit." *Philologus* 43: 430–48.

———. 1954. "Die Entstehung des Atomismus." In *Convivism: Beiträge zur Altertumswissenschaft (Conrat Ziegler zum 70. Geburtstag)*, Stuttgart, 14–40. Reprinted in *Studien zur antiken Literatur und ihrem Fortwirken*. Edited by Ernst Vogt, 228–46. Heidelberg: Carl Winter, 1967.

———. 1961. "*Sphragis*: Ichform und Namensiegel als Eingangs- und Schlussmotiv antiker Dichtung." *Rheinishces Museum* 104: 3–46; 97–124.

Krischer, Tilman. 1965. "Ἔτυμος und Ἀλήθεια." *Philologus* 109: 161–74.

Kuhn, Thomas S. 1996 [1962]. *The Structure of Scientific Revolutions*. 3rd ed. Chicago: University of Chicago Press.

Laks, André. 1983. *Diogène d'Apollonie: La Dernière Cosmologie Présocratique*. Lille: Presses Universitaires de Lille.

———. 1993. "Mind's Crisis: On Anaxagoras' Νοῦς." *Southern Journal of Philosophy* 31 (Supplement): 19–38.

———. 1999. "À propos du nouvel Empédocle: Les vers 267–290 du poème physique: Étayent-ils l'hypothèse d'une double zoogonie?" *Hyperboreus* 5: 15–21.

Laks, André, and Claire Louguet, eds. 2002. *Qu'est-ce que la Philosophie Préso-cratique?* Lille: Presses Universitaires du Septentrion.

Lassalle, Ferdinand. 1858. *Die Philosophie Herakleitos des Dunklen von Ephesos.* 2 vols. Berlin.

———. 1920. *Gesammelte Reden und Schriften.* Edited by Eduard Bernstein. Vol. 8. Berlin: Paul Cassirer.

Lebedev, Andrei. 1985. "The Cosmos as a Stadium: Agonistic Metaphors in Her-aclitus." *Phronesis* 30: 131–50.

Lesher, James H. 1983. "Heraclitus' Epistemological Vocabulary." *Hermes* 111: 155–70.

———. 1992. *Xenophanes of Colophon: Fragments.* Toronto: University of To-ronto Press.

———. 1995. "Mind's Knowledge and Powers of Control in Anaxagoras DK B12." *Phronesis* 40: 125–42.

Lewis, Eric. 2000. "Anaxagoras and the Seeds of a Physical Theory." *Apeiron* 33: 1–23.

Liepmann, Hugo Carl. 1885. *Die Mechanik der leucipp-democritschen Atome.* Berlin: Gustav Schade.

Lloyd, Alan B. 1976. *Herodotus, Book II. Commentary 1–98.* Leiden: E. J. Brill.

Lloyd, G.E.R. 1964. "The Hot and the Cold, the Dry and the Wet in Greek Philos-ophy." *Journal of Hellenic Studies* 84: 92–106.

———. 1966. *Polarity and Analogy.* Cambridge: Cambridge University Press.

———. 1967. "Popper Versus Kirk: A Controversy in the Interpretation of Greek Science." *British Journal for the Philosophy of Science* 18: 21–38.

———. 1970. *Early Greek Science: Thales to Aristotle.* New York: W. W. Norton.

———. 1975. "Greek Cosmologies." In *Ancient Cosmologies.* Edited by C. Blacker and M. Loewe, 198–224. London: George Allen & Unwin.

———. 1979. *Magic, Reason and Experience.* Cambridge: Cambridge University Press.

———. 1987. *The Revolutions of Wisdom.* Berkeley: University of California Press.

———. 2002. "Le pluralisme de la vie intellectuelle avant Platon." In Laks and Louguet, pp. 39–53.

Lloyd-Jones, Hugh. 1983 [1971]. *The Justice of Zeus.* 2nd ed. Berkeley: Univer-sity of California Press.

Long, A. A. 1963. "The Principles of Parmenides' Cosmogony." *Phronesis* 8: 90–107.

———. 1974. "Empedocles' Cosmic Cycle in the 'Sixties." In Mourelatos 1974, 397–425.

___, ed. 1999. *The Cambridge Companion to Early Greek Philosophy.* Cambridge: Cambridge University Press.

Longrigg, James. 1964. "A Note on Anaximenes' Fragment 2 (Diels/Kranz)." *Phronesis* 9: 1–4.

———. 1965. "Κρυσταλλοειδῶς." *Classical Quarterly* N.S. 15: 249–51.

———. 1993. *Greek Rational Science.* London: Routledge.

Lord, A. B. 1960. *The Singer of Tales.* Cambridge, MA: Harvard University Press.

Lüth, Johann Christoph. 1970. *Die Struktur des Wirklichen im empedokeischen System über die Natur*. Meisenheim am Glan: A. Hain.

Mackenzie, Mary Margaret. 1988. "Heraclitus and the Art of Paradox." *Oxford Studies in Ancient Philosophy* 6: 1–37.

Makin, Stephen. 1993. *Indifference Arguments*. Oxford: Blackwell.

Mann, Wolfgang-Reiner. 2000. *The Discovery of Things: Aristotle's Categories and Their Context*. Princeton, NJ: Princeton University Press.

Mansfeld, Jaap. 1964. *Die Offenbarung des Parmenides und die menschlichen Welt*. Assen: Van Gorcum.

———. 1979–1980. "The Chronology of Anaxagoras' Athenian Period and the Date of His Trial." *Mnemosyne*, Ser. 4, 32: 39–69; 33: 17–95.

———. 1980. "Anaxagoras' Other World." *Phronesis* 25: 1–4.

———. 1990. *Studies in the Historiography of Greek Philosophy*. Assen: Van Gorcum.

———. 1992. *Heresiography in Context: Hippolytus' Elenchos as a Source for Greek Philosophy*. Leiden: E. J. Brill.

Marcacci, Flavia. 2000. "Talete di Mileto: Tra filosofia e scienza." PhD diss., Universitá degli Studi di Perugia.

Marcovich, Miroslav. 1965. "Herakleitos." In *Paulys Encyclopädie der classischen Altertumswissenschaft*. Edited by G. Wissowa et al., Supplementband 10. Stuttgart: Druckenmüller.

———. 1967. *Heraclitus*. Mérida, Venezuela: University of the Andes Press.

Marino, Gordon Daniel. 1984. "An Analysis and Assessment of a Fragment from Jonathan Barnes' Reading of Heraclitus." *Apeiron* 18: 77–89.

Martin, Alain, and Oliver Primavesi. 1999. *L'Empédocle de Strasbourg*. Berlin: Walter de Gruyter.

Mathewson, R. 1958. "Aristotle and Anaxagoras: An Examination of F. M. Cornford's Interpretation." *Classical Quarterly* N.S. 8: 67–81.

Matson, Wallace I. 1963. "Democritus, Fragment 156." *Classical Quarterly* N.S. 13: 26–29.

McCabe, Mary Margaret. 2000. *Plato and His Predecessors*. Cambridge: Cambridge University Press.

McDiarmid, John B. 1953. "Theophrastus on the Presocratic Causes." *Harvard Studies in Classical Philology* 61: 85–156.

McGibbon, D. 1964. "The Atomists and Melissus." *Mnemosyne*, Ser. 4, 17: 249–55.

McKirahan, Richard D., Jr. 1994. *Philosophy Before Socrates*. Indianapolis: Hackett.

Meijer, P. A. 1997. *Parmenides Beyond the Gates: The Divine Revelation on Being, Thinking and the Doxa*. Amsterdam: J. C. Gieben.

Mejer, Jørgen. 1978. *Diogenes Laertius and His Hellenistic Background*. Hermes Einzelschriften, vol. 40. Wiesbaden: Franz Steiner.

Minar, Edwin L., Jr. 1939. "The *Logos* of Heraclitus." *Classical Philology* 34: 323–41.

———. 1963. "Cosmic Periods in the Philosophy of Empedocles." *Phronesis* 8: 127–45.

Mondolfo, Rudolfo. 1958. "Evidence of Plato and Aristotle Relating to the *Ekpurosis* in Heraclitus." *Phronesis* 3: 75–82.

———. 1960. "I Frammenti de fiume e il flusso universale in Eraclito." *Revista Critica di Storia della Filosofia* 15: 3–13.

Moorhouse, A. C. 1962. "Δέν in Classical Greek." *Classical Quarterly* N.S. 12: 235–38.

Moran, Jerome. 1973. "Ps-Plutarch's Account of the Heavenly Bodies in Anaximenes." *Mnemosyne* 26: 9–14.

———. 1975. "The Priority of Earth in the Cosmogony of Anaximenes." *Apeiron* 9: 17–19.

Moravcsic, Julius M. 1983. "Heraclitean Concepts and Explanations." In Robb 1983a, 134–52.

———. 1989. "Heraclitus at the Crossroads of Pre-Socratic Thought." In *Ionian Philosophy*. Edited by K. J. Boudouris, 255–69. Athens: International Association for Greek Philosophy.

Morel, Pierre-Marie. 1996. *Démocrite et la recherche des causes*. Paris: Klincksieck.

Morrison, J. S. 1955. "Parmenides and Er." *Journal of Hellenic Studies* 75: 59–68.

Mourelatos, Alexander P. D. 1970. *The Route of Parmenides*. New Haven, CT: Yale University Press.

———. 1973. "Heraclitus, Parmenides, and the Naive Metaphysics of Things." In *Exegesis and Argument*. Edited by E. N. Lee, A.P.D. Mourelatos, and R. Rorty, 16–48. Assen: Van Gorcum.

———, ed. 1974. *The Pre-Socratics: A Collection of Critical Essays*. Garden City, N.Y.: Anchor Press.

———. 1976. "Determinacy and Indeterminacy, Being and not-Being in the Fragments of Parmenides." *Canadian Journal of Philosophy*, suppl. vol. 2: 45–60.

———. 1979. "Some Alternatives in Interpreting Parmenides." *Monist* 62: 3–14.

———. 1987. "Quality, Structure, and Emergence in Later Pre-Socratic Philosophy." *Proceedings of the Boston Area Colloquium in Ancient Philosophy* 2: 127–94.

———. 1999. "Parmenides and the Pluralists." *Apeiron* 19: 117–29.

———. 2002. "La Terre et les étoiles dans la cosmologie de Xénophane." In Laks and Louguet 2002, 331–50.

Moyal, Georges J. D. 1990. "La rationalism inexprimé d'Héraclite." In *La Naissance de la raison en Grèce: Actes du Congrès de Nice, mai 1987*, Le rationalism. Paris: Presses Universitaires de France.

Mugler, Charles. 1956. "Le problème d'Anaxagore." *Revue des Études Grecques* 69: 314–76.

Nehamas, Alexander. 1981. "On Parmenides' Three Ways of Inquiry." Deucalion 33/34: 97–111.

———. 2002. "Parmenidean Being/Heraclitean Fire." In Caston and Graham 2002, 45–64.

Nestle, Wilhelm. 1942 [1940]. *Vom Mythos zum Logos*. 2nd ed. Stuttgart: Alfred Kröner Verlag.

Neugebauer, Otto. 1957. *The Exact Sciences in Antiquity*. 2nd ed. Providence, RI: Brown University Press.

Neugebauer, Otto. 1975. *A History of Ancient Mathematical Astronomy*. 3 vols. Berlin: Springer Verlag.

Nussbaum, Martha C. 1972. "Ψυχή in Heraclitus." *Phronesis* 17: 1–16; 153–70.

O'Brien, Denis. 1968. "The Relation of Anaxagoras and Empedocles." *Journal of Hellenic Studies* 88: 93–113.

———. 1969. *Empedocles' Cosmic Cycle*. Cambridge: Cambridge University Press.

———. 1981. *Theories of Weight in the Ancient World*. Vol. 1. Leiden: E. J. Brill.

———. 1990. "Héraclite et l'unité des opposés." *Revue de Métaphysique et de Morale* 95: 147–71.

———. 1995. "Empedocles Revisited." *Ancient Philosophy* 15: 403–70.

O'Grady, Patricia F. 2002. *Thales of Miletus: The Beginnings of Western Science and Philosophy*. Aldershot: Ashgate.

Osborne, Catherine. 1987. "Empedocles Recycled." *Classical Quarterly* 37: 24–50.

Owen, G.E.L. 1960. "Eleatic Questions." *Classical Quarterly* N.S. 10: 84–102.

———. 1966. "Plato and Parmenides on the Timeless Present." *Monist* 50: 317–40.

Palmer, John. 1999. *Plato's Reception of Parmenides*. Oxford: Clarendon Press.

———. 2004. "Melissus and Parmenides." *Oxford Studies in Ancient Philosophy* 26: 19–54.

Panchenko, Dmitri. 1997. "Anaxagoras' Argument Against the Sphericity of the Earth." *Hyperboreus* 3: 175–78.

Parry, Milman. 1971. *The Making of Homeric Verse: The Collected Papers of Milman Parry*. Edited by Adam Parry. Oxford: Clarendon Press.

Patin, Alois. 1899. *Parmenides im Kampfe gegen Heraklit*. Leipzig: B. G. Teubner.

Patzer, Andreas. 1986. *Der Sophist Hippias als Philosophiehistoriker*. Freiburg: Karl Alber.

Paxson, Thomas D., Jr. 1983. "The Holism of Anaxagoras." *Apeiron* 17: 85–91.

Peck, A. L. 1926. "Anaxagoras and the Parts." *Classical Quarterly* 20: 57–71.

Pellikaan-Engel, Maja E. 1974. *Hesiod and Parmenides: A New View on their Cosmologies and on Parmenides' Proem*. Amsterdam: Adolf M. Hakkert.

Pendrick, Gerard J. 2002. *Antiphon the Sophist: The Fragments*. Cambridge: Cambridge University Press.

Pepper, Stephen C. 1942. *World Hypotheses: A Study in Evidence*. Berkeley: University of California Press.

Pepple, John. 1996. "A Lost Fragment of Empedocles." *Journal of Value Inquiry* 30: 169–86.

Perilli, Lorenzo. 1988–1989. "Diog. Apoll. Fr. 5 D.-K." *Museum Criticum* 23–24: 293–98.

Petit, Alain. 1988. "Héraclite: La captation de la marge." *Études Philosophiques* 207–19.

Popper, Karl. 1958. "Back to the Presocratics." *Proceedings of the Aristotelian Society* 59.

———. 1963. "Kirk on Heraclitus, and on Fire as the Cause of Balance." *Mind* 72: 386–92.

———. 1968. *Conjectures and Refutations: The Growth of Scientific Knowledge*. New York: Harper & Row.

———. 1998. *The World of Parmenides: Essays on the Presocratic Enlightenment*. Edited by Arne F. Petersen and Jørgen Mejer. London: Routledge.

Potts, Ronald. 1984. "Anaxagoras' Cosmogony." *Apeiron* 18: 90–96.

Quine, Willard Van Orman. 1960. *Word and Object*. Cambridge, MA: MIT Press.

Rabinowitz, W. G., and W. I. Matson. 1956. "Heraclitus as Cosmologist." *Review of Metaphysics* 10: 244–57.

Raven, J. E. 1948. *Pythagoreans and Eleatics*. Cambridge: Cambridge University Press.

Reeve, C.D.C. 1982. "Ἐκπύρωσις and the Priority of Fire in Heraclitus." *Phronesis* 27: 299–305.

Reiner, E., and D. Pingree. 1975. *Babylonian Planetary Omens*. Part 1: *Enuma Anu Enlil, Tablet 63*. Bibliotheca Mesopotamica 2.1. Malibu.

Reinhardt, Karl. 1916. *Parmenides und die Geschichte der griechischen Philosophie*. Bonn: Friedrich Cohen.

———. 1942. "Heraklits Lehre vom Feuer." *Hermes* 77: 1–27.

Robb, Kevin, ed. 1983a. *Language and Thought in Early Greek Philosophy*. La Salle, IL: Hegeler Institute.

———. 1983b. "Preliterate Ages and the Linguistic Art of Heraclitus." In Robb 1983a, 153–206.

Robinson, T. M. 1987. *Heraclitus*. Toronto: University of Toronto Press.

Rösler, Wolfgang. 1971. "Ὁμοῦ χρήματα πάντα ἦν." *Hermes* 99: 246–48.

Ross, W. D. 1924. *Aristotle's Metaphysics*. 2 vols. Oxford: Clarendon Press.

Salem, Jean. 1996. *Démocrite: Grains de poussière dans un rayon de soleil*. Paris: J. Vrin.

Saussure, Ferdinand de. 1966 [1916]. *Course in General Linguistics*. 3rd ed. Translated by Wade Baskin. Edited by Charles Bally and Albert Sechehaye. New York: McGraw-Hill.

Schleiermacher, Friedrich Ernst Daniel. 1807. "Herakleitos der Dunkle von Ephesos." *Museum der Altertumswissenschaften* 1: 315–533.

———. 1838. *Friedrich Schleiermachers sämtliche Werke*. Vol. 2. Part 3. Berlin: G. Reimer.

Schmalzriedt, Egidius. 1970. Περὶ Φύσεως: *Zur Frühgeschichte der Buchtitel*. Munich: Wilhelm Fink.

Schofield, Malcolm. 1980. *An Essay on Anaxagoras*. Cambridge: Cambridge University Press.

———. 1997. "The Ionians." In Taylor 1997a, 47–87.

———. 2003a. "Leucippus, Democritus and the Οὐ μᾶλλον Principle: An Examination of Theophrastus *Phys. Op.* Fr. 8." *Phronesis* 48: 253–63.

———. 2003b. "The Presocatics." In *The Cambridge Companion to Greek and Roman Philosophy*. Edited by David Sedley, 42–72. Cambridge: Cambridge University Press.

Schwabe, Wilhelm. 1975. "Welches sind die materiellen Elemente bei Anaxagoras?" *Phronesis* 20: 1–10.

Schwabl, Hans. 1953. "Sein und Doxa bei Parmenides." *Wiener Studien* 66: 50–75.

Schwabl, Hans. 1963a. "Aufbau und Struktur des Prooimions des hesiodischen Theogonie." *Hermes* 91: 385–415.

———. 1963b. "Hesiod und Parmenides: Zur Formung des parmenideischen Prooimions." *Rheinisches Museum* 106: 134–42.

———. 1966. "Anaximenes und die Gestirne." *Wiener Studien* 79: 33–38.

Scolnicov, Samuel. 1983. "Eraclito e la preistoria del principio di non-contradizzione." In *Atti del Symposium Heracliteum*. Edited by L. Rossetti, vol. 1, 97–110. Rome: Edizioni dell'Ateneo.

Sedley, David. 1981. "Two Conceptions of a Vacuum." *Phronesis* 27: 175–93.

Seligman, Paul. 1962. *The Apeiron of Anaximander.* London: Athlone Press.

Shelley, Cameron. 2000. "The Influence of Folk Meteorology in the Anaximander Fragment." *Journal of the History of Ideas* 61: 1–17.

Sider, David. 1973. "Anaxagoras on the Size of the Sun." *Classical Philology* 68: 128–29.

———. 1981. *The Fragments of Anaxagoras.* Meisenheim am Glan: Verlag Anton Hain.

Sinnige, Theo Gerard. 1971. *Matter and Infinity in the Presocratic Schools and Plato.* 2nd ed. Assen: Van Gorcum.

Sisko, John E. 2003. "Anaxagoras' Parmenidean Cosmology: Worlds within Worlds within the One." *Apeiron* 36: 87–114.

Smyth, Herbert Weir. 1956 [1920]. *Greek Grammar.* Edited by Gordon M. Messing. Cambridge, MA: Harvard University Press.

Snell, Bruno. 1926. "Die Sprache Heraklits." *Hermes* 61: 353–81.

———. 1944. "Die Nachrichten über die Lehren des Thales und die Anfänge der griechischen Philosophie- und Literaturgeschichte." *Philologus* 96: 170–82.

———. 1978. "Die Entwicklung des Wahrheitsbegriffs bei den Griechen." In *Der Weg zum Denken und zur Wahrheit*, 91–104. Göttingen: Vandenhoeck & Ruprecht.

Snodgrass, A. M. 1971. *The Dark Age of Greece.* New York: Routledge.

Solmsen, Friedrich. 1962. "Anaximander's Infinite: Traces and Influences." *Archiv für Geschichte der Philosophie* 44: 109–31.

———. 1965. "Love and Strife in Empedocles' Cosmology." *Phronesis* 10: 109–48.

———. 1971. "The Tradition about Zeno of Elea Re-Examined." *Phronesis* 16: 116–41.

———. 1988. "Abdera's Arguments for Atomic Theory." *Greek, Roman and Byzantine Studies* 29: 59–73.

Souilhé, Joseph. 1919. *Étude sur le Terme Dunamis dans les Dialogues de Platon.* Paris: Félix Alcan.

Spanu, Helene. 1987–1988. "Inhalt und Form der Theorie von Anaxagoras." *Archaiognosia* 5: 11–19.

Stannard, Jerry. 1965. "The Presocratic Origin of Explanatory Method." *Philosophical Quarterly* 15: 193–206.

Stevenson, J. G. 1974. "Aristotle as Historian of Philosophy." *Journal of Hellenic Studies* 94: 138–43.

Stokes, Michael C. 1962–1963. "Hesiod and Milesian Cosmogonies." *Phronesis* 7: 1–37; 8:1–34.

————. 1965. "On Anaxagoras." *Archiv für Geschichte der Philosophie* 47: 1–19; 217–50.

————. 1971. *One and Many in Presocratic Philosophy.* Washington: Center for Hellenic Studies.

————. 1976. "Anaximander's Argument." *Canadian Journal of Philosophy*, suppl. vol. 2: 1–22.

Strang, Colin. 1963. "The Physical Theory of Anaxagoras." *Archiv für Geschichte der Philosophie* 45: 101–18.

Strawson, P. F. 1963 [1959]. *Individuals.* Garden City, N.Y.: Doubleday.

Sweeney, Leo. 1972. *Infinity in the Presocratics.* The Hague: Martinus Nijhoff.

Tannery, Paul. 1887. "La Cosmogonie d'Empédocle." *Revue Philosophique de la France et de l'Étranger* 24: 285–300.

————. 1930 [1887]. *Pour l'Histoire de la Science Hellène.* 2nd ed. Edited by A. Diès. Paris: Gauthiers-Villars.

Tarán, Leonardo. 1965. *Parmenides.* Princeton, NJ: Princeton University Press.

————. 1999. "Heraclitus: The River-Fragments and Their Implications." *Elenchos* 20: 9–52.

Taylor, A. E. 1917. "On the Date of the Trial of Anaxagoras." *Classical Quarterly* 11: 81–87.

————. 1928. *A Commentary of Plato's Timaeus.* Oxford: Clarendon Press.

Taylor, C.C.W., ed. 1997a. *Routledge History of Philosophy.* Vol. 1, *From the Beginning to Plato.* London: Routledge.

————. 1997b. "Anaxagoras and the Atomists." In Taylor 1997a, 208–43.

————. 1999a. "The Atomists." In Long 1999, 181–204.

————. 1999b. *The Atomists: Leucippus and Democritus.* Toronto: University of Toronto Press.

Teodorsson, Sven-Tage. 1982. *Anaxagoras' Theory of Matter.* Göteborg: Acta Universitatis Gothoburgensis.

Theiler, Willy. 1965 [1925]. *Zur Geschichte der teleologischen Naturbetrachtung bis auf Aristoteles.* 2nd ed. Berlin: Walter de Gruyter.

Trépanier, Simon. 2003. "Empedocles on the Ultimate Symmetry of the World." *Oxford Studies in Ancient Philosophy* 24: 1–57.

————. 2004. *Empedocles: An Interpretation.* London: Routledge.

Triplett, Timm. 1986. "Barnes on Heraclitus and the Unity of Opposites." *Ancient Philosophy* 6: 15–23.

Ueberweg, Friedrich. 1920. *Grundriss der Geschichte der Philosophie.* Edited by Karl Praechter. 1st Part. 11th ed. Berlin: Ernst Friedrich Siegfried und Sohn.

Verdenius, W. J. 1942. *Parmenides: Some Comments on His Poem.* Groningen: J. B. Wolters.

————. 1966–1967. "Der Logos begriff bei Heraklit und Parmenides." *Phronesis* 11: 81–98; 12: 99–117.

Vlastos, Gregory. 1946. "Parmenides' Theory of Knowledge." *Transactions and Proceedings of the American Philological Association* 77: 66–77.

————. 1947. "Equality and Justice in Early Greek Cosmologies." *Classical Philology* 42: 156–78.

————. 1950. "The Physical Theory of Anaxagoras." *Philosophical Review* 59: 31–57.

Vlastos, Gregory. 1952. "Theology and Philosophy in Early Greek Thought." *Philosophical Quarterly* 2: 97–123.

———. 1953a. "Isonomia." *American Journal of Philology* 74: 337–66.

———. 1953b. Review of *Pythagoreans and Eleatics*, by J. E. Raven. *Gnomon* 25: 29–35.

———. 1955a. "On Heraclitus." *American Journal of Philology* 76: 337–78.

———. 1955b. Review of *Principium Sapientiae* by F. M. Cornford. *Gnomon* 27: 65–76.

———. 1959. Review of *Wege und Formen frügriechischen Denkens* by Hermann Fränkel. *Gnomon* 31: 193–204.

———. 1967. "Zeno of Elea." In *Encyclopedia of Philosophy*. Edited by Paul Edwards. Vol. 8, 369–79. New York: Macmillan.

———. 1975. *Plato's Universe*. Seattle: University of Washington Press.

Wardy, R.B.B. 1988. "Eleatic Pluralism." *Archiv für Geschichte der Philosophie* 70: 125–46.

Wehrli, Fritz. 1967–1969. *Die Schule des Aristoteles*. Vol. 6, *Lykon und Ariston von Keos*. Basel: Schwabe.

Weidemann, Alfred. 1890. *Herodots zweites Buch mit sachlichen Erläuterungen*. Reprint, Milan: Cisalpino-Goliardica, 1971.

West, M. L. 1960. "Anaxagoras and the Meteorite of 467 BC" *Journal of the British Astronomical Association* 70: 368–69.

———. 1966. *Hesiod: Theogony*. Oxford: Clarendon Press.

———. 1971. *Early Greek Philosophy and the Orient*. Oxford: Clarendon Press.

White, Stephen. 2002. "Thales and the Stars." In Caston and Graham 2002, 3–18.

Wiesner, Jürgen. 1989. "Theophrastos und der Beginn des Archereferats von Simplikios' Physikkommentar." *Hermes* 117: 288–303.

———. 1996. *Parmenides: Der Beginn der Aletheia*. Berlin: Walter de Gruyter.

Wiggins, David. 1982. "Heraclitus' Conceptions of Flux, Fire and Material Persistence." In *Language and Logos: Studies Presented to G.E.L. Owen*. Edited by M. Schofield and M. C. Nussbaum, 1–32. Cambridge: Cambridge University Press.

Wilamowitz-Moellendorff, Ulrich von. 1959 [1931–1932]. *Der Glaube der Hellenen*. 2 vols. 3rd ed. Basel: Benno Schwabe.

Windelband, Wilhelm. 1894. *Geschichte der alten Philosophie*. Munich: C. H. Beck.

Wittgenstein, Ludwig. 1953. *Philosophical Investigations*. Translated by G.E.M. Anscombe. Oxford: Basil Blackwell.

Wöhrle, Georg. 1993. *Anaximenes aus Milet; die Fragmente zu seiner Lehre*. Stuttgart: Franz Steiner Verlag.

Woodbury, Leonard. 1958. "Parmenides on Names." *Harvard Studies in Classical Philology* 63: 145–60.

———. 1981. "Anaxagoras and Athens." *Phoenix* 35: 295–315.

Wright, M. R. 1981. *Empedocles: The Extant Fragments*. New Haven, CT: Yale University Press.

———. 1995. *Cosmology in Antiquity*. London: Routledge.

Zafiropulo, Jean. 1948. *Anaxagore de Clazomène*. Paris: Société d'Édition "Les Belles Lettres."

Zeller, Eduard. 1919–1920. *Die Philosophie der Griechen in ihrer geschichtliche Entwicklung*. Edited by Wilhelm Nestle. Part 1, *Vorsokratische Philosophie*. 2 vols. Leipzig: O. R. Reisland.

Zhmud, Leonid. 1997. *Wissenschaft, Philosophie und Religion im frühen Pytha-goreismus*. Berlin: Akademie Verlag.

INDEX LOCORUM

1.6.1: 107n.40
1.6.1–2: 28n.1
1.6.2: 31n.12, 109n.51, 184n.103
1.6.3–5: 7n.16
1.6.6: 8n.19
1.7.1: 53, 57
1.7.2: 83n.93
1.7.3: 46
1.7.4–5: 59
1.7.5: 79nn.88 and 90, 222n.84
1.8.1: 25n.66
1.8.3: 48n.6
1.8.6: 79n.89
1.8.6–10: 181n.97
1.8.8: 221 and n.81
1.8.9: 221
1.13.2: 275n.37
1.14.2–6: 38n.38
1.14.5–6: 71, 175n.78
1.14.6: 135n.55
1.16: 279n.10

Hippon
A3: 279n.10
A5: 279n.10

Homer
Iliad
1.396–406: 13n.36
1.57ff: 100n.27
1.70: 54n.22
2.5ff: 69n.61
2.484–92: 97n.22
2.484–93: 12n.29
8.18–27: 13n.36
15.189–93; oage 95n.20
18.497–508: 38n.36
18.509–40: 36n.30
Odyssey
18.219: 179n.88

Isocrates
Antidosis
268: 24n.62, 51n.18
Helen
3: 24n.62

Leucippus
A1: 14n.42, 79n.89, 222n.85, 272–74
A7: 260–61n.15
A19: 252n.7
A24: 273–74

A25: 15n.46
B1: 169n.62

Lucretius
1.231: 61
1.422–25: 260n.14
1.665–71: 125
1.693–700: 260n.14
1.709–10: 72n.67
1.713: 72n.67
1.1090: 61 and n.43
5.621–36: 181n.95

Melissus
A5: 283n.17
B1–B6: 251n.4
B7: 251–52, 263
B8: 121n.27, 198n.27, 253, 257
B8.3: 79

Parmenides
A9: 191n.12
A10: 191n.13
A24: 173n.68
A40a: 182n.101
A42: 179n.87, 180
B1: 94n.18, 152
B1.8–11: 202
B1.9: 94n.18
B1.11: 94n.18
B1.31–32: 178
B2: 156 and n.29, 158, 162, 183, 266, 267
B2.5: 158
B2.6–8: 158
B2.7: 158
B3: 167n.57
B4: 160–61n.41, 164n.47
B4.1: 151, 160n.40
B4.2: 211
B6: 156–57, 159, 164n.47, 266, 267
B6.2: 263n.20
B6.3–9: 246
B6.4: 156–57n.29
B6.4–7: 159
B6.8–9: 158, 159, 160–61n.41
B6.9: 160 and n.40
B7: 164n.47, 246
B7.3–5: 257
B8: 162, 164n.47, 166, 169, 183, 203, 204, 206, 266, 267
B8.1–4: 162

GENERAL INDEX